Risk, Society and Policy Series

The Social Contours of Risk: Volume II:

Risk Analysis, Corporations and the Globalization of Risk

Risk, Society and Policy Series

The Social Contours of Risk: Volume II:

Risk Analysis, Corporations and the Globalization of Risk

Jeanne X. Kasperson and Roger E. Kasperson, with contributors

with the assistance of
Mimi Berberian and Lu Ann Pacenka

from Routledge

First published by Earthscan in the UK and USA in 2005

For a full list of publications please contact:
Earthscan
2 Park Square, Milton Park, Abingdon, Oxon OX14 4RN
Simultaneously published in the USA and Canada by Earthscan
711 Third Avenue, New York, NY 10017

Earthscan is an imprint of the Taylor & Francis Group, an informa business

Typesetting by Mapset Ltd, Gateshead, UK
Cover design by Yvonne Booth

ISBN 13: 978-1-844-07175-3 (pbk)

A catalogue record for this book is available from the British Library

Library of Congress Cataloging-in-Publication Data

The social contours of risk / by Jeanne X. Kasperson and Roger E. Kasperson,
with contributors, with the assistance of Mimi Berberian and Lu Ann Pacenka.
        p. cm. – (Risk, society, and policy series)
    Includes bibliographical references and index.
    ISBN 1-84407-073-5 (v. 1 : pbk.) – ISBN 1-84407-072-7 (v. 1 : hardback) –
ISBN 1-84407-175-8 (v. 2 : pbk.) – ISBN 1-84407-176-6 (v. 2 : hardback)
    1. Risk–Sociological aspects. 2. Environmental risk assessment. I. Kasperson,
Jeanne X. II. Kasperson, Roger E. III. Series.
HM1101.S635 2005
302'12–dc22

2005003307

Contents

PART 1 RISK AND SOCIETY: FRAMING THE ISSUES

PART 2 CORPORATIONS AND RISK

PART 3 THE GLOBALIZATION OF RISK

List of Figures

List of Tables

List of Boxes

Acknowledgements

The intellectual debt of this book to colleagues, collaborators and friends over the past several decades is unusually large; in the 'Introduction and Overview' we try to note some of the more important of these. Beginning with graduate studies at the University of Chicago, Norton Ginsburg, Marvin Mikesell and Gilbert White were exemplary mentors. Our long professional collaboration and friendship with Robert Kates stimulated our interest in hazards research, led Jeanne to work on world hunger for ten years and enriched many of the studies reported on in this volume. Others at the George Perkins Marsh Institute at Clark University who have been important colleagues and collaborators over the years include Christoph Hohenemser, Rob Goble, Billie Turner, Patrick Derr, Dominic Golding, Sam Ratick, Halina Brown and Ortwin Renn. Outside of Clark, we have benefited greatly from our collaborators and friends Kirstin Dow, Paul Slovic, James Flynn, Bill Clark, Bert Bolin and Howard Kunreuther.

Mimi Berberian and Lu Ann Pacenka of the George Perkins Marsh Institute played central roles in the preparation of this book, and it would not have been possible without their generous effort and customary excellence in the multitude of tasks involved in the preparation of a volume for publication. Our good fortune to work with them over the past several decades is something we have valued greatly. Others who contributed significantly to the preparation of the book are Teresa Ogenstad and Erik Willis of the Stockholm Environment Institute.

The authors are also indebted to a variety of funding sources that supported the work in this volume over the years. We wish to note, in particular, the National Science Foundation, the Nevada Nuclear Waste Project (and particularly Joe Strolin), the United Nations University, the National Oceanic and Atmospheric Administration and the US Environmental Protection Agency. The George Perkins Marsh Institute of Clark University and the Stockholm Environment Institute supported the preparation of the book in a variety of ways and we acknowledge our appreciation to them.

Finally, we note the continuing support and joy we have received from our children, Demetri and Kyra.

Roger E. Kasperson, on behalf of Jeanne X. Kasperson and Roger E. Kasperson
Stockholm, Sweden
1 June 2004

Acronyms and Abbreviations

ACGIH	American Conference of Governmental Industrial Hygienists
AE	architect-engineer
AIDS	acquired immune deficiency syndrome
AIR	*All India Reporter*
ALARA	as low as reasonable achievable
AOSIS	Alliance of Small Island States
ASARCO	American Smelting and Refining Company
ASSOCHAM	Associated Chambers of Commerce and Industry of India
BBC	British Broadcasting Corporation
BGH	bovine growth hormone
BRWM	Board on Radioactive Waste Management
BSE	bovine spongiform encephalopathy
BST	bovine somatotrophin
CAER	Community Awareness and Emergency Response
CEGB	Central Electricity Generating Board
CENTED	Center for Technology, Environment and Development
CEO	chief executive officer
CFC	chlorofluorocarbon
CIA	Central Intelligence Agency
CIIT	Chemical Industry Institute of Toxicology
CJD	Creutzfeldt–Jakob disease
CMA	Chemical Manufacturers Association
CPSC	Consumer Product Safety Commission
DAD	decide-announce-defend
DDT	dichlorodiphenyl-trichloroethane
DEA	data envelopment analysis
DFID	Department for International Development
DHHS	Department of Health and Human Services
DNA	deoxyribose nucleic acid
DOE	Department of Energy
EDB	ethylene dibromide
EIA	environmental impact assessment
EK-A	Energikommissionens Expertgrupp für Sakerhat och Miljo
ENSO	El Niño Southern Oscillation

EPA	Environmental Protection Agency
EPRI	Electric Power Research Institute
ERDA	Energy Research and Development Agency
ESRC	Economic and Social Research Council
EU	European Union
EVIST	Ethics and Values in Science and Technology
FAO	Food and Agriculture Organization
FEMA	Federal Emergency Management Agency
FEWS	famine early warning system
FIFRA	Federal Insecticide, Fungicide, Rodenticide Act
FIVIMS	food insecurity and vulnerability mapping system
GACGC	German Advisory Council on Global Change
GAO	General Accounting Office
GATT	General Agreement on Tariffs and Trade
GCMs	general circulation models
GM	genetically modified
GMO	genetically modified organism
GSS	general social survey
HIV	human immunodeficiency virus
HLNW	high-level nuclear wastes
IAEA	International Atomic Energy Agency
ICPS	International Trade Unions of Chemical, Oil and Allied Workers
ICSU	International Council of Scientific Unions
IEO	industry and environment
IFRC	International Federation of Red Cross and Red Crescent Societies
IGU	International Geographical Union
IHDP	International Human Dimensions Programme on Global Environmental Change
IIASA	International Institute for Applied Systems Analysis
IIED	International Institute for Environment and Development
ILO	International Labour Office
INPO	Institute of Nuclear Power Operations
IPAT	impact of population, affluence and technology
IPCC	Intergovernmental Panel on Climate Change
KBS	Kärn-Bränsle-Sakerhet
LISREL	linear structural relationships
LMO	living modified organism
LULUs	locally unwanted land uses
MEF	Ministry of Environment and Forests
MIC	methyl isocyanate
MRS	monitored retrievable storage
MSDSs	material safety data sheets
MTHM	metric tonnes of heavy metal

NACEC	North American Commission for Environmental Cooperation
NBC	National Broadcasting Company
NE NIGEC	Northeast Regional Center of the National Institutes for Global Environmental Change
NEPA	National Environmental Policy Act
NGO	non-governmental organization
NIMBY	not in my backyard
NIMTOF	not in my term of office
NNC	National Nuclear Corporation
NOAA	National Oceanographic and Atmospheric Administration
NORC	National Opinion Research Center
NRC	National Research Council
NRC	Nuclear Regulatory Commission
NSSS	nuclear steam system supplier
NWPA	Nuclear Waste Policy Act
OCRWM	Office of Civilian Radioactive Waste Management
OECD	Organisation for Economic Co-operation and Development
OFDA	Office of US Foreign Disaster Assistance
OMB	Office of Management and Budget
OPEC	Organization of Petroleum Exporting Countries
ORNL	Oak Ridge National Laboratory
OSHA	Occupational Safety and Health Administration
OTA	Office of Technology Assessment
PCB	polychlorinated biphenyl
PRA	probabilistic risk analysis
psi	pounds per square inch
PWR	pressurized-water reactor
RAINS	Regional Acidification INformation and Simulation
rem	roentgen equivalent in man
R.I.P.	rest in peace
SARA	Superfund Amendments and Reauthorization Act
SARF	social amplification of risk framework
SEI	Stockholm Environment Institute
SEK	Swedish krona
SFFI	Shriram Food and Fertilizer Industries (India)
SRA	Society for Risk Analysis
SSI	Swedish National Radiation Protection Institute
START	SysTem for Analysis, Research and Training
SUPRA	Scottish Universities Policy Research and Advice Network
TLV	threshold limit value
TMI	Three Mile Island
TRIS	flame retardant, chemical structure is (2, 3 dibromolpropyl) phosphate

TSCA	Toxic Substances Control Act
UCIL	Union Carbide of India Limited
UCS	Union of Concerned Scientists
UN	United Nations
UNCED	United Nations Conference on Environment and Development
UNDP	United Nations Development Programme
UNDRO	United Nations Disaster Relief Organization
UNEP	United Nations Environment Programme
UNFCCC	United Nations Framework Convention on Climate Change
UNICEF	United Nations International Children Emergency Fund
UNISDR	United Nations International Strategy for Disaster Reduction
UNLV	University of Nevada, Las Vegas
USNUREG	United States Nuclear Regulatory Commission
USSR	Union of Soviet Socialist Republics
VDT	visual display terminal
WBGU	Wissenschaftlicher Beirat Globale Umweltveränderungen
WCED	World Commission on Environment and Development
WHO	World Health Organization
WICEM	World Industry Conference on Environmental Management
WIPP	Waste Isolation Pilot Plant
WTO	World Trade Organization

Introduction and Overview[1]

HISTORY

Roger Kasperson and Jeanne Xanthakos met as first year students at Clark University in the autumn of 1955, beginning a college romance that resulted in their marriage in 1959 and what eventually also became a long scholarly collaboration. Loading their few personal belongings and a family cat, as it turned out, into a small U-Haul trailer, they set off for the University of Chicago, where Jeanne commenced graduate studies in English (receiving an MA in 1961), while also working in the University of Chicago Education Library and Roger began his graduate programme in the Geography Department (receiving his PhD in 1966). Geography at Chicago was a hotbed at that time, under the direction of Gilbert F. White, for the creation of a field of natural hazards research and cross-disciplinary research on environmental issues. While Roger was pursuing a programme focused on political geography, the tutelage of Marvin Mikesell, Norton S. Ginsburg and Gilbert White had an important lasting influence on his interest in human–nature studies, and introduced both Roger and Jeanne to a talented cadre of fellow graduate students (e.g. Robert Kates, Ian Burton and Tom Saarinen) who were pushing the boundaries of environmental geography. Meanwhile, Jeanne was also extending the library skills that eventually became a professional career track.

After stints at the Massachusetts State College at Bridgewater, the University of Connecticut and Michigan State University, where Jeanne held library and Roger teaching positions, and a brief stint at the University of Puerto Rico, where both served as researchers, Roger accepted a joint position in government and geography at Clark University and Jeanne spent the next six years at home with two young children. It was at Clark that both Roger and Jeanne developed professional interests in risk research.

ENTRANCE TO RISK WORK

It all began with an educational innovation. In the early 1970s, many colleges and universities instituted a brief term – January term – between the two semesters during which innovative teaching and learning experiences would occur. At Clark, this promised such inviting things as field research on environmental problems in Puerto Rico and the Virgin Islands (geographers know where to go during New England winters!). On this occasion, in 1972, Chris Hohenemser had challenged Robert Kates and Roger Kasperson to apply their natural hazards frameworks and thinking to 'a really complicated problem – the nuclear fuel cycle'. The result was an extremely rich dialogue between geographers and physicists that highlighted a host of risk issues and questions. When the Ford Foundation shortly thereafter in 1973 announced a new research initiative centred on important interdisciplinary problems, the Clark group applied, won one of the several grants awarded, and a new collaboration on technological risk that was to last for several decades was off the ground.

While the nuclear fuel cycle work led to an important publication interpreting 'The distrust of nuclear power' in *Science* (Hohenemser et al, 1977), it more importantly revealed the need for stronger theoretical and analytical underpinnings to the young field of risk analysis. During the late 1970s, as Jeanne was joining the Clark risk group as its bibliographer and librarian, we applied to the new risk programme in the National Science Foundation and secured several grants to work on foundational ideas and approaches to the study of risk.

Using a style of working based on an interdisciplinary team, bridging the natural and social sciences, that was to become a hallmark of Clark risk research, we undertook analyses treating the causal structure of risk, a framework for analysing risk management, a taxonomy of technological risk and a large number of case studies to which these analytical structures were applied. *Perilous Progress: Technology as Hazard* (Kates et al, 1985), with Jeanne as co-editor and both of us as contributors to many of the individual chapters, brought together these results. In the two volumes of this work, several chapters emerge from these foundational efforts of the late 1970s and early 1980s. We took, for example, the pervasive question of the time: how safe is safe enough? We argued (Chapter 1, this volume) that this is a normative question to which no simple answer suffices. What is 'acceptable' or 'tolerable' is, essentially, a decision problem in which broad considerations of risk, benefits, industrial structure, equity and technological development enter. Such problems are better treated from the standpoint of process than by such 'fixes' as '1 in 1 million risk' or 'best practicable technology'. With David Pijawka, we drew upon hazard frameworks to compare natural and technological hazards, especially the shifting patterns of risk apparent in both developed and developing societies, the greater difficulties that technological risks pose for management, and the potential role of the therapeutic community during risk emergencies (Chapter 2, this volume). This chapter also explored the possibility for using common analytical approaches in some depth.

From this basic work on technological risks, we identified a number of new research directions. One that resulted in sustained work involved the intertwining of risk management with value and ethical questions, exploration of issues that were to continue to occupy our research attention to the current time and helped inform our later work, treated in detail in Volume I, on the social amplification of risk.

RISK, ETHICS AND VALUES

Our work on nuclear risks brought home the realization that the most difficult risk problems are nearly always heavily value laden. This even involves the definition of risk, something about which people tend not to agree. Early in its history the Society of Risk Analysis sponsored an effort to reach a consensual definition of 'risk', and commissioned a committee for that purpose. After a year of work it was apparent that defining risk was essentially a political act and the committee gave up on its effort. Indeed, risk controversies suggest that it is often not the magnitude or probabilities of specific risks upon which interested parties disagree, but, more basically, what the risk of concern in the panoply of effects should be and what sign (positive or negative) should be associated with certain postulated changes.

Work on the nuclear fuel cycle highlighted the problem of nuclear waste, an issue largely underestimated and neglected in the US and elsewhere. Alvin Weinberg (1977) was not alone in designating this problem as the one in the nuclear fuel cycle that he had most underestimated. Our work suggested strongly the need to tackle the underlying problems of what we termed 'locus' and 'legacy': how the value problems involved in putting waste in someone's backyard or exporting the risks and burdens of management to future generations could be overcome. These questions stimulated what was to be a decade of work in a series of projects supported by a new National Science Foundation programme, Ethics and Values in Science and Technology (EVIST). Specifically, the 'locus and legacy' effort addressed the issue of how wastes could be equitably stored at one place for the benefit of a large and diffuse population at many other places, as well as how the burdens and risks of waste management could be equitably distributed over this generation, which has made the decisions and reaped most of the benefits, and future generations, which have had no voice in decisions and will bear the long-term burdens.

This early foray into risk and equity issues led in two directions. Firstly, together with the earlier work on the nuclear fuel cycle, we began an extended project on the complex of social, equity and risk questions contained in radioactive waste management. This resulted in a book on *Equity Issues in Radioactive Waste Management* (Kasperson, 1983a) in which we and others identified and assessed a number of equity problems, as well as explored alternative principles of social justice that could be brought to bear on them. One of us (R.E.K.) also during this period served two terms on the National Research Council's Board on Radioactive Waste Management and chaired the panel that produced the report *Social and Economic Aspects of Radioactive Waste Disposal* (USNRC, 1984b). Eventually, we and others at Clark participated in a remarkable team of social scientists assembled by the state of Nevada to assess the social and economic impacts of the proposed Yucca Mountain nuclear waste repository (see Chapter 8, Volume I). Chapter 14 in Volume I summarizes the history of the nuclear waste adventure and the social and value issues that have pervaded the siting process. Specifically, it criticizes the 'tunnel vision' that has been apparent in the repository developmental effort and the focus of this vision on technical issues to the exclusion of the equity and trust issues that have driven societal concern in Nevada and elsewhere. Chapter 15 (Volume I) then generalizes these equity issues to a much broader array of siting undertakings, both in the US and other countries. The latter chapter assesses the underlying problems that have stalemated siting ventures in many countries, contrasts the major approaches that have evolved, and prescribes a series of innovations in process and substantive equity aimed at getting the process through 'roadblocks' and securing greater success in outcomes.

The second line of work involved uncovering new risk equity problems that needed consideration. While analysing the difference in radiation standards used for workers and publics, for example, we discovered that this was only one example of what we came to term the 'double standard' in risk protection, namely, that it was legally permissible to expose workers to much higher levels of risk than publics. 'What ethical systems supported such a position?' we asked. Chapter 12 in Volume I reports on the results of a National Science Foundation project that identified the moral arguments used to support differential protection and analyses and then tested the validity of each of these arguments and their assumptions. The chapter concludes that the assumptions supporting many of the propositions are flawed and based on incomplete or erroneous evidence, such as the view that workers are compensated for the higher risk and voluntarily undertake them, and argues that steps should be taken to narrow or eliminate the divergence in risk standards and to afford workers greater protection.

In the various case studies of different equity problems, it became apparent in this work that a framework for analysing equity issues was badly needed. As a result, in 1991, working with Kirstin Dow in the context of global environmental problems, we formulated such a framework (see Chapter 13, Volume I). It allows the treatment of a number of equity problems, including both geographical equity (the not-in-my-backyard, or NIMBY, issue) and temporal equity (fairness over current and future generations). Since it combines an empirical analysis of the 'facts' of existing or projected inequities with the ability to apply different normative principles to the distributions, analysts or decision makers could use the framework, drawing upon their own definitions of social justice. Finally, the framework provides for application of different management systems for responding to the risk equity problems, including means for building greater procedural equity.

Finally, as part of a larger effort during the second half of the 1990s that concentrated on the most vulnerable peoples and places, we examined the global social justice problems that climate change may pose for international efforts to address global warming. In Chapter 16, Volume I, we argue that the current Kyoto process fails to treat the range of equity issues involved and, as a result, is unlikely to be successful until such questions are internalized. This includes not only differential past and future greenhouse gas emissions, but also the distribution of impacts that have already begun and the differing abilities to deal with such impacts. We set forth in this chapter principles of social justice that should be recognized and a 'resilience strategy' to supplement initiatives aimed at reducing emissions worldwide.

CORPORATIONS AS RISK MANAGERS

The studies of risk management during the late 1970s and early 1980s had convinced us and others in the Clark group that those closest to the technologies and production processes were in the best position to be able to manage risks if only they could be trusted with the social mandate to protect the public and the environment. A study of the Bhopal accident, undertaken with B. Bowonder of the Administrative Staff College in Hyderabad, provided a dramatic and poignant case of how industrial management of risk can go wrong. The results of the post-mortem study of the accident are set forth in Chapter 5 of this volume. Bhopal revealed how essential management attention to both safety design and ongoing management systems and attention to risk 'vigilance' was, but how important safety culture, auditing and high-level management priority to risk were, as well. Furthermore, Bhopal suggested the need for public authorities to exercise monitoring and control over what happens in individual corporations and plants. In a follow-up assessment (see Chapter 8, this volume) some decades later with B. Bowonder, we assessed what had been learned from the accident and the extent to which India was better prepared in the future to deal with industrial risks and another possible Bhopal. As might be expected, the progress we saw was greatest at the level of policy and standards and weakest at the level of monitoring, implementation and enforcement.

Bhopal was not the only case of industrial disaster that drew our attention. Some years earlier the accident at the Three Mile Island nuclear plant in Pennsylvania had produced major concerns over the adequacy of emergency planning around nuclear plants in the US. With the aid of a grant from the public fund established in the aftermath of the Three Mile

Island plant accident, we broadly assessed the state of emergency planning for nuclear plants in the US, the major results of which appear in Chapter 6 in this volume. The review was not reassuring as to management plans for emergency response. The general approach was to try to 'engineer' local responses to accidents and, as much as possible, to use 'command-and-control' procedures built upon military models of communication and organization. In Chapter 6 of this volume we argue, by contrast, that emergency management designs should build upon the adaptive behaviour of people at risk and intentionally seek to create rich information environments that allow people to exercise informed judgements and adaptive behaviour as they select protective strategies. In short, the chapter is a call for a major overhaul of approaches to managing serious industrial accidents.

The Bhopal and Three Mile Island accidents, taken together with the many case studies of risk management appearing in *Perilous Progress*, pointed to the need for a deeper understanding of how corporations actually went about the task of managing risks. And so we initiated an effort during the mid 1980s to survey what was known about the corporate management of product and occupational risks. Chapter 4 in this volume, taken from *Corporate Management of Health and Safety Hazards: A Comparison of Current Practice* (Kasperson et al, 1988a), argues that corporate risk management is 'terra incognita.' Despite an extensive industrial trade literature and numerous exposés of particular failures, little was known systematically about the structures, practices and resources employed by corporations in their risk work. But the chapter does report on a number of findings emerging from case studies and the secondary literature: that corporations vary widely in the types and effectiveness of their risk management efforts, that a variety of exogenous and endogenous factors helped to explain the variance, and that much more systematic and in-depth research was needed to fill out the many empty spaces in our current knowledge. The analysis also called attention to what might be termed the 'shadow' regulatory system that exists in corporations of processes that assess risk, set standards, audit performance and seek to ensure implementation of decisions. But, again, research is needed to build upon the existing sketchy state of empirical understanding of what actually happens in corporations and how decisions are really made.

This general study of corporate risk management highlighted the role that corporate culture, sometimes termed 'safety culture', played in risk management. In a project focused on the risk and ethical issues involved in the transfer of technology and the location of plants by multinational corporations in developing countries, we undertook a case study of DuPont Corporation, a firm noted for having one of the most advanced corporate safety cultures, in its location of a new plant in Thailand. The study, reported in Chapter 7 of this volume, examined the literature on corporate culture and then analysed the extent to which corporate culture considerations actually entered into the location and development of the DuPont plant and how conflicting objectives were negotiated and resolved. While the seriousness of replicating DuPont safety goals and performance in Thailand was quite apparent, the case also indicates the challenges arising from cross-cultural contexts and the ways in which corporate culture may interfere with other goals involved in technology transfer. This is an issue of considerable importance to sustainable development programmes globally.

RISK COMMUNICATION AND PUBLIC PARTICIPATION

As the recognition of the limits to risk regulation grew during the 1980s, interest increased in the possibilities for managing risk through better communication of risks to the public and through greater involvement of publics in risk decisions. This impetus received considerable motive force through the support of Williams Ruckelshaus in his second term as administrator of the US Environmental Protection Agency, and his use of risk communication as a central ingredient in a regulatory decision involving the release of arsenic into the air by an American Smelting and Refining Company (ASARCO) smelter in Tacoma, Washington, in 1983. While the results of this process were ambiguous and controversial, risk communication became a central topic of management interest, not only in the EPA, but in private corporations and state environmental agencies as well. Risk communication rapidly became a dominant topic at annual meetings of the Society for Risk Analysis and the risk communication group became one of the society's largest specialty groups. In 1986, the EPA sponsored the first National Conference on Risk Communication, attended by 500 people. Later, in a searching overall assessment of literature and experience, the National Research Council published *Improving Risk Communication* (USNRC, 1989).

In this first generation of risk communication studies during the 1980s, the orientation was both simplistic and specific. Drawing from advertising and public relations, the risk communication task was conceived of as identifying 'target audiences', designing the 'right messages' and using the 'right channels'. Public relations people were viewed to be the relevant experts for the risk communication job and for the task of facilitating discourse about complex risks as being intrinsically, no different to selling soap. Engineering risk communication 'targets', 'channels' and 'messages' were the approach.

Predictably, these first generation efforts produced meagre encouraging results; by the end of the 1980s, the limits of risk communication programmes were becoming painfully evident. Obviously, risk communication, heretofore the domain of advertising and public relations firms, needed to be informed by psychometric and cultural studies of risk perception and, equally important, communication needed to be integrated with empowering those at risk and with more democratic procedures in risk decision making.

A recognition also grew that the primary risk communication problem involved the failure of risk managers to listen to those who were bearing the risks and to act upon their feedback. So the decade of infatuation with risk communication as a 'fix' to the management problem ran its course and more sophisticated approaches began to evolve, essentially, the 'second generation' of risk studies.

The Clark group made several modest contributions to this evolution. First, with Ingar Palmlund, we proposed a framework for evaluating risk communication programmes (see Chapter 3, Volume I). This format argued that evaluation needed to begin as soon, or even before, the communication programme began. And it needed to be collaborative, so that publics participated in defining what the programme goals would be and what outcomes should be pursued. Properly designed, we argued, evaluation should be integrated with the substance and procedures of the communication programme, so that mid-course corrections could be undertaken to continue to develop the programme as it moved forward.

But we were also concerned that risk communication should be integrated more generally with efforts to empower publics and to enhance their participation in risk decision making. Much earlier, we had prepared a resource paper for the Association of American Geographers dealing with public participation and advocacy planning (Kasperson and Breitbart, 1974). We went back into the extensive literature that had developed during the 1970s and early 1980s to glean major propositions that were consensual findings from previous analyses and experience. Chapter 1 of Volume I provides the results of that foray, which were published in *Risk Analysis*.

Interestingly, as experience with risk communication grew during the second half of the 1980s and the 1990s, the appeal of risk communication as an answer to how the various problems of risk management might be resolved using non-regulatory approaches began to fade. The mixed success of risk communication experience contributed to this, as well as critiques of 'message engineering' from various precincts of the social sciences. And the pendulum swung to strategies aimed at what came to be termed 'stakeholder involvement,' which was increasingly seen both as a principal mechanism for informing risk managers of public concerns and values and as a means of winning the support of various participants in the process. And soon, through various federal and state agencies, stakeholder involvement had replaced risk communication as the required ingredient of any risk management effort and 'focus groups' had become the preferred tool. As with risk communication a decade earlier, little substantive engagement with the conflicting purposes and complexities in achieving effective public participation occurred, leading us to prepare a short cautionary statement on the euphoria for the 'stakeholder express', provided in Chapter 5 of Volume I. Nonetheless, the uncritical embrace of stakeholder involvement continues at the time of writing, not only in the US and Europe, but in developing countries, as well, where it is supplemented by mandatory efforts at 'capacity building'.

It is still important, of course, to draw together what has been learned from a wide range of experience and scholarly analysis. We were asked in 1997 to take stock of the risk communication efforts over two decades as they bore upon industrial accidents and emergency planning. In Chapter 4, Volume I, we reviewed the various approaches to risk communication that had developed over time, pointing to the strengths and limits of each. We then proposed, drawing upon our social amplification research discussed in the following section, that an integrative approach needed to be taken for designing effective programmes. Using that as a base, we then proposed a series of practical guides or advice for corporate and government officials charged with developing and implementing risk communication with various publics.

THE SOCIAL AMPLIFICATION OF RISK

By the mid 1980s, risk studies had gone through a period of rapid foment and development. The Society for Risk Analysis had been created and enjoyed rapid growth, a new risk and decision programme had been established in the National Science Foundation and the National Research Council had published an influential study, *Risk Assessment in the Federal Government: Managing the Process* that encouraged the use of risk analysis in support of regulatory decision making (USNRC, 1983). And yet, progress in conceptual approaches appeared limited due to the separation of natural and technological hazards, each of which had its own journals, professional societies and annual meetings; social and technical analyses of risks that proceeded largely independently from one another; and the preoccupation among social scientists with quarrels over which risk approach (the psychometric model, cultural theory or economic analysis) should be preferred. The risk field, in short, appeared hamstrung by fragmented thinking and the lack of overarching analysis.

It was also at this time (1986, to be exact) that we were drawn with other risk scholars into what would become a remarkable chapter in social science research. The state of Nevada was a candidate site for the development of a high-level nuclear waste repository and had decided to embark on a programme of social and economic studies to identify and assess what impacts the state might experience as a result of both the consideration process and then the development of the facility itself. The study team assembled boasted diverse experience and talents, and included such prominent risk scholars as Paul Slovic, Jim Flynn, Howard Kunreuther, Bill Freudenburg, Alvin Mushkatel and David Pijawka. But the state also assembled a remarkable technical review committee chaired by Gilbert F. White that included the likes of the economist Allen Kneese, the sociologist Kai Erikson, and the anthropologist Roy Rappaport. The research group interacted intensely with these advisers over a decade of collaborative work, producing not only several hundred articles and technical reports, but breaking new ground on a number of risk issues and constructs.

One of these was the social amplification of risk framework. In 1986, as part of our review of risk analyses performed in support of the US Department of Energy's (DOE's) high-level radioactive waste management programme, we were pondering what effects small accidents in the transportation or operational system at the repository, or perhaps even significant mishaps in management without major radiation releases, might have on the programme. From past work, we recognized that such risk events would likely receive high levels of media attention and close public scrutiny. Paul Slovic (1987) was developing the concept of 'risk signals', risk events or occurrences that suggest to the public that the risk is more serious or difficult to manage than had been previously assumed. So we began a discussion with Paul Slovic and his colleagues at Decision Research about how we might analyse such issues, and together we started to try to describe, in schematic or simple conceptual form, what processes would be likely to emerge where things go wrong concerning risks that are widely feared. The team at Clark that worked intensively on this was made up of the Kaspersons, Rob Goble, Sam Ratick and Ortwin Renn. Alternative schematic frameworks flowed back and forth between Clark and Decision Research until the two groups reached agreement in the form of an article for *Risk Analysis*. This piece, provided as Chapter 6 of Volume I, stimulated lively debate from those who commented on it in the journal and also a vigorous exploration of the conceptual framework by not only the original authors, but many others over the next 15 years. Indeed, in terms of our own work, we subsequently wrote some 17 published articles or book chapters expanding on the framework and applying it to a broad array of risk issues and situations.

'The social amplification of risk' sought to produce an integrative framework that could be used to integrate both technical and social aspects of risk, but would also bring under one roof findings from a variety of theoretical and social science perspectives. The intention was to examine the major structures of society that enter into the processing of risk and risk events. This processing can be either proactive, in anticipation of risk events, or reactive to them. We define 'social stations' that are active in processing or augmenting the flow of 'risk signals' and interpreting their social meaning. The actions of these social stations may either dampen the flow of signals, as in risk *attenuation*, or amplify them, as in risk *amplification*. Both affect the rippling of consequences in time and space. Highly socially amplified risks have ripples that extend beyond the immediately affected persons or institutions, and may have large effects upon distant actors. Highly attenuated risks, by contrast, typically have low visibility and concern, and impacts are restricted to those most directly affected.

The social amplification framework proved to be highly useful not only for the various studies conducted in the Nevada project on high-level radioactive waste management, but for other risk issues as well. In Chapter 7 of Volume I, for example, we ask the question: 'How is it, given all the ongoing attention to risks in the media and elsewhere, that certain risks pass unnoticed or unattended, growing in size until they have taken a serious toll?' Using notions from the social amplification framework, we explored the phenomena of 'hidden hazards' and found a variety of causes and explanations. Some hazards, such as what we term 'global elusive hazards', remain hidden because of the nature of the risks themselves. Elusive global hazards, such as acid rain and global warming, for example, have widely diffused effects that are difficult to pinpoint in particular times or places. Some risks, by contrast, are widely attenuated due to society's ideological structures, and so in the US occupational risks, for example, have often generated little concern. Other risks are concentrated among marginal peoples who have little access to political power or the media; therefore, hunger and famines often grow in severity and scope before they are 'discovered' by society's watchdogs or monitoring institutions. Meanwhile, socially amplified and value-threatening hazards, such as genetically modified foods, spark social clashes and conflicts while attenuated hazards generate yawns over the breakfast table.

Risk signals occupy an important place in social amplification thinking. But except for Paul Slovic's hypothesis that particular risk events have high signal value, little attention had been given to the types of signals that exist or the effects they might have. Therefore, working with Betty Jean Perkins, Ortwin Renn and Allen White, as recounted in Chapter 8 of Volume I, we undertook an analysis of the flow of risk signals related to the nuclear waste repository issue in the major newspaper of Las Vegas. This necessitated the development of methodology for identifying and analysing risk signals. We also created a taxonomy of risk signals for classifying the kinds of interpretation and inferences involved. The analysis of the flow of signals in the major Las Vegas newspaper pointed clearly to a major shift in the focus of the Nevada nuclear waste repository debate, from one initially centred on traditional risk and benefit issues, to one almost wholly preoccupied with equity, social trust and the use of political power. Risk signal analyses are a promising new approach to understanding the social contours of risk, but have yet to be fully developed by us or other social risk analysts.

Another aspect of risk emerging from the Nevada studies and one closely related to the social amplification of risk is the concept of stigma. In the nuclear waste context, the intense media coverage of repository-related risks and the social conflict surrounding the repository siting process has a large potential to stigmatize places under consideration for repository sites (invariably termed as 'dumps') and, perhaps, the waste disposal technology itself. These issues are explored in depth in *Risk, Media and Stigma: Understanding Public Challenges to Modern Science and Technology* (Flynn et al, 2001). As part of their efforts on the stigma issue, we were asked to apply the social amplification framework to the development of stigma. This analysis, reported in Chapter 9 of Volume I, modified the social amplification framework to focus it directly on stigma evolution and effects. We give particular attention, drawing both on the stigma and amplification literatures, to how people, places or technologies come to be 'marked', how this marking, over time, changes the identity by which people view themselves and, finally, how identity changes in the perceptions of others. Clearly, social amplification of risks can constitute a powerful process in marking and identity changes. And stigma becomes not only a consequence and part of the 'rippling' of effects, but also a causal factor in future amplification processes.

During the late 1990s, the UK Health and Safety Executive became interested in the social amplification of risk framework and instituted a research programme on that theme. In 1999, with Nick Pidgeon and Paul Slovic, we convened a workshop in the UK that brought together researchers who had been working on social amplification concepts, themes and applications, and a series of papers was subsequently published in *The Social Amplification of Risk* (Pidgeon et al, 2003). For the workshop and volume, we joined with Pidgeon and Slovic in the stocktaking on social amplification work over the past 15 years, as reported in Chapter 11 of Volume I. This overview reviews the principal scholarly debates that have emerged around the social amplification framework, major findings from empirical research on the concepts, diverse applications that have been undertaken and unresolved issues that remain to be addressed. Sessions at the World Congress on Risk in 2003 showcased yet a new array of empirical and theoretical studies, indicating that the framework continues to be useful for a range of social studies of risk and policy applications.

THE GLOBALIZATION OF RISK

Global environmental risk, we have noted elsewhere (Kasperson and Kasperson, 2001b), is the ultimate threat. What is at stake is the survival of the planet itself, and the life support systems it provides for humans and other species. At the same time, the risks are highly uncertain, and only partially knowable and manageable. The risk portfolios of individual countries and places are also becoming progressively more global in their sources. And so the increasing globalization of risk confronts humans with some of their most challenging and perplexing risk problems.

For some time our interests in global risks had been growing, as reflected in our writings with B. Bowonder (Chapters 5, 8 and 9 of this volume), our international studies of comparative management of the nuclear fuel cycle (Chapter 3 of this volume) and the decade that one of us (J.X.K. spent at the World Hunger Program at Brown University during the 1980s and 1990s (see Newman, 1990). But two events – one international and one local – accelerated our attention to the global arena. The World Commission on Environment and Development (WCED, 1987) report clearly demarcated the need for concerted attention to changes in the basic biogeochemical cycles of the Earth and the threats that human activities posed for the long-term security and well-being of the planet. It also called for an international programme of risk research to build the necessary knowledge base. At about the same time, Clark University hosted a landmark international conference entitled 'The Earth as Transformed by Human Action' (Turner et al, 1990a) that documented and assessed human-driven changes in the planet over the last 300 years. At Clark, this led us to join forces with Bill Turner and his research group who were examining human transformation of the planet, particularly in terms of land use, land cover and agricultural systems.

The concrete result of this new collaboration was a major international project entitled Critical Environmental Zones, which was to occupy a decade of effort analysing nine high-risk regions of the world (the Olgallala aquifer, the Basin of Mexico, Amazonia, the North Sea, the Aral Sea, Ukambani in Kenya, the Ordos Plateau of China, the Middle Hills of Nepal, and the Sundaland area of Southeast Asia). The goals were unabashedly ambitious – to understand the principal human driving forces in each of these regions, the human and ecosystem vulnerabilities, and the patterns of human response that arose to deal with the risks and their effects. And, as with the social amplification work, we sought integrative frameworks crossing the natural and social sciences in each of the regions. The work was also avowedly collaborative in that scientists indigenous to the regions were involved in each of the regional assessments.

The project taught us a great deal, both about the challenges of integrative work and how comparative studies can actually be achieved when everyone has their own pet interests and differing theoretical perspectives. We particularly came to believe in the importance of process – the need to formulate team-oriented questions and formats, frequent project meetings where research designs and initial findings are presented and defended, and successive 'approximations' of what the emerging comparative findings are. Eventually, over the ten years of effort, a major comparative volume (Kasperson et al, 1995) and five regional books (see Chapter 11 of this volume) appeared. Although the regions had been selected because they were seriously environmentally threatened areas, the results were, nonetheless, sobering: each region had distinctive arrays of human driving forces; state policies and globalization processes were assuming increased importance; high vulnerability was apparent in many subgroups and marginal areas; and policy interventions aimed at controlling driving forces and mitigating impacts and vulnerabilities were lagging seriously behind the pace of environmental degradation. So, while the Earth Summit in Johannesburg in 2002 could point to many good works and scattered successes in the first generation of sustainability work, *Regions at Risk* details that, however commendable these efforts, evidence abounds that we are losing ground. Chapter 11 in this volume details the major findings at some length.

As we worked on regional patterns of environmental change in different parts of the globe, vulnerability issues were assuming a growing prominence in our analyses. And, of course, our sustained work on equity issues had invariably involved questions of differential vulnerability to risk. During the mid 1990s, vulnerability research showed many of the same patterns and characteristics that the risk field as a whole had a decade earlier. Despite the fact that vulnerability is an integral part of the risk – threat is an interaction between stresses and perturbation and the degree of vulnerability that exists among receptor systems to them – the social science community researching the risk field had not given concerted attention to the assessment of vulnerability. Such attention as had occurred had come largely from analysts working on natural hazards and climate impacts; but they were heavily divided between the ecological and social science communities, while, in addition, the social scientists themselves were fragmented into competing ideological and theoretical camps. Indeed, a common view during the 1990s was that 'social vulnerability' was what really mattered, and thus linkages and interactions with ecosystems and ecosystem services could be left to others. Meanwhile, a variety of international efforts, such as those on the Intergovernmental Panel on Climate Change, the International Human Dimensions Programme, the emerging Millennium Ecosystem Assessment and agencies such as the United Nations Environment Programme and the United Nations Development Programme had identified vulnerability as a priority issue. Expectations were high for what vulnerability assessment could deliver. And so the time seemed right to convene some of the leaders in vulnerability research to take stock of the state of theory and research and to explore whether broad-based, more integrative approaches could be identified.

So, with support from the International Human Dimensions Programme and the Land Use, Land Cover Research Program, we convened a workshop at the Stockholm Environment Institute to assess research and practice in this field. The discussions indicated that, while many differences remained among competing approaches, it might indeed be possible to find a middle ground. Many agreed that a common conceptual framework, such as that presented by Clark and the Stockholm Environment Institute (Kasperson and Kasperson, 2001a, p16), was essential for greater cumulative progress in the field. But it was also apparent that two requirements, in particular, stood out: (1) whatever framework emerged needed to treat the basic receptor as a social-ecological system as researchers in the Resilience Alliance had been doing (Berkes and Folke, 1998), and (2) that empirical applications and validation of any conceptual framework was a high priority.

Earlier we (along with George Clark and others) had been part of an effort to assess human vulnerability to severe storms along the northeast coast of the US (see Chapter 12 of this volume). This analysis sought to assess multiple dimensions of vulnerability in the coastal community of Revere, Massachusetts, and to integrate physical dimensions of risk with social vulnerabilities to them, drawing upon census data in particular for the latter. Data envelope analysis was then used to analyse the results and maps were prepared that captured the interactions between physical and social risk. Drawing upon that study and the workshop results, and collaborating with Bill Turner, Wen Hsieh and Andrew Schiller, we developed a lengthy analysis of the fundamental issues involved in vulnerability, with a particular focus upon creating a sound conceptual framework that captured the essential elements and dynamics of vulnerability (see Chapter 14 in this volume). This work also benefited greatly from interactions with a research team convened by Bill Clark at Harvard University and two intensive workshops held at Airlie House in the US that reviewed and contributed to our thinking in Chapter 14. Vulnerability analysis poses a major challenge in global change assessments and the need for sustained efforts on this in future research topics remains clear.

HUSBAND–WIFE COLLABORATION

Over the past 25 years, as a husband and wife team, we have worked together on some 40–50 research projects, written or co-edited eight books and monographs, and co-authored some 50 articles, chapters and technical reports related to the subject of this book, *The Social Contours of Risk*. The question has been often posed to us – how do you work together so much and still manage to have dinner together every night? So some comment on this may be of interest to some of the readers of this volume.

Typically, we were both heavily involved, often along with other colleagues at Clark or elsewhere, in designing research projects. Since we almost invariably worked in a context of interdisciplinary collaborative research, we often sought to have something like a 'mini-seminar' to talk through a research issue with colleagues in order to determine what the central questions should be and how the research would be focused. Roger often played a role in helping to organize and structure these discussions; Jeanne was typically the expert on literature and bibliography and was always the meeting scribe. Jeanne also always sat quietly in a corner of the room, talking little but listening carefully and thus was always the best source as to who has said what and why (a precious skill; we know how good most men are at the listening function!).

As for collaboration in writing, we would always talk through what the central questions for a particular piece would be, what the argument was that would run through the work, and how it would be organized. We would then prepare an outline upon which we agreed. Roger would usually write the first draft. Jeanne would review the draft and then, armed with a host of questions and issues, discuss what she saw as the necessary revisions. After these had been talked through, Jeanne would write the revised version. Roger would review this, raise remaining issues, and Jeanne would subsequently write the final and polished draft for publication. Over time, we got quite good at this style of working and could be reasonably productive in our joint writing.

While we were both at the Stockholm Environment Institute, on leave from Clark in 2002, Jeanne died unexpectedly while this book was in process. The book is lovingly dedicated to her.

Roger E. Kasperson
Stockholm
June 2004

NOTE

1   This 'Introduction and Overview', slightly modified, also appears in Volume I.

Part 1

Risk and Society:
Framing the Issues

1 Acceptability of Human Risk

Roger E. Kasperson

INTRODUCTION

This chapter has three objectives: to explore the nature of the problem implicit in the term 'risk acceptability'; to examine the possible contributions of scientific information to risk standard-setting; and to argue that societal response is best guided by considerations of process rather than formal methods of analysis.

Most technological risks are not accepted but are imposed. There is also little reason to expect consensus among individuals on their tolerance of risk. Moreover, debates about risk levels are often debates over the adequacy of the institutions which manage the risks. Scientific information can contribute three broad types of analyses to risk-setting deliberations: contextual analysis, equity assessment and public preference analysis.

More effective risk-setting decisions will involve attention to the process used, particularly with regard to the requirements of procedural justice and democratic responsibility.

Recent events have conspired to intensify the societal discussion of the level of risk appropriate for the control of technological hazards. The US Supreme Court decision on the Occupational Safety and Health Administration (OSHA) regulation to control the exposure of workers to benzene, for example, called for a determination of the presence of significant risks that could be reduced by the new standard, but not that the work environment be made risk free (Industrial Union Department, 1980). The recent Reagan administration proposal to relax or eliminate 35 air quality or safety regulations (Farnesworth, 1981) has raised anew the conflicts between an ailing economy and life-saving measures. The Three Mile Island accident of 1979 has provoked increased calls for a safety goal to which all nuclear regulation and licensing should aspire. Even the popular press has joined the fray, airing issues usually restricted to scientists and regulators.

Note: Reprinted from *Environmental Health Perspectives*, vol 52, Kasperson, R. E., 'Acceptability of human risk', © (1983), US Government Printing Office

These concerns have generally been characterized by the question: how safe is safe enough? In fact, this posing of the question has done much to muddle serious discussion of a complex problem. There are even those who assume that this question can and should be answered. Yet setting risk standards involves thorny choices between enlarging benefits and reducing risks, healthier workplaces versus increased unemployment, and present versus future well-being. It is a situation ideally designed to breed indecision among beleaguered public officials whose training and experience are remote from the skills needed to resolve such issues.

This chapter has three objectives:

1   to explore the nature of the problem implicit in the term 'risk acceptability';
2   to examine the possible contributions of scientific information to risk standard-setting;
3   to argue that societal response is best guided by considerations of process rather than formal methods of analysis.

THE NATURE OF THE PROBLEM

The term 'risk acceptability' conveys the impression that society purposely accepts risks as the reasonable price for some beneficial technology or activity. For some special cases this may approach reality. Hang-gliding, racing car driving, mountain climbing and even adultery are all high-risk activities in which the benefits are intrinsically entwined with the risks. These activities are exhilarating because they are dangerous. But most risks of concern are the undesired and oft unforeseen by-products of otherwise beneficial activities or technologies.

Acceptability is the concept that underlies judgements of safety. Lowrance, for example, argues that 'a thing is safe if its attendant risks are judged acceptable' (Lowrance, 1976). Setting aside, for the moment, the important questions as to how and by whom such judgements are made, probably no risk is acceptable if it can be readily reduced still further. To suggest otherwise is to invoke moral justification for trading practical constraints against human lives, a position that most risk guardians will wisely evade. The marketplace, then, is a poor guide to what risks are acceptable, as attested to by the century-long struggle by workers to reduce workplace risks. The occurrence of a past risk suggests more about the balance of political forces which pertained at that time than its acceptability to those who bore the risk.

What does it mean to accept a risk? Does the daily commuter who disdains seat belts accept the risks of automobile driving? Do the workers in textile plants accept the risk of cotton dust exposure? At this individual level, guidance can be found in the practices of informed consent formulated to protect subjects in human experiments. Here, risk acceptance involves several important ingredients: the provision of full information concerning all potential risks; evidence that the subject understands the information; genuine freedom of choice for entering into the experiment; and the option to terminate one's participation at any time.

On the basis of the informational criterion alone, it is apparent that few risks meet the test of acceptance. Whereas some classes of risk (e.g. high probability/acute consequences) are undoubtedly better understood than others (e.g. low probability/chronic consequences), it is only a minority of risks for which the public approaches anything like full information and understanding. Nor is this irrational: given the relentless parade of risks that confront the individual, limited information is undoubtedly a prerequisite for warding off hypochondria, if not despair. There are also large classes of risks, including many of those most feared by the public, that are involuntary in nature. Even risks – such as workplace risks – traditionally labelled voluntary in nature, probably fail to have much actual breadth of choice; workers, for example, tend to have restricted occupational and residential mobility and are not free to seek out less hazardous jobs, particularly when jobs are scarce. Suffice it to conclude, then, that most technological risks are not accepted; they are imposed, often without warning, information or means of redress.

Since most risks are imposed on a less than fully informed risk bearer, the response is more properly thought of as tolerance or acquiescence rather than acceptance. With limited choice and imperfect knowledge, the individual does not resist the imposition of the risk. As knowledge of the risk and range of choice grow, the individual will usually become more risk averse and the degree of risk acceptance will also increase. The area between the tolerated and the accepted risk is the latitude available to the risk guardian for standard-setting (see Figure 1.1). This structure of risk response is, of course, specific to a particular point of time and may be expected to change.

At the societal level, the issue is still more complicated. There is little reason to expect consensus among individuals in their thresholds of tolerability and acceptability. In fact, some of the most difficult risks are those in which individual structures of risk tolerance tend to be divergent rather than convergent. Such appears to be the case, for example, with nuclear power where there are notable sex differences in the response to the hazard (Kasperson, 1980; Nelkin, 1981). In such cases, the current tendency is to set the standard at the level deemed appropriate by the expert with an adjustment to reflect what Bill Rowe would term the risk's 'squawk' factor (Rowe, 1977). As often or not, this adjustment for public response fails to resolve the issue, leaving the risk guardian perplexed, frustrated and irritated. And the means for communicating between anxious publics and well-intentioned experts fail, leaving the public distrustful of the expert and the expert convinced of the public's irrationality.

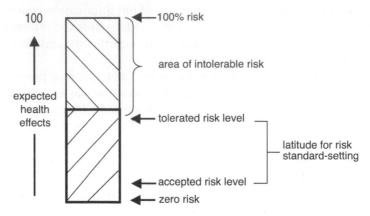

Figure 1.1 *Schematic diagram of individual response to risk*

Increasingly, it is apparent that judgements of appropriate risk levels are inherently problems of ethics and politics. Debates over risk are often, at root, debates over the adequacy and credibility of the institutions that manage the risk and not debates over the actual level of risk. Within the latitude available for risk setting, the risk manager must weigh and trade off multiple objectives and conflicting values. In such decision situations, the preferred choices will not always be those with the lowest risk. Above all, the public wants to be assured that these decisions are made fairly and with a strong commitment to the safety and well-being of the public. With this in mind, it is appropriate to consider current interest in a comprehensive risk standard and the possible contributions of scientific knowledge.

THE SEARCH FOR AN ANALYTIC FIX

There is a great temptation to tidy up this confusing mishmash of risk decisions through some common yardstick and consistent standard for risk imposition. Several means have been proposed. Some see in the historical pattern of risk occurrence evidence that society has arrived at a recurring balance between risk and benefit (Starr, 1969). Others see the need for a consistent quantitative level of risk to serve as the basis for all regulation (Reissland and Harries, 1979). Still others would have cost effectiveness serve as the guiding principle in responding to risk across technologies (Wilson, 1975).

This search for an analytic fix for the risk acceptability problem is misguided. Worse still, it reveals a profound misunderstanding of the nature of the problem. First, it wrongly assumes that one risk is like any other, whereas it is patently clear that risks are multidimensional phenomena that fall into complex groupings. Death by cancer is not the same as death by accident; catastrophic risks are more feared and carry greater social toll than smaller fatality risks; imposed risks are unlike accepted risks. Research at Clark University over the past five years has identified 19 hazard attributes which, when factor-analysed for some 93 technological hazards, fall into five major factors that differentiate such hazards (Hohenemser et al, 1982). When compared with studies of public response to the same hazards conducted by Paul Slovic and colleagues at Decision Research (Slovic et al, 1982), a remarkably close correspondence emerges between the structure of technological hazards and the nature of public response, providing hope that an overall taxonomy of such hazards is possible that will have strong public policy relevance. But it makes clear that regulatory approaches to risk will need to be plural, taking account of major important differences among risks.

Second, decisions on risk levels do not occur in isolation from other social objectives and constraints. Each risk decision, then, tends to be technology – or even situation – specific (Fischhoff et al, 1980). The particular set of values, scientific information, cost considerations and safety opportunities differ from one risk to another and from one time to another for the same risk. Moreover, different regulatory agencies have different legislative mandates and programme priorities for the same risks. Sound decisions on risk levels and distribution, therefore, will and should show substantial variation among even similar risks. However untidy that may appear to some, it is an inescapable reality of responsible and rational risk management; a conclusion, by the way, shared by two recent appraisals of the risk acceptability issue (Fischhoff et al, 1980; Salem et al, 1980).

THE CONTRIBUTION OF SCIENTIFIC INFORMATION

The role of the expert in judgements on risk levels and distributions is to provide information and analyses to inform the decision process. Such formal analyses, however, should not pre-empt the established process that provides participatory or consultative roles for interested parties. The purpose of such analyses is to support an expert judgement for a draft standard, which will then be tested in the political process. Three broad types of analyses – contextual analysis, equity analysis and public preference analysis – are essential for sound decisions (see Table 1.1).

Contextual analysis

The risk under consideration should be placed in appropriate contexts in order to shed light on its social meaning. Five contexts or comparisons are paramount: the risk compared with natural background; the risk compared with other risks prevalent in society; the risk in the context of the magnitude of associated benefits; the costs of risk reduction; and the risks of available substitutes (if needed).

In the first, the risk is compared with natural background levels of exposure. This comparison suggests the increment to risk afforded by the

Table 1.1 *Expert assessment for judging risk intolerability*

Contextual analysis	Equity analysis	Public preference analysis

Risk in the context of:
• natural background
  levels
• other extant risks
• magnitude of benefits
• costs of control
• risks of available
  substitutes
	Distribution of risks,
	benefits and control
	costs over:
	• workers and publics
	• generations
	• backyards
	• social groups
		Public risk reduction
		preferences as indicated
		by:
		• incurred risk inference
		• legal legacy inference
		• expressed values

use of some technology or activity. Large departures are obvious sources of concern, whereas those undetectable in variations in natural exposure merit much less concern. Such analyses have been helpful in the field of radiation control.

The second context is the risk compared with other risks prevalent in society, often with the assumption that risk levels should be balanced. Typically, the comparisons are with similar technologies, other stages of a fuel or production cycle, or risks previously determined to be tolerable by a particular risk guardian. In the UK chemical industry, for example, if a given risk contributes more than 4.0 to the fatal accident frequency rate – the number of fatal accidents in a group of 1000 men in a working lifetime (100 million hours) – risk reduction is undertaken. Two well-known comparisons in the energy area are those of the Rasmussen report (USNUREG, 1975) and the Inhaber analysis (Inhaber, 1979). The issue in such comparisons is whether or not the choice of contextual risks clarifies or obfuscates the risk under consideration. To compare reactor risks with the chance of being hit by a meteor, radioactive waste repository risks with the danger of lightning, or nuclear power with automobile fatalities does little to clarify public choice considerations. There is, however, considerable value, as Gori (1982) has argued, to risk comparisons within a functional class of products needed to sustain a modern society, where benefit levels and uses tend to be similar.

Perhaps the most common form of contextual analysis is a comparison of risks and benefits. This method recognizes some level of risk above zero as necessary and balances the benefits of the activity or technology against the risk to determine how much risk reduction should be undertaken. The quality of such analyses varies widely, depending upon such factors as the messiness of the problem, the skill of the analyst, the way in which the analytic question is posed, the existence of appropriate techniques and the analyst's ability to fashion new techniques (Fischhoff, 1977). Benefit analysis is particularly underdeveloped, with considerable difficulty often apparent in determining even whether a specific outcome is beneficial or harmful (e.g. increased electricity consumption). While useful as one type

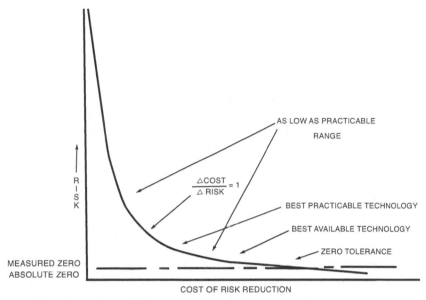

R
I
S
K

AS LOW AS PRACTICABLE
RANGE

$$\frac{\triangle COST}{\triangle RISK} = 1$$

BEST PRACTICABLE TECHNOLOGY

BEST AVAILABLE TECHNOLOGY

ZERO TOLERANCE

MEASURED ZERO
ABSOLUTE ZERO

COST OF RISK REDUCTION

Source: adapted from Rowe (1977)

Figure 1.2 *Social criteria for risk reduction*

of decision information, by itself risk–benefit analysis cannot solve, and may obscure, important policy and ethical issues (e.g. distributive justice). A third type of contextual analysis is the cost effectiveness of risk reduction. The question at stake is how much society wishes to spend in order to avoid a particular consequence. It is well known that such expenditures vary widely; in 1972, for example, Sinclair estimated that in the UK, US$2000 was spent to save an employee's life in agriculture, US$200,000 in steel handling and US$5 million in the pharmaceuticals industry (Sinclair et al, 1972). In a search for an analytic fix, Wilson has suggested that a 'risk tax' of US$1 million per life be used to achieve maximum overall reduction in the array of risks facing society (Wilson, 1979). But risk levels can also be set by changes in the slope of the curve in risk reduction efficiency for any given risk (see Figure 1.2).

A final form of contextual analysis involves the examination of risks of available substitutes. Actions designed to reduce risks sometimes create new and perhaps larger risks, such as increased coal burning when nuclear plants are shut down for safety reasons or the use of TRIS-treated pyjamas as a substitute for flammable materials. Judgements on tolerable levels of risk imposition must consider the risks likely to accrue from increased use of substitute products or technologies.

Equity analysis

Equity analysis is a second – and oft-neglected – need in scientific analysis to support risk decision making. Characteristically, those who enjoy the benefits of a technology are not the same as those who bear the risks. Risks are rarely distributed evenly throughout society and they are sometimes exported to future generations. Attempts to control risks may benefit different groups than those who pay the control costs. While scientific analysis cannot and should not, of course, resolve equity issues, it can lay bare the distributional inequities and value problems.

Four types of inequity merit analysis. First is potential inequity between workers and publics. Higher exposure standards for workers are tolerated in this society and others than those permitted for publics on the basis of questionable moral assumptions. The societal risk from some problems – the toxic waste clean-up comes to mind – can easily be displaced to workers who need jobs. Second is the inequity over generations. Increasingly, there is concern over the export of risks to the future, particularly where the effects may be irreversible. Ozone depletion and radioactive waste disposal are prominent examples. Third is the geographical inequity often referred to as the backyard problem. Traditionally, our society has located noxious facilities and hazardous activities in the backyards of vulnerable and politically powerless peoples. Finally, a more general analysis is required to assess impacts across social groups, including native peoples, minorities and social classes.

Public preference analysis

The third major type of scientific information needed for judgements on risk levels is an assessment of public preferences for risk reduction. The purpose of such enquiry is not, it should be emphasized, to substitute for the direct expression by the public of its wishes, but rather to anticipate what preferences are likely to be and to indicate where there are large departures in expert and lay public assessments of risk.

Three major types of information concerning public preferences are useful. Inferences from incurred risks, commonly described as 'revealed preferences', use statistical risk and economics data for risk–benefits trade-offs acceptable to the public. The assumptions are that, by trial and error, society has arrived at a nearly optimal balance between risks and benefits, and that prevailing social and economic relationships are just and consonant with public values. Both are in doubt, so the results need to be compared with other public preference indicators. Inference from legal legacy looks to the past accumulation of regulatory decisions and court cases for guidance to appropriate standard-setting. Finally, expressed preferences involve directly eliciting risk reduction preferences from the public itself. Considerable progress has occurred in this research, suggesting that lay publics are basically rational on risk questions, order risks similarly to experts but systematically overestimate well-publicized and dramatic risks and underestimate chronic, dispersed risks, and are very risk averse for risks with catastrophic potential.

By putting risks into appropriate contexts, assessing equity issues and anticipating likely public preferences, the scientist can provide information needed for risk decision making. However, the key to effective risk decisions lies in the process and institutions responsible for the judgements that emerge, an issue to which we now turn.

TOWARDS AN EFFECTIVE DECISION PROCESS

Viable decisions on risk levels and distribution require a process consistent with Western democratic theory and directive to the risk guardian. Since the public cannot hope to inform itself and participate in the innumerable decisions on risk, it delegates discretion to the legislators who pass laws and to the regulators who implement them. Doubts as to the credibility of these institutions and processes have provoked much of the current debate over risk decision making. If and when that credibility is recovered, 'how safe is safe enough' will cease to be the subject of societal debate. In the meantime, extraordinary efforts will be required for the recovery of trust and for socially acceptable decisions on risks.

In a democracy, what the risk guardian ideally wants to know to make value-laden decisions is what would public preferences be in the context of informed consent, where interests have become clear, issues dissected and debated, opposing views confronted, and individuals free to choose. It is a hypothetical state, of course: a modern democracy cannot realize such requirements for risk or for any other public issue. All such decisions will inevitably involve choices between the interests of society and the prerogatives of the individual. Public officials make such decisions not in grand master strokes, but in piecemeal decisions that emerge in a series of moves made over time. Customarily we think of the goal of such decisions as lying in realization of the 'public interest', a concept itself the subject of disagreement and confusion (Cochran, 1974).

A credible process finds its starting point in the recognition that power relationships in risk decision arenas are asymmetrical. The creators of risk nearly always have superior knowledge and resources to promote the expansion of potentially hazardous technologies. A variety of forces incline many scientists and risk guardians in the direction of risk creators. Those who fear the risk, by contrast, have few resources, and limited and usually tardy access to the decision process. It is this structure of decision making that is most at stake in the current wars over risk tolerability.

A viable process for risk decisions is one that recognizes the requirements of procedural justice and democratic responsibility. The details of such a process are the subject for a lengthier discussion; suffice it to note here five major considerations for such a process:

1   Decisions on risk are rarely made in isolation, but are part of broader societal choices on the use and expansion of particular technologies

and activities. 'Best solutions' involve choices that take account of competing social values and multiple goals. The appropriate role of the scientist lies in the estimation and measurement of risk and the creation of information needed to assess its meaning, but not in determining its preferred level or distribution.

2   Attempts to find an analytic fix for the risk tolerability problem are misguided. Risk standard-setting should begin with the recognition that such standards should be plural in nature, varying in level and distribution with magnitudes of benefits, equity consideration, opportunities for risk reduction, availability of less risky alternatives, public preferences for risk reduction, and other considerations.

3   Risks cannot be made fully voluntary if society is to realize the potential good associated with existing and new technologies. The emphasis in risk management should be on avoiding rather than mitigating risk, in making unavoidable risks as voluntary as is feasible, and in compensating the bearers of unavoidable risks from beneficiaries where possible.

4   Since risks tend to be imposed rather than accepted, the burden of proof should be on the risk creator to demonstrate the need for the technology and the absence of the risk.

5   Fairness in risk imposition is best achieved by the active participation of risk bearers on their own behalf in decisions regarding the tolerability of particular risk levels and allocations. Risk bearers should not be dependencies in the decision process, but require their own technical capability, right to negotiation and legitimacy in the process. They also have the right to full information as to the risks that will be imposed upon them.

2 Societal Response to Hazards and Major Hazard Events: Comparing Natural and Technological Hazards

Roger E. Kasperson and K. David Pijawka

INTRODUCTION

Danger is an inherent part of human existence. On some occasions we court it for the exhilaration of a particular experience. Sometimes this involves human confrontation of nature, as with mountain climbing or white watering; at other times it is human confrontation of technology, as with racing car driving or test piloting. Usually, however, it is interaction not confrontation, and the danger is unwanted. Involved are the threatening processes of nature over which we have limited control or the adverse prices of a technology that otherwise adds to our health, wealth and well-being.

For most of human experience, the events of nature have exacted the highest toll and caused the greatest concern (Nash, 1976). Throughout history, floods, and drought have been the scourge of mankind, registering such tolls as over 1 million dead in the 1899–1901 drought in India and in the 1931 Hwang-Ho flood in China. The bubonic plague in Europe from 1348 to 1666 is estimated to have killed some 25 million people, roughly one third of the population of the continent. Influenza during 1917–1919 claimed 13 million victims in India, over 500,000 in North America, and millions in Africa and Europe.

In developing countries, natural hazards remain as major problems. The losses from geophysical hazards (floods, droughts, earthquakes and tropical cyclones) alone total an annual average of 250,000 deaths and US$15 billion in damage and costs of prevention and mitigation (Burton

Note: Reprinted from *Public Administration Review*, vol 45, Kasperson, R. E. and Pijawka, K. D., 'Societal response to hazards and major hazard events: Comparing natural and technological hazards', pp15–20, with permission from *Public Administration Review* © (1983) by the American Society for Public Administration (ASPA), 1120 G Street NW, Suite 700, Washington DC, All rights reserved

et al, 1978), while infectious disease still accounts for 10–25 per cent of human mortality (Harriss et al, 1978). But in developed societies, major gains have been made on this broad class of hazards. Geophysical hazards, for example, now result in fewer than 1000 fatalities per annum in the US, a figure that pales by comparison with the 40,000–50,000 annual fatalities from automobile accidents. Infectious disease, with the notable current exception of acquired immune deficiency syndrome (AIDS), has shrunk to a tiny fraction of its earlier mortality toll. All of this has contributed to dramatic increases in life expectancy – from 47 years in 1900 to 74 years in 1979 in the US (USNRC, 1982). And technology has often been the handmaiden in reducing ancient hazards and extending the lifespan.

Yet, technology has emerged as the major source of hazard for modern society. The accumulated exposure of 8 million to 11 million workers to asbestos since the beginning of World War II is expected to result in as many as 67,000 workers dying prematurely each year over the next two decades, with cancer rates among the heavily exposed rising to 35–44 per cent. The chemical revolution of the 20th century has produced widespread exposure of workers and publics to a number of known carcinogens and a larger number whose toxicity remains un-assessed. One recent estimate places the burden of technological hazards at 20–30 per cent of all male deaths and 10–20 per cent of all female deaths, with overall expenditures and losses at 10–15 per cent of gross national product (Harriss et al, 1978).

The hazards of technology pose different managerial problems than those arising from nature. Natural hazards are familiar, and substantial accumulated trial-and-error responses exist to guide management; technological hazards are often unfamiliar and lack precedents in efforts at control. Natural hazards tend to have relatively well-understood 'hazard chains' (see the following section), making opportunities for control intervention relatively clear; the hazard chains for technological hazards, by comparison, are often poorly understood, particularly when the consequences are chronic and the sources of exposure are multiple. Natural hazards tend to provide only limited potential for preventing events; thus, management tends to occur 'late' in the hazard chain. Moreover, technological hazards show a wide variation in loci for control intervention.

Members of the public tend to see natural hazards as acts of God whose effects can only be mitigated; technological hazards, especially those associated with new technologies or those that are imposed, are assumed to be amenable to 'fixes' of various kinds and amenable to substantial reduction. Managing technological hazards requires the simultaneous goals of enlarging social benefits and reducing risks and, where those at benefit and those at risk diverge, action to reduce inequity. Managing natural hazards requires judging the proper allocation of societal effort and the appropriate types of intervention to be undertaken in risk reduction.

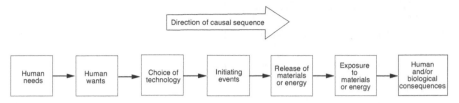

Figure 2.1 *The chain of hazard evolution as applied to technological hazards*

This chapter enquires into the range of problems encountered by society as it attempts to avoid and respond to the hazardous events rooted in technology. Two tasks, in particular, are recognized: first, to characterize the hazard management process and to highlight the particularly difficult problems encountered, and, second, to assess how the major hazardous events arising from technology affect people, their communities and their institutions.

HAZARD EVOLUTION

Hazards may be broadly defined as threats to humans and what they value – life, well-being, material goods and environment (Hohenemser et al, 1982). *Risk*, as differentiated from hazard, may be thought of as the probability that a particular technology or activity (automobile driving) will lead to a specified consequence (death from crashes) over time or activity unit (one driving year). Traditionally, research on natural hazards has envisioned hazards as comprising events → consequences suggesting three broad classes of hazard management: preventing events, preventing consequences and mitigating consequences after they have occurred (Burton et al, 1978).

More recently, researchers have developed a more complex hazard 'chain' or model, based upon hazard evolution (Hohenemser et al, 1982). This model elaborates the events → consequences chain into a multi-stage structure (see Figure 2.1). The model begins with an 'upstream' component of the hazard in which basic human needs (e.g. food, shelter, security) are converted into human wants. Still in the upstream portion of the hazard chain is the choice of technology, involving considerations of realizing benefits and minimizing risks. Thus, in Figure 2.2 the need for food results in human objectives to reduce crop damage due to insects through the use of insecticides, which represents only one technological option. Inevitably, initiating events (e.g. failure of brakes) trigger a release of materials or energy (e.g. collision). The 'downstream' portion of the hazard chain consists of the exposure of humans and ecosystems to these releases, leading to adverse consequences.

Even this elaborated model is a very simplified structure of hazard, for feedback occurs among the stages. Yet, the model is useful in that it

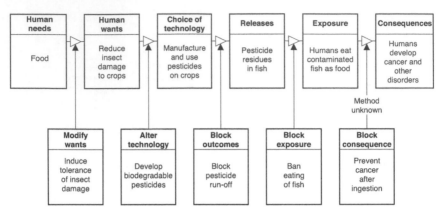

Figure 2.2 *Expansion of the model of hazard evolution into the full range of stages extending from human needs to consequences*

provides a standardized means for structuring hazards and for identifying systematic opportunities for hazard control. Each stage in the hazard evolution is connected by links, each of which represents an opportunity for blocking the hazard. Figure 2.2 provides an illustration of how the hazard chain may be used for identifying a set of managerial control (blocking) opportunities for pesticide hazards. Not shown is a parallel benefit chain that also results from the choice of technology.

Using this notion of a hazard chain, the following section characterizes the management process.

A FLOW CHART OF HAZARD MANAGEMENT

Hazard management is the purposeful activity by which society informs itself about hazards, decides what to do about them, and implements measures to control them or to mitigate their consequences. This activity has two essential functions: intelligence and control. Intelligence provides the information needed to determine whether a problem exists, to define choices and, retrospectively, to determine whether success has been achieved. The control function consists of the design and implementation of measures aimed at preventing, reducing or redistributing the hazard, and/or mitigating its consequences.

Figure 2.3 depicts the management process as a loop of activity. In the centre of the diagram is the hazard chain through which the deployment of technology may cause harmful consequences for human beings and their communities. In 'Major hazard events: Impacts and social response', later in this chapter we enquire at length into the nature of these impacts. Four major managerial activities – hazard assessment, control analysis, control strategy, and implementation and evaluation – surround the chain. Each, as we shall see, characteristically involves both normative and scientific

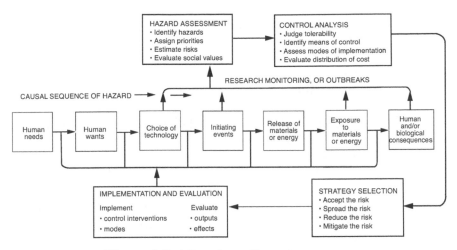

Figure 2.3 *A flow chart of hazard management*

judgements. The depicted sequence, of course, is an idealization and simplification of a process that is often not linear or which jumps over stages. It is not unusual, for example, for control actions to be instituted prior to a thorough hazard assessment.

Hazard assessment

This process involves four major steps: hazard identification, assignment of priorities, risk estimation and social evaluation. A variety of methods – research, engineering analysis, screening and monitoring, and diagnosis – exists for identifying hazards. The scientific capability for these methods has improved enormously over the past several decades. That progress is Janus-faced, however, for it threatens to overwhelm the more limited societal capability to act upon the vast new stores of information and to proliferate low-level hazards.

Somehow these large new hazard domains must be ordered and priorities must be established for the many competing candidates for managerial attention. And the problem is formidable. The US Consumer Product Safety Commission, for example, oversees a hazard domain that includes some 2.5 million firms, more than 10,000 products, and some 30,000 consumer deaths and 20 million consumer injuries. The Toxic Substances Control Act mandates that the US Environmental Protection Agency screen the 70,000 or more chemical substances in commerce and the 1000 or so entering the market each year. The task is not made easier by the fact that establishing priorities receives precious little congressional guidance, yet is laden with value considerations. Is the aggregate risk or the distribution of the risk more important? Should ecological risk receive lower priority than human health risks? Should children enjoy higher priority than adults? How should effects in future generations be valued?

Once a hazard is identified and priority assigned, it is necessary to:

- characterize and, where possible, quantify the stages and linkages in the hazard chain; and
- evaluate this characterization in social terms.

These tasks are often viewed as separate and distinct. Public officials look to scientific experts for advice and evidence on the former while searching their consciences or deferring to the political process for guidance on the latter. Complicating this portion of the assessment process is the fact, addressed at length later in this discussion, that no simple relationship exists between scientific estimates of risk and public response to it. A great deal of confusion and social conflict in hazard management arises from this departure.

Control analysis

Following the hazard assessment, control analysis judges the tolerability of the risk and rationalizes the effort that is made in preventing, reducing and mitigating a hazard. The first of these – judging the tolerability of the hazard – is one of the most perplexing issues in hazard management (see Chapter 1; see also Kasperson and Kasperson, 1983). Fischhoff et al (1982) distinguish four principal methods for judging the tolerability of hazards: risk–benefit analysis, revealed preferences, expressed preferences and natural standards. According to these methods, a technology is judged tolerable if, respectively, its benefits outweigh its risks; its risks do not exceed those of historically tolerated technologies of equivalent benefit; people's expressed opinions indicate that the risks are tolerable; or the risks do not exceed those fixed by nature through the process of evolution. These methods, it should be noted, are often in conflict and no consensus exists to indicate which are preferred in alternative hazard situations.

Whatever the methods, four different criteria tend to be employed in reaching risk tolerability judgements. According to the first, risk aversion, any level of risk is considered intolerable because of the nature of the product, its use or its consequences. Thus, we ban biological weapons, dichlorodiphenyl-trichloroethane (DDT) and chlorofluorocarbon (CFC) aerosols. The other three criteria all involve some form of comparison. Thus, risks may be compared with other risks, often with the assumption that they should be balanced. A well-known, and often criticized, set of risk comparisons is the *Reactor Safety Study* (USNUREG, 1975). Another criterion, cost effectiveness, compares the efficiency involved in various opportunities for risk reduction. Controls for forestalling a fatality in automobile accidents, for example, range in cost from US$500 for enforcing mandatory seat-belt usage to US$7.6 million for road alignment and gradient change. Finally, risk–benefit comparison seeks to balance the benefits of an activity or technology against the risk, to determine how much risk reduction should be undertaken. The quality of such analyses, as with cost–benefit analysis, depends upon such factors as the messiness of the problem, the skill of the analyst, the way in which the analytic question is posed, the existence of appropriate techniques and the analyst's ability to fashion new ones (Fischhoff, 1979).

The second task in a control analysis is to identify the means of control. Here a scientific analysis of the hazard chain is essential in identifying opportunities, ranging from altering our wants and choice of technology (upstream), to preventing or mitigating consequences (downstream) for blocking the evolution of the hazard. Generalizing from the hazard chain, some seven major control interventions are possible:

1   modify wants;
2   choose alternative technology;
3   prevent initiating events;
4   prevent releases;
5   restrict exposure;
6   block consequences; and
7   mitigate consequences.

Complex cases may require a full fault- or event-tree analysis, as in the *Reactor Safety Study* (USNUREG, 1975).

Each control action can be realized through different modes of implementation, which may be grouped into three major classes:

1   Society can mandate the action by law, regulation or court order and, thereby, ban or regulate the product.
2   Managers can encourage the action through persuasion, incentives or penalties.
3   Managers can inform those creating or bearing the risk and allow them voluntarily to reduce or tolerate the hazard. Box 2.1 provides subclasses for each of these major modes.

Finally, there must, of course, be a cost analysis of various control options. Known as cost-effectiveness analysis, this approach permits the hazard manager to select the most efficient action from the available candidates. It is known that existing control measures reflect very different requirements to invest in 'life-saving'. Wilson (1979), for example, has estimated that in 1975 the US expended US$1000 for avoiding a death in the liquefied natural gas industry, compared with US$750,000 for nuclear power.

Selecting a management strategy

Equipped with a hazard assessment and a control analysis, the manager is positioned to select a hazard management strategy. Such a strategy is posited to consist of an overall management goal and a 'package' of control

---

Box 2.1 Modes of implementation

Mandate
Ban the product or process
Regulate the product or process (e.g. performance and design standards;
use and dissemination restrictions)

Encourage
Seek voluntary compliance
Provide incentives (e.g. credits or subsidies)
Penalize through indemnifying those harmed:

- via the market (wages)
- via the courts (award damages)
- via transfer payments (taxes)

Provide insurance

Inform
Inform hazard makers (by monitoring and screening)
Inform those at risk (e.g. by labelling and advertising)

---

measures designed to achieve the goal. The control package will specifically include both control interventions (oriented toward intervention in the hazard chain) and modes of implementation (oriented toward institutional means of control). The management goal will comprise four major possibilities (see Figure 2.3):

1   Risk acceptance seeks to achieve voluntary willingness to tolerate risk, usually through risk compensation or increased information. The former is common in occupational settings where workers are compensated, albeit only partially, for risky work by higher wages or workmen's compensation. Similarly, financial incentives may be offered to communities to accept a nuclear power plant or a hazardous waste facility. In other cases, such as the warning labels on cigarette packages or the patient package inserts for oral contraceptives, information on risk enables those exposed to do their own weighing of benefits against risks.
2   Risk spreading seeks to make a risk distribution more equitable by redistributing it over social groups, geographic regions or generations. The new distribution may aim at equalizing the risk or, alternatively, allocating risk in relation to benefits or to a differential ability to bear the risk. A notable example was the introduction of tall stacks in coal-burning plants to reduce local pollution, resulting in long-distance transport of pollutants and a new regional inequity in risk.

3   Risk reduction has already been discussed at length. Suffice it to note here that efforts to reduce risk often involve the curtailment of benefits or the creation of new, and sometimes unsuspected, hazards. A classic example from the early 1970s was the introduction of the flame retardant TRIS into children's pyjamas and the subsequent discovery that TRIS is a carcinogen.

4   Risk mitigation does not attempt to prevent consequences, but rather to mitigate their effects once they have occurred. Typical actions include disaster relief, medical intervention and family assistance. A number of these will be discussed later in this chapter when we address the impacts of technological disasters (see 'Technological disasters and their long-term effects').

Although presented here as a rational choice process, hazard management strategies tend to develop in piecemeal fashion, are a result of trial and error and build upon previous precedents. They also result, of course, from the mutual partisan adjustments of which Lindblom (1959) writes. However they develop, they nearly always must steer a course between the realization of the benefits of a technology and a minimization of its risks. When the hazard chain is poorly understood scientifically or when the control analysis is incomplete or highly uncertain, society tends to respond by mitigating consequences. A prominent example is occupationally induced cancer, where exposure sources are multiple and the causal agents largely unknown. By contrast, a 'mature' hazard management strategy will tend to employ a complex system of interventions along the hazard chain and a rich set of modes of implementation. Such a system evolves both from growing scientific knowledge and from trial and error in attempts at control.

Implementation and evaluation

Implementation is a crucial and problem-prone stage of hazard management. A lengthy review by the National Research Council (USNRC, 1977) of the Environmental Protection Agency indicates why control actions often fail in implementation:

*   Administrative resources are often inadequate, particularly in a decentralized system in which lower administrative levels face large enforcement burdens but lack resources.
*   Those charged with implementing health and safety control actions are often reluctant to do so because it conflicts with other organizational and political interests.
*   Implementation always contains implicit notions regarding how hazard managers can be induced to accept mandated control actions. If these assumptions are incorrect (and they often are), implementation fails.

- Where managers lack monitoring and surveillance resources in their intelligence function, implementation becomes dependent upon data furnished by hazard makers.

Many of these problems have pervaded the Occupational Safety and Health Administration (OSHA). Even before extensive cutbacks during the Reagan administration, OSHA had very limited inspection and enforcement programmes. Only about 10 per cent of OSHA inspections of places dealt with health issues; yet safety hazards, the easiest violations to identify, were often trivial in their overall impact on health and safety. Meanwhile, the average fine for most violations was a few dollars, and even the small number of serious violations carried penalties of only several hundred dollars (Mendeloff, 1980). Compliance, in short, was heavily dependent upon the voluntary cooperation of the regulated firms, a situation which has increased and has been formalized in recent years.

Hazard management is not complete, even with the implementation of control measures. There must be some evaluation of the accomplishments of hazard management and an assessment of the broad consequences of managerial outputs (i.e. control measures). This evaluation involves the application of social criteria to determine whether success has been achieved. Four criteria may be proposed for such retrospective assessment. First, the managerial actions must be effective: the degree of risk reduction, redistribution or acceptance actually achieved must be measured. Second, management must be efficient: the two relevant measures of efficiency are minimal interference with technological benefits and the choice of the most cost-effective measures for risk control. Third, management must be timely: managers should move through assessment and control activities with a minimum of delay. Finally, management should be equitable: risks and costs to those not benefiting from the technology should be minimized.

This flow chart of hazard management identifies major activities which must be undertaken and issues that will arise. No matter how competent the management process, hazardous events will still occur, harm people, disrupt communities and endanger institutions. We turn next to such occurrences, their effects and potential means of response.

MAJOR HAZARD EVENTS:
Impacts and social response

Several conceptual frameworks exist for assessing public response to technological hazards and emergencies (see Lindell and Perry, 1980; Quarantelli, 1981; Stallen and Tomas, 1981; Baum et al, 1983; Perry et al, 1983; Sorensen et al, 1983). In comparing the response to the threat of a flood with behaviour during a derailment that posed the risk of releasing radioactive material, Perry noted that every disaster situation has the same phases of social response – threat detection, threat evaluation and information dissemination (Lindell and Perry, 1980). Thus, although characteristics of disaster agents may vary, natural and technological disasters may be examined within the same conceptual and analytical framework. Nevertheless, differences in characteristics of threat between natural and technological hazards do result in different management problems (noted above), as well as different patterns of response (Quarantelli, 1981). These contrasts are important to understand in order to develop effective means of controlling and managing the hazards of technological origin. This section highlights some of these differences by focusing on emergency events resulting from failures in technology, and sociopsychological impacts of recovery from technological disasters.

It is important to distinguish among three generic types of technological hazard events – mostly releases in terms of the hazard chain. The first, the ubiquitous (or routine) hazard events of technology, involves exposure over a substantial period of time to low emissions of chemical contamination or other hazardous activity that poses chronic and, perhaps, unacceptable consequences. These hazards do not represent major failures and are typically addressed by established management structures and processes. The second class of hazard event involves the failure of a technology, resulting in release or potential release of hazardous material and necessitating emergency response. The third group, technological disasters, is characterized by exposure to harmful substances for a particular population and locale, resulting in major loss of life or injury with long latency periods, social disruption and relocation. The Love Canal and Times Beach situations represent this class. In such events, the hazards are often identified through scientific efforts and formal risk assessments. Emergency responses to disasters are less important than interrupting the causal chain of the hazard and thereby averting adverse consequences. The failure of a technology may result in disaster if loss or impact is substantial. The distinction between the two latter classes of hazard events is based on the greater likelihood of occurrence of small random events that can usually be contained or mitigated through emergency responses. If technological failure results in adverse and long-term secondary effects with attendant social disruption, then it can be defined as a disaster.

The first class of hazard events, routine releases of hazardous technology, has been discussed above in the context of the flow chart of hazard management. We turn, therefore, to the two classes of hazard events that represent major failures and releases.

Failure of technology requiring emergency response

This class of hazard events includes accidents at nuclear facilities, transportation incidents with potential release of hazardous substances and explosions at fixed facilities that pose both immediate and chronic

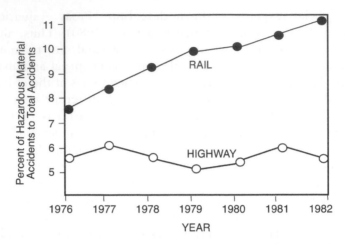

Figure 2.4 *Hazardous material transit accidents, rail and highway, as percentage of total accidents*

threats to health. Growing public concern over this class of hazard events may reflect its increasing incidence. Figures 2.4 and 2.5 provide time series data on the incidence and magnitude of damages from hazardous materials transit accidents. As Figure 2.4 indicates, the percentage of rail hazardous material accidents to the total number of rail accidents has increased from less than 8 per cent to 11 per cent over the 1976 to 1982 period. This is significant because 35 per cent of all freight trains have been estimated to carry hazardous material. Highway accidents involving hazardous material have remained fairly steady, between 5 and 6 per cent of all commercial highway accidents. In addition, as Figure 2.5 shows, property damages per accident for hazardous material carriers are also increasing and are severe when compared to damages of non-hazardous material carriers. A recent survey of 300 community and organization officials in 19 cities found a very high level of perception associated with the probability of a chemical accident. Although local emergency organizations assigned even higher probabilities to the occurrence of such accidents, communities were generally found to be inadequately prepared for serious chemical emergencies (Helms, 1981).

Community vulnerability to chemical threats is a function of the magnitude and nature of the risks and level of preparedness. With natural hazards, communities are generally familiar with the threat, have past experience, and may have previously instituted controls to prevent or reduce damage. Floods, for example, can be forecast, and rise in water level can be monitored as a basis for an evacuation warning. In coping with natural hazards, individuals and communities enjoy control of the situation in terms of the adjustment choice; experience and familiarity with the hazard are critical decision factors in response, and protection measures are understood.

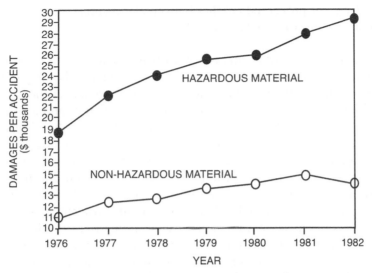

Figure 2.5 *Transit damages per accident, hazardous material and non-hazardous material accidents*

By contrast, familiarity with the hazard is relatively low in communities suddenly faced with technological emergencies (Zeigler et al, 1983). First, the random nature of occurrences of serious chemical releases means that few communities have experienced these hazards. Thus, predictive knowledge is low and effects of release may be unknown in small communities. Even where such experience has occurred, serious problems exist in the effectiveness of response and emergency management, particularly with respect to threat identification and coordination of response. Quarantelli (1981) has suggested that first on-scene responders typically lack knowledge of the full range of possible responses to the variety of chemical hazards and that the identification of chemical hazards during emergencies has often been a problem. Problems in hazard identification, lack of experience with chemical emergencies and fear of secondary impacts have resulted in both over-response and delays in emergency management. Small communities also tend to rely upon outside resources for aid in response to serious chemical incidents. Such vertical dependencies for response at the local level have resulted in low levels of community preparedness to chemical hazards (Tierney, 1981). Such dependencies are also not free of problems. In Arizona the state supports six hazardous material response teams whose responsibility it is to coordinate emergency response to local communities. However, an ongoing assessment of the role of these teams suggests that:

• Their dispersed locations have resulted in arrival at the scene of an emergency only after initial actions were taken by local response organizations.

- They have not prevented conflict between local and state emergency planning organizations (Pijawka, 1984).

Prevention of such hazard events is very difficult. Rerouting of hazardous cargo is possible; but this implies knowledge, which may be lacking, of the type and volume of material and the jurisdictional authority of communities to take action. Prevention consists of activities and decisions that reduce the probability of occurrence of technological failure. Most of these activities occur at the national level through promulgation of engineering containment standards, packaging standards and enforcement of regulations, such as those of the Department of Transportation under the Hazardous Materials Act. Hazard prevention is less efficacious at the local level. At the community level, actions to reduce risk through various activities – zoning, screening of hazardous industries, disclosure of materials stored or processed by firms, and inspections by fire organizations – are contingent upon community norms and values regarding the relationship of roles between private and public sectors.

Unlike most natural disasters, technological failures can occur quickly and without warning. Perry (1985) found that quick onset of technological accidents places an enormous burden, particularly for evacuation decision making, on emergency managers. He argues that in contrast to natural disasters, technological emergencies are compressed in time; that is, the time between hazard awareness, problem identification, risk assessment and the decision to evacuate may be extremely short. Chemical release into the environment may also be hidden and identified only after considerable lapse of time. Thus, unlike natural hazards, technological hazards may not be readily observable and may require a specialized analytical capability, lacking in most communities, for identifying, estimating and evaluating the risk. Lack of familiarity with technological hazards, generally low levels of community awareness and preparedness, the rapid onset of the hazard event, and the potential for larger secondary consequences present critical problems for emergency managers.

Lack of familiarity with technological hazards and the perception of these hazards are important factors in differentiating response to technological and natural disasters. Although no formal evacuation order was given at Three Mile Island, approximately 40 per cent of the population within 24km of the disabled plant evacuated (Flynn, 1982). The extent and voluntary nature of the evacuation contrast with the general reluctance in natural disasters to evacuate until the event is perceived as extreme and impending. Moreover, when presented with a hypothetical nuclear accident scenario, people express evacuation intentions that exceed official expectations for which evacuation plans were prepared (Johnson and Zeigler, 1984). The high level of concern over the risk of nuclear power and chemical release may be attributable to the public's perceiving the risks as more threatening and catastrophic than natural disasters (Fischhoff, 1979).

Biases are associated with perceived threat and consequences of technological hazard. Risk consequences perceived as dreaded (feared), uncontrollable, irreversible or catastrophic are generally overestimated. Public fears may be intense for particular hazards, especially those with low probability, high consequences (such as those presented by nuclear reactor accidents). Imagery of threat in terms of fear or irreversibility of effects heightens the public perception of risk. Recent studies on judgemental biases in risk perception indicate that members of the public tend to overestimate rare causes of death and those with high imagery and to underestimate common causes of death. Accordingly, the risks of accidents, floods, botulism, fire and homicides are overestimated; risks of death from diabetes, stroke and emphysema are underestimated (Fischhoff et al, 1982). More recently, a proposed taxonomy of technological hazards has identified major biophysical attributes of hazards that may have considerable potential for predicting public response. While such fears may well seem exaggerated compared with 'objective' risk probabilities, it is more appropriate to view the assessment of risk by members of the public as proceeding on a different, and probably broader, basis than those incorporated in quantitative risk assessment by experts. In any event, the fear of the public is an objective reality that contributes to stress and emotional disruption.

The complexity of technological threats, as augmented by heightened perceptions of risk, results in response patterns that differ from those associated with natural disasters. One study found that during a derailment that had the potential to release chemical substances, a large proportion of the threatened population complied with the evacuation warning even though the risk was perceived to be low (Perry et al, 1983). In addition, as the Three Mile Island accident has shown, the technical and scientific aspects of man-made threats, the uncertainty surrounding stages of the hazard chain and lack of familiarity with the hazard result in greater public dependence upon governmental authorities and a reduced reliance upon social networks. Thus, emergency managers face an extremely difficult situation when confronted with technological hazard events. The special nature of the threat may result in over-response by the public and increased dependence upon local authorities for information (Johnson and Zeigler, 1984). In this light, Sorensen has argued that emergency response plans for nuclear facilities have been overly mechanistic and have failed to incorporate sufficiently a knowledge of behaviour into their design (Sorensen, 1984).

Technological disasters and their long-term effects

Although a substantial body of sociopsychological research has addressed the effects of disasters on individuals and families (Helms, 1981; Quarantelli, 1981), a number of issues have surfaced that require expanded research. Recent reviews on psychological and emotional effects of natural disaster have identified serious inconsistencies in earlier findings (Perry, 1983). The work by Erikson (1976), for example, shows that disasters can result in adverse and long-lasting psychological disruptive effects. By contrast, several studies suggest that adverse emotional effects following natural disasters have not been pervasive or have occurred only during the immediate post-impact period (Quarantelli, 1979). Perry (1983) has argued that this disagreement stems from definitional problems over the meaning of psychological impacts, methodological differences among studies and a lack of attention to explanatory theory.

Except for the few major studies on the Love Canal disaster, the Three Mile Island accident and the Buffalo Creek catastrophe, knowledge of long-term impacts of technological disasters is scant. Behavioural assessments of the effects of Three Mile Island, however, provide evidence of distress during a relatively short period of extreme threat to safety, and a longer term due to impending threats of restart and periodic low-level releases of radioactive material from the disabled reactor unit. The Three Mile Island data point to a high level of distress during the crisis period; but it was relatively short in duration. One study concluded that approximately 10 per cent of the surrounding population experienced severe distress equivalent to chronic mental disorder; but such levels declined soon after the accident was under control. In fact, one month subsequent to the accident, only 15 per cent of the nearby population was experiencing severe distress compared to 26 per cent during the crisis (Bromet, 1980). Retrospective assessments of stress six months after the accident found that during the emergency period, 48 per cent perceived the threat to safety as very serious and 22 per cent were extremely upset. However, at the time of the survey, only 12 per cent continued to perceive the threat as very serious (Flynn, 1982). A survey of residents revealed that perceived effects on the physical health decreased from 14.4 per cent during the accident to 7.6 per cent one year later (Cutter, 1984).

Although the incidence of serious distress declined rapidly for a large segment of the population, for others psychological impacts persisted. Immediately after the accident, 24 per cent of nearby residents indicated personal consequences of the accident on mental health; one year later, 25 per cent reported effects on mental health (Cutter, 1984). In 1983, self-reporting workshops held with neighbours of the Three Mile Island reactor found the following effects: 'psychic numbing', hopelessness, anxiety, feelings of being trapped by the situation and lack of peace of mind. These emotional effects were manifest four years after the accident, whereas concern over radiation releases and other threats from the impaired reactor had significantly diminished within one year following the accident (Sorensen, 1984). For these individuals, recurring reminders of the accident, secondary stressful events such as venting of radioactive substances, and fear of disaster during subsequent operations have resulted in apparently serious continued psychological and emotional effects.

A study of long-term family recovery from natural disaster found the degree of economic loss and lengthy return factors associated with a slower pace of emotional recovery (Bolin, 1982). The Three Mile Island accident, however, did not result in higher levels of permanent unemployment, declines in property values, long-term evacuation and out-migration, or significant economic loss for most residents, including evacuees (Flynn and Chalmers, 1980). In fact, family activities resumed shortly after control of the accident was under way. Thus, reporting on mental health effects years after the accident has to be explained by factors other than delay in material recovery. The fact that radioactive materials were, except for a small amount, contained suggests that the health consequence factor would also not emerge as a critical variable in explaining the prolonged state of emotional upset. Three years after the accident, those reporting concern for physical health effects because of Three Mile Island constituted about 8 per cent of the surrounding population, a figure which represents a decrease of 50 per cent since the accident (Cutter, 1984).

Speculation on factors influencing prolonged mental health impacts related to Three Mile Island has centred on the impact of continuous media attention to the issue, perceived risks of intermittent venting of radioactive gases, release of tritium-contaminated water, the threat of restarting the undamaged facility, and the visible presence of the facility as a reminder of potential threat. The perceived threat posed by possible restart of Three Mile Island may present an unacceptable risk situation for some residents. The fear of another accident or a catastrophic accident following restart may be a significant factor. Three years following the accident, over 25 per cent of the nearby population believes that the frequency of a major nuclear power plant accident is one in ten years, whereas in 1979 this figure was only 15 per cent. The perceived increased likelihood of an accident within the population is reinforced by the substantial growth in distrust of government regulatory agencies and the perceived lack of personal control in managing the risk.

Threat may be defined as constituting two dimensions – degree of danger (perceived or real) and degree of control. Often with technological hazard, as we have noted above, the hazard chain is poorly understood and the management process is fraught with conflict and ill-defined trade-offs. In the few cases of technological disasters, governmental actions have been delayed, ineffectual and conflict ridden. Individual or group actions by victims to remedy the situation have been generally ineffective. In technological disasters individuals see few opportunities to reduce exposure through physical adjustments because of the pervasiveness of the threat and the 'no-threshold' level of effects for carcinogens. The perceived loss of control over the hazard event and the inability to undertake adjustments tend to result in anxiety and emotional disturbances.

Technological disasters, unlike natural disasters, pose different risk situations and present coping problems that differ from most natural disasters. Attempts to reduce exposure from technological hazards are often difficult because of the pervasiveness of chemical substances and the inability to establish a safety threshold for exposure to carcinogens. There is, furthermore, little experience in defining such threats, forecasting their occurrence and magnitude, and predicting the chronic health consequences. Because of these characteristics, managerial responses intended to produce 'blocking' controls or to reduce stress have been ineffective in gaining public acceptance at Three Mile Island (Holt, 1982).

THE 'THERAPEUTIC' COMMUNITY:
Natural and technological disasters

The emergence of a 'therapeutic' post-disaster community – the spontaneous altruism by community victims and non-victims in aid and sheltering activities – has been suggested as one powerful means of reducing disaster-related stress (Tierney, 1981). The therapeutic community emerges in the immediate post-disaster period because it replaces institutional and organizational resources that are limited or inoperative (Perry, 1985). The degree of success and the extent of community reliance upon such social innovation require further research. The therapeutic community has not prevented the most vulnerable groups from suffering proportionately the greatest hardships in recovery; familial stress due to uncertainties prevalent in the early post-impact period; and inequities among social groups in later phases of reconstruction (Bolin, 1982).

Why does the therapeutic community emerge? In areas struck by natural disasters, populations characteristically are differentially affected. Yet, since natural disasters are often perceived as acts of God, victims and non-victims share a commonality. Loss is directly observable and thus therapeutic response can be focused, defined and targeted. Additionally, damage caused by the disaster has immediate and, perhaps, long-term effects on the community's economy and social structure. Erikson (1976) suggests that euphoria often exists in the immediate post-disaster environment, which is due to the realization that the community has survived and to a sense of strength in rebuilding. Except for droughts, the onset of most natural disasters occurs relatively quickly and tends to be short-lived. Therefore, energies are focused on recovery; mutual aid systems enhance the rebuilding process.

By contrast with natural disasters, technological disasters, particularly those characterized by prolonged exposure to harmful chemicals, result in attention to activities that will accelerate the decision for permanent relocation and a heightened sense that the community has been destroyed and cannot be rebuilt. The therapeutic community will not emerge when the number of victims exceeds the number of non-victims in the stricken area. For technological disasters, that is typically the case. Additionally, because the problem is caused by humans and the degree of victimization and health consequences are uncertain, the loss is not physically observable, and normal support patterns of care and shelter are not appropriate. Thus, traditional patterns of social support may not develop in technological disasters.

Aid, rescue and rebuilding are familiar activities by which communities recover from natural disasters. In such cases, the public recognizes and responds to emotional disruption and physical recovery. Chemical disasters, by contrast, are a new societal phenomenon, and because of the technical nature of the threat, substantial dependence rests on scientific and regulatory institutions, rather than on individual family or community efforts. Recovery efforts from natural disasters have revealed some problems in relationships between victims and disaster aid/recovery agencies due to formal agency rules and structures, institutional insensitivity and delay. In technological disasters, the few situations for which data are available suggest that victim–institutional relationships are often in conflict. At Love Canal, recommendations and actions by the Environment Protection Agency awaited a number of scientific assessments of risks. The apparent slow response by regulators, the uncertainty of scientific assessments and the forced reliance on the media for information fostered deep resentments among victims. If exposure poses risk to only part of a community and there are scientific uncertainties about the nature of the hazard, then the potential adverse impacts on the economy of the area, prolonged debate, political activity among victims and intense media attention may induce substantial community conflict. Research on technological disaster reveals that the non-victimized community may develop sharp resentment against the disaster victims (Baum et al, 1983).

The sustained threat from chemical hazards, the inappropriateness of traditional coping mechanisms, the dependence of victims upon governmental regulatory actions to ameliorate the hazard, and the tendency to blame individuals or institutions as a cause for the suffering have stimulated intense political activity on the part of victims of technological disaster. Political conflict, therefore, should be expected in technological disasters. Whereas such political participation may provide some form of emotional support, the awareness of threat may, nonetheless, increase over time. The inability of the individual to control the evolving situation may be expected to result in continued or even increased levels of distress (Holt, 1982). Sorensen and colleagues argue that a number of factors promote harmony during natural hazard events: the conception of the external threat as an act of God; the ability to identify and understand the threat and its impacts; the community identification with loss; and the tendency for there to be community consensus on disaster-related problems and solutions (Sorensen et al, 1983). In technological disasters, a number of factors promote disharmony: exposure to contaminants is pervasive; environmental clean-up measures are often only temporary solutions; and relocation becomes a substitute for recovery.

CONCLUSIONS

From this overview and analysis of technological hazards and major hazard events, we reach a number of conclusions:

- The major burden of hazard management in developed societies has shifted from risk associated with natural processes to risk arising from technological development and application.
- Technological hazards pose different, and often more difficult, management problems than do natural hazards. Contributing factors to this greater difficulty are the unfamiliarity and newness of technological hazards; the lack of accumulated experience with control or coping measures; the less understood hazard chains; the broader opportunities for control intervention; the perceived amenability of technological hazards to fixes; and the simultaneous need to enlarge benefits and reduce risks in judging the tolerability of technological hazards and instituting control strategies.
- Hazards may be conceived as comprising a series of linked stages, beginning with human needs and ending in adverse consequences. Each stage presents an opportunity for managerial intervention to block or control the emergence of the hazard.
- In idealized form, hazard management consists of a sequence of four major activities – assessment, control analysis, selection of management strategy, and implementation and evaluation. Two major functions – intelligence and control – are involved and both necessitate normative, as well as empirical, judgements.
- Technological disasters tend to elicit a different pattern of public response than do natural disasters. Whereas publics tend to be reluctant to evacuate in natural disasters, evacuation from technological disasters tends to exceed official expectations. Factors contributing to this difference are the lack of familiarity and greater perception of threat associated with the latter. Technological disasters, unlike natural disasters, result in a greater reliance upon governmental authorities and a reduced use of community and family social networks.
- Although the knowledge of long-term impacts arising from technological disasters is scant, experience following the Three Mile Island accident suggests that although severe distress has been short-lived, other psychological impacts have been persistent. Continual media attention, public perception of risks and threats associated with the restart of the undamaged reactor, and the visible presence of the facility have all contributed to continuing stress.

- The emergence of a therapeutic community to ameliorate effects during the post-disaster period appears substantially less likely for technological than natural disasters. Reasons for this include the lack of a sense of rebuilding, priority to relocation as a means of mitigation, conflict between victims and non-victims, and delayed response by government authorities.

3 Large-scale Nuclear Risk Analysis: Its Impacts and Future

Roger E. Kasperson, James E. Dooley, Bengt Hansson,
Jeanne X. Kasperson, Timothy O'Riordan and
Herbert Paschen

INTRODUCTION

The international emergence of the large-scale nuclear risk study has stimulated both common and divergent responses and societal impacts. The commonalities doubtlessly owe much to the nature of this 'technique' as an innovation in technology assessment and safety assurance institutions and practices. Divergences arise from a number of sources – the emphasis placed upon nuclear power as part of a national energy strategy, the scale and political effectiveness of anti-nuclear sentiment, the degree to which safety regulation is a public or relatively closed exercise, the particular mix of nuclear generation systems adopted, and, above all, the configuration of the particular political culture. It is striking that the large-scale risk study has emerged in those countries (the US, West Germany and Sweden) where public scrutiny of government is extensive, where the law and the constitutional norms encourage citizen use of the courts to test the legitimacy and administrative fairness of regulatory decisions (not applicable to Sweden), and where local government (state, land, commune) plays an important role in licensing the construction of a plant. It also emerged in countries where public (notably, *informed* public) suspicion of nuclear power was growing and where new developments or an acceleration in the existing nuclear programme were in process or proposed.

The large-scale risk study did not, by contrast, develop in those countries (the UK and Canada) where a more consensual (internal) form of accountability prevails, where the indigenous nuclear industry was regarded as competent and able to complete its own safety analyses (drawing only, when necessary, for support from the major US and West

Note: Reprinted from *Nuclear Risk Analysis in Comparative Perspective*, Kasperson, R. E. and Kasperson, J. X. (eds), 'Large Scale Nuclear Risk Analysis: US impacts and failure', Kasperson, R. E., Dooley, J. E., Hansson, B., Kasperson, J. X., O'Riordan, T. and Paschen, H., pp219–236, © (1987), with permission from Taylor and Francis

German studies), and where public opposition, whether directly through political channels or indirectly through the courts, was subdued, obstructed or simply not allowed to voice criticisms. Again, Sweden occupies a middle position in this spectrum of adoption.

Risk analysis, used broadly, refers to a broad miscellany of techniques and analytic methods. Probabilistic risk analysis is but one form of the more generic activity, and the plant-specific analysis is a recent form of this type of analysis. The comprehensive risk study addresses an ambitious scale of broad issues, usually intended to characterize the risks of a technology or programme. In this sense, it functions rather like a generic environmental impact statement. These various species of risk analysis coexist in considerable tension with one another.

The purposes of generic risk analysis have been several – to increase knowledge about risk; to inform licensing and regulation; to improve plant design, construction and management; and to provide a base for energy policy decisions. The purpose of the *comprehensive risk study*, by contrast, has been largely political – it has been initiated to solve or serve as background for overtly political questions and its emergence, therefore, shows a clear connection to the general political 'style' in the countries studied. It is therefore valuable to summarize the effects of the risk studies on society, at large, and on public attitude toward nuclear power, in particular. These impacts will include those that are *inward looking* or on the *private dimension* of the risk studies – effects on the nuclear industry and institutional complex – as well as those that are *outward looking* or on the *public dimension* – effects on national politicians, the scientific community, environmental groups, the media and the general public.

STANDARD-SETTING, SAFETY IMPROVEMENTS AND PLANT DESIGN

Large-scale risk studies have deeply and extensively influenced safety work in nuclear power. Their methods penetrate safety thinking. They have substantially affected, and improved, the design of the plants planned and built during the latter part of the 1970s and the early 1980s. In addition, extensive changes have been retrofitted to plants already in operation. One major example is the filtered vent containment (to control iodine release) to be built at the Swedish Bärseback plant, which was a direct result of the 'Swedish reactor safety study' (Hansson, 1987). Another is the introduction of automatic control devices for emergency cooling as a result of the 'German risk study' (Paschen et al, 1987). Still another is the improvement of control room instrumentation called for in the Kemeny report (Kasperson, 1987). More important than even such conspicuous safety improvements is the extensive penetration into safety philosophy and approach of the methodology of the probabilistic risk analysis (PRA). Thus, safety questions have for the first time come to be regarded as an integral part of nuclear engineering, with integrated assessment of system design, training of operators, and formulation of administrative rules and guidance.

The significance of risk studies for nuclear plant design is quite apparent in the Federal Republic of Germany. Given the finding in the 'German risk study' of the important contribution of operator error to serious accidents, improvements in plants have centred upon the installation of automatic control devices for cooling down the plant at Biblis B. Similarly, results pointing to stuck-open pressurizer relief valves have led to design and monitoring improvements designed to reduce significantly the probability of this accident sequence.

Common lessons from risk studies are apparent in all five countries. Safety managers are giving renewed attention to quality control, particularly with respect to pressure vessels, pipes, valves, pumps and motors. The importance of designing control rooms and operator panels in order to aid operator understanding of what is going on inside the reactor, thereby reducing the risk of operator error and increasing the probability that the operator will be able to handle situations that have not been explicitly covered or foreseen in the instruction, is widely recognized. Whereas the need for such measures was, of course, quite obvious after the Three Mile Island accident, a subcommittee to the Swedish Energy Commission (EK-A, 1978) and the Lewis Commission (US Risk Assessment Review Group, 1978) in the US had earlier emphasized these issues.

A recent review of plant safety enhancements attributable to plant-specific probabilistic risk analysis found no shortage of examples, including:

- six design changes and one procedural change associated with the Big Rock Point risk study;
- modification of the design of viewing windows on containment hatches so that their ultimate strength matched those of other structures in the containment;
- the preference of a diesel-driven containment spray pump modified to be independent of AC power over other proposed design changes at the Zion plant;
- modifications to increase the gap between the control room building and an adjoining structure and the installation of rubber bumpers to reduce the potential for loss of the control room as the result of an earthquake at Indian Point 2;
- procedural changes to reduce the probability of a core melt at Arkansas Nuclear 1 (Joksimovich, 1984, pp263–264).

Perhaps most significantly, the risk studies have deeply influenced general attitudes toward safety questions. First is the recognition, particularly in the aftermath of the 1986 disaster at Chernobyl, that the unthinkable can happen – namely, that serious accidents involving releases are possible (albeit at very low probabilities). This is a healthy realization, for it directs energy to the constant need to assess and reduce further highly unlikely combinations of events and accident sequences. Second, it is now understood that the combination of several commonplace failures can be as, or even more, important than the risk of large pipe breaks or leaks. Third, whereas safety philosophy previously focused almost exclusively on the prevention of major accidents, more attention now goes to mitigating the consequences. Part of this has produced new efforts on emergency preparedness and evacuation. These efforts, in turn, have led to new arguments about public communication and evacuation procedures, to the question of siting in proximity to population centres, to consideration of secondary containment, and to the merits and demerits of providing potassium iodide tablets. All of these issues will receive heightened attention over the next several years as a result of the Chernobyl accident.

Within these commonalities lie divergences. Many industrial practitioners continue to believe that the primary stimulus for specific safety improvements comes from feedback from operational experience. Those close to policy tend to be advocates of the large-scale risk studies. Those in hands-on positions, however, like to stress the importance of assiduous work with small improvements, based on collected operational experience, and often view the comprehensive study recommendations as theoretical or somewhat out of touch with reality. The tension between these different viewpoints, also apparent among the regulators, appears to decrease as risk studies become more detailed and more site specific, although even there a tension remains between the contributions of plant-level versus sub-plant-level assessments, between the engineering and the conceptual approach, and between the divergent viewpoints of licensees and regulators.

NUCLEAR INSTITUTIONS

The most apparent outward indication of new safety emphasis in nuclear power in the five countries was the considerable internal reorganization of nuclear institutions that occurred over the past eight or so years.

In Sweden, for example, the State Radiation Protection Institute and the State Nuclear Power Inspectorate received increased funds, the State Power Board set up a permanent reactor safety section, and two new bodies – the Council for Nuclear Safety (established by industry) and the Board for the Handling of Spent Nuclear Fuels (a state organization) – were created.

In the US, despite recommendations by the Kemeny Commission for a radical restructuring of the Nuclear Regulatory Commission (NRC) and a strengthening of the Advisory Committee on Reactor Safeguards, no major organizational changes took place in the regulatory bodies or other federal departments and agencies. The one new public institution – the Nuclear Safety Oversight Committee – lived a short life before its termination by the Reagan administration. The NRC, to be sure, established a new Office for Analysis and Evaluation of Operational Data, aimed at learning from past reactor incidents and malfunctions. In addition, its Office of Inspection and Enforcement was reorganized and strengthened (to a staff of 1000), with resident inspectors established at each plant site (Phung, 1984, p10).

By contrast, the private sphere reacted much more swiftly and extensively to the Three Mile Island accident. Within a month, the US nuclear industry had established the Nuclear Safety Analysis Center to conduct technical studies of the accident, to interpret the lessons to be learned, to develop strategies to prevent future accidents of that kind, and to consider generic safety issues. The industry also established an Institute of Nuclear Power Operations (INPO) to look initially at operator training and capabilities on a plant-by-plant basis. The emergence of INPO as a major new force in American nuclear institutional structure may well alter the traditional regulator–regulatee relationship, as indicated by the role that highly critical INPO reviews placed in the closing of nuclear plants in the Tennessee Valley Authority System.

The UK, Canada and West Germany show a slightly different picture. In the UK, the House of Commons Select Committee on Energy made a number of recommendations dealing with both the supervision of safety and the institutional organization of safety. The committee sought a technically strengthened Nuclear Installations Inspectorate, a safety monitoring group in the Department of Energy, and a more effective organization of safety responsibility between the National Nuclear Corporation (NNC) and the Central Electricity Generating Board (CEGB), with the latter becoming less involved in the specifics of plant design. But the government reply indicated that it was broadly satisfied with existing arrangements, though it did add a specialist technical expert on nuclear matters to the staff of the government chief scientist in the Department of Energy.

Subsequently, however, the government agreed to a reorganization of the project management arrangement for the design and construction of pressurized-water reactors (PWRs). Enter the Project Management Board, a separate organization within the CEGB but incorporating the NNC. In addition, Westinghouse and the NNC each own half of a PWR plant project, responsible for designing and constructing the nuclear island of all future PWRS in the UK. The project will coordinate export as well as domestic contracts. These changes strengthen the CEGB's hold on the design and construction of nuclear plants and lock in an international component for the British PWR.

In Canada, the Porter Commission in Ontario suggested a number of organizational changes; but since the commission was provincially created and advisory, its policy-related recommendations produced little institutional response at the federal level or by the nuclear industry generally (although the safety division at Ontario Hydro was upgraded).

This lack of response probably reflects the secure position of nuclear authority, as well as the distribution of political power in Canada. In the Federal Republic of Germany, although substantial societal conflict over nuclear power continues, the 'German risk study' elicited little executive or parliamentary response and no major institutional initiatives.

To the extent that a generalization is possible, it is that industry has probably undergone the greatest institutional change. Beyond that, it is apparent that a constellation of political forces is required to overcome inertia, built-in basic institutional structures of management and existing allocations of political stakes. The Three Mile Island accident in the US and the Swedish referendum were key events, not replicated in the other three countries, which permitted the development of a momentum for change. It is too early to assess the impacts of a major event outside these countries – the Chernobyl accident – upon the array of contending forces within the five countries studied.

IMPACTS OF THE SCIENTIFIC COMMUNITY

Since only the Swedish case addresses in depth the response of the scientific community to the major nuclear risk studies, the evidence on this subject draws on that case, with more limited indications and perspectives from the UK, the US, Canada and West Germany. Generally, the studies have not, except perhaps for the two Kärn-Bränsle-Säkerhet (KBS) (1977, 1979) studies, contributed substantially to new scientific knowledge. In the scientific flow of information, the scientists have generally been the donors, not the recipients.

Nevertheless, the risk studies have wrought a rather profound effect on the engineering sciences. This impact manifests itself mainly through more frequent contacts, both among different research areas and in relation to other groups in society, through new ways of posing problems in the individual sciences and through the emergence of new areas of research (often on the border of two old ones). Meanwhile, there has also been a stir in the scientific community: people have been forced to reconsider their work and positions, and to become more versatile and more open in their ways of thinking. These effects are most pronounced for those who have taken part in the *process* of the risk study, rather than from lessons imparted by the final product.

A notable difference between Sweden and the other countries is that a sizeable percentage of all Swedish scientists in nuclear-related disciplines has participated more or less actively in a major nuclear risk study. Considering the comparatively large number of Swedish risk studies and the small size of the Swedish scientific community, practically all scientific voices have had a chance to be heard. By and large, the studies have tapped what Swedish science has to offer on the applied side. The rub is that many scientists have been so busy that fundamental research in some areas may have suffered. In the other countries, where the scientific communities are so large that the risk studies have not to any discernible degree increased competition for the services of qualified scientists, these effects are certainly less manifest. In such settings it should be easier to find scientists for peer review who have not previously been involved in that particular study.

To be sure, such scientists are in scarce supply in Sweden. Given the vigour and high quality of Swedish science, this hegemonistic situation – which also exists in the UK – carries an alert for small countries moving to large-scale risk analysis in safety work and national policy making. Interestingly, the Swedish scientific community harbours a certain disquiet over the quality of the risk studies. The unease turns on suspicion that the competence of the scientists could have been better used, both in the assessment itself and in the dissemination of risk information to politicians and the public. At the same time, scientists are uncertain, often perplexed, about how to remedy the situation. Across countries, scientists are disenchanted over the performance of the media and the apparent inability of the public to deal with risks rationally (that is, as scientists do!). It may well be that PRAs and major risk studies serve to increase the scope of public dissatisfaction and lack of understanding.

IMPACTS ON POLITICIANS AND POLICY MAKERS

The success of risk studies in the public arena is less apparent – and certainly much more uneven – than that within the internal nuclear complex. Yet it is probably the politicians who have been most impressed and reassured by the risk studies. This is not to say that anti-nuclear politicians do not abound, only that risk studies appear to have done more to assuage or reassure than to spark official opposition. Basically, and not surprisingly, the impressive amount of work and technical resources put into the risk studies has, on balance, reached conclusions generally favourable for nuclear power. Comparable risk studies of other industries also serve the nuclear option. And nuclear fares extremely well in comparative, albeit controversial, risk analyses of competing energy technologies (see, for example, Inhaber, 1982). If politicians have not been convinced that nuclear power is desirable, at least they have not concluded that reactor safety is an intractable problem.

The US appears to be an exception to this generalization. There the risk study results appear to have disturbed politicians rather than to reassure them. This is not to suggest that US politicians have a more critical attitude toward nuclear power than politicians in other countries; rather, it may be that the original vision in the US involved greater illusion, or at least commitment, than in other countries. The dismantling of the American nuclear dream owes more to the overextension of the technology, the inaccuracy of utility economic projections, and the presence of a committed and vigilant opposition than to the results of nuclear risk studies.

Indeed, it seems doubtful that the substantive work and findings of the large-scale risk studies, despite their imposing bulk and prestige, have been major political determinants anywhere. In the Federal Republic of Germany, parliament largely ignored the 'German risk study'. In Canada, the Porter Commission report met a similar fate. In the US, political debates over the 'Reactor safety study' (USNUREG, 1975) the Lewis report (US Risk Assessment Review Group, 1978), and the Kemeny Commission report (US President's Commission on the Accident at Three Mile Island, 1979) were often acrimonious and generally inconclusive. In all three countries, the summaries of studies largely dominated the general picture of their results among influential politicians.

Not so in Sweden, the smallest country, where neither the studies *nor* their written summaries were influential among national politicians. Rather, the important sources of information were verbal contacts with those who performed the studies. Many of the committees that conducted the studies in Sweden were of a parliamentary character, and those that were not had close links with leading parliamentarians. Since all the major parties have assigned the task of looking after energy questions to a trusted member of the inner circles (who then sits on the parliamentary committees and serves as the intermediary between the committees and his fellow party members), the bulk of politicians' risk information comes through a single channel. Since the 'party specialist' filters out much of the information, most members of a political party tend to regard the risk studies in a uniform way, and a wealth of information in the study reports remains untapped and of little political consequence.

A party's political stance on overall energy policy is likely to colour political response to particular risk studies. In the UK, for example, only the ruling Conservative administration is strongly pro-nuclear. Other parties are adopting a variety of pro-coal and pro-conservation positions and placing greater emphasis on renewable energy sources and co-generation schemes, thereby distancing themselves from a nuclear commitment. In this post-Chernobyl era, nuclear power generation is a major electoral issue in the UK. This political jockeying is far more influential in political contexts than the results of particular studies or public inquiries.

IMPACTS ON THE NUCLEAR DEBATE

The debate in the scientific community and within society as a whole is the implicit, but pervasive, issue that underlies the major nuclear risk studies. Whereas risk analysis addresses many goals, it is doubtful that the large-scale, comprehensive risk study would have arisen in the absence of the active controversy during the 1970s over nuclear power. This is, perhaps, most striking in the case of the US *Reactor Safety Study*, which sought to end once and for all the debate over reactor risks. The Swedish risk studies typically aimed to shape a broader societal consensus for energy choices. In Canada, the Ontario Royal Commission on Electric Power Planning came to view its role as a vehicle for public scrutiny and education.

The motivation for the risk study, in this context, is to clarify the risks, to narrow the debate to the axes of 'real' disagreement and to develop, in the parlance of the expert, a more 'rational' approach to risk by policy makers and publics. The results in the five countries show conclusively that the risk studies have enlarged, not narrowed, the debate. Instead of shaping an accepted discourse for the discussion of risk, they have intensified the conflict over the proper range of consequences to be treated in the assessment and whose values should prevail in their weighting. Instead of greater acceptance of the bases upon which numerical calculations rest, they have heightened suspicions over the motivations for the studies and those who conduct them.

This conflict is directly apparent in the concept of risk itself, as discussed in the Canadian context. For the practitioner and the professional risk assessor in the nuclear arena, risk is typically the estimated probability of expected consequences, nearly always restricted to radiation-induced fatalities and other health effects. The means for calculating the risks are formal analytic procedures relating to events and faults that become accidents. For many outside this professional community, risk means something quite different: specifically, it refers to a broad array of technology impacts upon society as well as health, including qualitative attributes of the risks involved and the decision processes that produced the risk in the first place. The narrower treatment of risk excludes many of the broader concerns that are crucial for the non-professionals. This divergence has been manifest in risk perception research, which has emphasized the differences between expert and lay assessments of risk (Slovic et al, 1985). So the first-order political conflict over risk assessment, characteristically fought out in debates over the study methods and results, concerns the conception of risk that should guide public policy and management efforts.

The comprehensive risk study typically has something for everyone. Differing groups and perspectives invariably find evidence to support their claims, as was evident in the different messages drawn in the adversarial press to the Kemeny report or the contrasting interpretation of the major Swedish risk studies in the Swedish mass media. Whereas individual risk areas rarely are 'settled' or are dropped off the public agenda by the risk study, new risk issues often appear. Thus, the characteristic net effect is to broaden the arena of debate.

A second way by which risk studies spawn new issues or intensify the debate over existing questions is that they often lay bare problems in risk management that previously have been contained within the authority structure and hidden from the public view. This is particularly the case in political cultures where the public decision process is opaque. The Sizewell Inquiry has permitted a public view of the nuclear risk assessment and management process that is quite remarkable in the UK (O'Riordan, 1984). Similarly, the Ontario Royal Commission on Electric Power Planning provided extensive information on the workings of the major governmental institutions and Ontario Hydro. But even in more open governmental systems, the revelations of management deficiencies and problems can add to the debate and revitalize dormant issues (as happened with the Kemeny report).

RESPONSES OF THE MEDIA AND THE GENERAL PUBLIC

The response of the mass media shows considerable variation across countries. On the whole, the press appears to have treated the risk studies with reasonable accuracy. Whereas factual mistakes were not unusual (and a source of much concern to scientists), almost all were obvious misunderstandings by the journalist of complex technical matters and did not distort the overall picture. The publication of the risk studies, as opposed to risk events, was treated as ordinary news items. In Sweden, detailed studies of media response indicate that most newspapers relied upon the material delivered by the national news agency, which in turn relied heavily upon summaries by the risk study committees themselves. West German newspapers relied heavily upon the press conference materials summarizing the 'German risk study' and paid little attention to the report itself. In Sweden, when the newspapers used their own correspondents, their reports were, nonetheless, remarkably similar. The striking uniformity of news reporting in the Swedish press finds parallels in the US and West German experiences.

Despite the general factual accuracy of information, the press was not always impartial. Rather, there was often selective reference to risk reports or selective choice of risk topics to be addressed. A typical case was *Dagens Nyheter*, a major Swedish morning newspaper with an editorial policy pronouncedly anti-nuclear, which referred to a number of otherwise largely ignored sources, most of which were critical of nuclear power.

Several other characteristics of press coverage of the reports are notable. As was quite evident with the Kemeny report, news coverage often tended to dwell on the process that created the report, on pre-report 'leaks' of upcoming issues and recommendations, and on the likely political ramifications of the results. Such aspects received more attention than the substantive risk issues in the studies. Once the report appeared, there was typically little sustained attention to the results. Furthermore, experience from Sweden, West Germany and the US points to the tendency to stress certain risk issues in the reports and to ignore others, even though the neglected items often appear no less, or even more, important. In the UK, a generally anti-nuclear and anti-industry bias in journalists has resulted in a corresponding selection bias in most media coverage.

Television shows several notable differences from the press. Typically, because of characteristics of the medium, perhaps, the coverage was considerably more narrow and superficial. In Sweden, this included an emphasis on spectacular scenarios and very few references to the actual risk studies. In West Germany, television news reporting appeared to be biased against nuclear power, whereas British television viewers received most of their information from impartial documentaries rather than from news reporting.

It may be concluded, then, that media coverage was generally balanced but uneven, factually accurate, but in many cases unsophisticated and superficial in treatment. Unfortunately, the media tended to regard the risk reports as isolated news items, with little context as to how society grapples over time with the issues of nuclear safety, and with a lack of sustained follow-through on the eventual results wrought by the study findings.

As to the impact of risk studies on the general public, the general conclusion shows little evidence of any significant direct effect, either on levels of knowledge or attitudes. This is not surprising insofar as reports in the media only touch upon the detailed findings of the studies, and attitudes to nuclear power involve basic value, as well as factual, considerations. Furthermore, since various viewpoints readily draw supporting messages from the studies, the general public probably does not receive sharply delineated messages. In short, the risk study has done little to draw the public into the fray. If there are effects, they are probably primarily indirect and cumulative in nature. As suggested in a study on West Germany (Kasperson and Kasperson, 1987b, Chapter 4), the risk study undisputedly becomes a datum in the continuing public discussion, which has to be considered in all future deliberations of nuclear power. To the extent that the risk studies over time have a cumulative impact on the ability of the public to deal with probability and uncertainty, to place risks in broader context, or to engage in comparisons among diverse risks, then a certain shift or evolution in the centre of gravity of public response may occur.

CURRENT DEVELOPMENTS AND FUTURE PROSPECTS

Risk or reliability analysis is still very much in the heyday of its development, most notably in nuclear power risk management, but present in other areas (e.g. chemical plants, oil drilling rigs and aircraft safety) as well. Considering that risk analysis techniques were only modestly employed during the 1970s following the pioneering work of Farmer (1967) in the UK, the Swedish risk assessments and the US 'Reactor safety study', the growth since the Three Mile Island accident in

1979 has been quite remarkable. In the US alone, where the embrace of these techniques has been greatest, some 13 'full-scope' (level 3)[1] probabilistic risk assessments (see Table 3.1) and at least nine level 1 or level 2 assessments had been completed by 1984.

The large-scale (i.e. level 3) studies have been sponsored by the utilities of the Electric Power Research Institute (EPRI), a utility research organization. The motivations for the studies have been diverse: several were initiated in response to regulatory requests (e.g. concern over proximity to metropolitan areas), one (Big Rock Point) was viewed as a means of evaluating the cost effectiveness of risk reduction measures proposed by the US Nuclear Regulatory Commission, and most of the remaining were part of long-range utility risk management programmes (USNUREG, 1984, pp58–59). In addition to these assessments, four other major comprehensive nuclear risk studies – the 'German risk study', the US Department of Energy risk assessment for a high-temperature gas-cooled reactor, an assessment of the Clinch River Breeder Reactor in the US, and the Sizewell B study – have been completed. Other major studies are now under way, including the broad-scale assessment of Ontario Hydro in Canada, as well as a host of smaller nuclear risk assessments in Italy, Japan, Spain, Switzerland and Taiwan, and the continuing sub-generic studies that characterize the risk analysis effort in the UK.

Interestingly, this growth has emphasized research in the private dimension of risk. With the exception of the Sizewell B Inquiry, the political motivations for the studies have receded into the background. The Sizewell B Inquiry was very unusual in that the inspector took it upon himself to assess safety management by interpreting the terms of reference very broadly and by following up on specific issues raised in the proof of evidence of the CEGB.

When placed in the context of the overall effort in probabilistic risk analysis over the past five years, it is evident that the many assessments have enhanced our understanding of nuclear power plant risks and are beginning to contribute to the regulatory process. Notable contributions include the following:

- The estimated frequency of core melts is generally higher and also covers a broader range of core-damage/frequency point estimates (about 10^{-6} to 10^{-3} per year) than had been previously believed possible (see Table 3.2). The small fraction of accidents that might result in off-site consequences generally involves either an early failure of containment or a containment bypass.
- Estimated risks of early fatalities and injuries are highly sensitive to source-term magnitudes and the timing of releases and emergency responses.
- Accidents beyond the design base (including those initiated by earthquakes) are the principal contributors to public risk.

Table 3.1 *Completed full-scope (level 3) US probabilistic risk analyses (as of 1984)*

Plant	Issuance	Operating licence	Rating (MWe)	NSSS/AE[a]	Containment	Sponsor	Report
Surry 1	1975	1972	788	W/S&W	Dry cylinder	NRC	NUREG-75/014 (WASH-1400)
Peach Bottom 2	1975	1973	1065	GE/Bechtel	Mark I	NRC	NUREG-75/014 (WASH-1400)
Big Rock Point	1981	1962	71	GE/Bechtel	Dry sphere	utility	USNRC docket 50-155
Zion 1 & 2	1981	1973	1040	W/S&L	Dry cylinder	utility	USNRC dockets 50-295 and 50-304
Indian Point 2 & 3	1982	1973	873	W/UE&C	Dry cylinder	utility	USNRC dockets 50-247 and 50-286
Yankee Rowe	1982	1960	175	W/S&W	Dry sphere	utility	USNRC docket 50-29
Limerick 1 & 2	1983	–	1055	GE/Bechtel	Mark II	utility	USNRC dockets 50-352 and 50-353
Shoreham	1983	–	819	GE/S&W	Mark II	utility	USNRC docket 50-322
Millstone 3	1983	1986	1150	W/S&W	Dry cylinder	utility	controlled document
Susquehanna 1	1983	1982	1050	GE/Bechtel	Mark II	utility	draft
Oconee 3	1983	1974	860	B&W/Duke	Dry cylinder	EPRI/NSAC	draft
Seabrook	1984	–	1150	W/UE&C	Dry cylinder	utility	draft

Note: a NSSS, nuclear steam system supplier; AE, architect-engineer

Source: USNUREG, 1984, p58

Table 3.2 *Probabilistic risk analysis (PRA) study results: Estimated core-melt frequencies (per reactor year)*[a]

Plant	Rating in megawatts (MWe)	Type, nuclear steam system supplier (NSSS)	Core-melt probability (years)[b]
Arkansas 1	836	PWR, B&W	5×10^{-5}
Biblis B	1240	PWR, KWU	4×10^{-5}
Big Rock Point	71	BWR, GE	1×10^{-3}
Browns Ferry 1	1065	BWR, GE	2×10^{-4}
Calvert Cliffs 1	845	PWR, CE	2×10^{-3}
Crystal River 3	797	PWR, B&W	4×10^{-4}
Grand Gulf 1	1250	BWR, GE	2.9×10^{-5}
Indian Point 2	873	PWR, W	4×10^{-4}
Indian Point 3	965	PWR, W	9×10^{-5}
Limerick	1055	BWR, GE	3×10^{-5}
Millstone 1	652	BWR, GE	3×10^{-4}
Millstone 3	1150	PWR, W	1×10^{-4}
Oconee 3	860	PWR, B&W	8×10^{-5}
Peach Bottom 2	1065	BWR, GE	8.2×10^{-6}
Ringhals 2	800	PWR, W	4×10^{-6}
Seabrook	1150	PWR, W	2×10^{-4}
Sequoyah 1	1148	PWR, W	1.0×10^{-4}
Shoreham	819	BWR, GE	4×10^{-5}
Sizewell B	1200	PWR, W	1×10^{-6}
Surry 1	788	PWR, W	2.6×10^{5}
Yankee Rowe	175	PWR, W	2×10^{-6}
Zion	1040	PWR, W	1.5×10^{-4}

Note: a Includes external event contribution, where appropriate. Comparisons of values listed should be made with extreme caution. Different models, assumptions and degree of sophistication were employed.

b Median values; otherwise point estimates are listed.

Source: USNUREG (1985, 1987)

- Human interactions are extremely important contributors to the safety and reliability of plants.
- Small loss-of-coolant accidents and transients are dominant contributors to core-melt frequency.
- The loss of off-site power is a key risk event.
- Certain common-mode failures – such as earthquakes, internal fires and floods – appear to play an important role in risk, but may be highly plant specific (Budnitz, 1984; Garrick, 1984; USNUREG, 1984, pp63–65).

Beyond these specific findings, it has become clear that probabilistic risk analysis has a number of values and uses in nuclear risk management. First, plant risk assessments can become 'living' mathematical representations of the system as it evolves over time, with constant updating from operating experience and integration of modifications. Such a 'living' system has potential for overcoming the traditional split between safety assurance and the operations/reliability segments in the organizational structures of utilities and reactor vendors. Second, because of its integrated nature, probabilistic risk analysis can be very helpful in assigning priorities to either generic or plant-level safety issues, or in allocating resources among various safety or regulatory needs. Much the same use is possible in allocating resources for inspection and enforcement. Third, risk assessment may be used to guide decision making: where should regulations or standards be strengthened or relaxed; where is there confirmation for existing approaches; where are there gaps in the regulatory regime? Finally, the assessments at the plant level can be very helpful in identifying design or operational deficiencies. Despite these virtues and despite the continuing evolution of risk assessment methodology, significant uncertainties and limitations remain. It is difficult to know whether a given risk assessment underestimates or overestimates risk, so that such studies need to be used with great caution and a healthy dose of scepticism. The US General Accounting Office (GAO), in a review of probabilistic risk assessment, concluded that these methods entail four sources of high uncertainty:

1   *Completeness of analysis.* It is patently impossible to ensure the identification of all events and combinations of events that could initiate or direct the course of an accident. Logic diagrams or initiating-event/mitigating systems analysis may not succeed in identifying events that are outside of historical operating experience.
2   *Sufficiency and reliability of data.* Lack of historical experience or lack of understanding usually renders scarce the data requisite to quantify the systems analysis. In the absence of historical data, subjective judgement enters into the selection of appropriate data, thereby introducing additional uncertainty.
3   *Assumptions made by study analysts.* Inevitably, the analyst must make assumptions for purposes of simplification or to fill gaps in data or understanding. One basic assumption attendant on nearly all assessments is that the nuclear plant was actually built according to the design specifications. Invalid assumptions increase uncertainties.
4   *Relationship of computer models to reality.* Risk assessment invariably relies upon abstract models to portray plant systems, accident phenomena within the containment building, human interactions and accident consequences. Such models cannot be subject to strict validation in real-life experience. Moreover, the extent of conservatism – which

impinges upon both design and theory – introduced by the analyst often defies quantification (USGAO, 1985, pp16–17).

Research programmes are under way in various countries to reduce the uncertainties in these (and other) areas. But the uncertainties cannot be completely eliminated; indeed, they may remain large because they are intrinsic to the methods themselves and to inherent limits on the data that can be made available.

Another broad limitation is that such risk studies cannot answer some of the questions of greatest concern to the policy makers and the public. Risk assessment, viewed narrowly, has little to say about the level of risk that should prevail, the so-called 'acceptable risk' problem. Thus, although the mobilization of various results and a comparative perspective may enhance that effort, the assessments themselves provide no intrinsic guidance. Furthermore, the assessments cannot provide assurance that safety goals are met. In the face of some temptation to judge the overall safety achieved by the results of a plant-specific risk assessment, then, it is important to remember that significant model components and the overall system model are unvalidated and that such use is sure to encourage mischief in the assessment process.

The rapid development of risk analysis over the past five years has not yet stimulated substantial discussion of the social impacts of this new technique upon the institutions and people who use it or upon the society in which it is embedded. Yet, as illustrated by the cases of programme planning budget review and systems analysis during the 1960s, such methodologies can disburse far-reaching impacts. Environmental impact statements have also certainly altered the planning and decision processes of federal (and often state) agencies, as well as the roles of citizen groups and members of the public. Probabilistic risk analysis is considerably more complex and resource-demanding than other methodologies that have been introduced. Somehow (no easy feat!), it must be incorporated within the institutions responsible for managing nuclear safety. The institutional and social impacts associated with this integration need to be anticipated and, where adverse, perhaps ameliorated. Numerous significant social impacts seem likely.

Whereas risk analysis carries the potential for clarifying major trade-offs and safety decisions, as well as for placing a particular plant or risk issue in broader context, few groups or individuals outside the technical community now boast or are likely in the near future to possess the requisite expertise and resources to penetrate a large-scale nuclear risk assessment. Widespread use of such assessments in licensing and regulation could have the net effect of greatly narrowing the range of outside scrutiny, review and public participation in a technology that has drawn fire for the centralization of its decision making and the limited opportunities for citizen participation in licensing, rule making and management.

The general problem could find specific representation in several problems:

- In societies with more open and participatory political cultures, large-scale probabilistic risk analyses are likely to produce increased difficulty for potential interveners to participate in formal licensing and regulatory procedures. In addition to existing obstacles associated with the need for high technical expertise and sizeable financial resources, additional capability is required to penetrate and review the methodology, analysis and documentation of a massive nuclear risk assessment. Such independent review is difficult even for governmental regulators who have access to extensive expertise and substantial resources. It is noteworthy, for example, that the US Nuclear Regulatory Commission had, by January 1984, reviewed four major full-scope risk assessments at a per-study cost of US$200,000–$600,000 and a time commitment of 9 to 18 months (USGAO, 1985, p74). Such review demands may exceed the resources available to regulators in small countries and are certainly beyond the capability of nearly all potential critics outside the industry and its regulators.
- Members of the general public should be less directly affected since they usually already lack the capability to understand highly technical issues. But the licensing of nuclear facilities and the management of nuclear safety will likely become even more inscrutable.
- Communication of risk and safety issues to public officials, the mass media and members of the public, already difficult, will face substantially enlarged burdens of explaining probabilistic risk analysis, its contributions to a technical understanding of risk, and why its results should provide added assurance and public confidence.
- As indicated most dramatically by experience in Sweden, large-scale risk assessments consume much of the existing expertise in their creation. This leaves few highly qualified technical experts available for peer review and continuing criticism. Such hegemony could, particularly in smaller countries, contribute to a further closure of nuclear power management to diverse viewpoints and social criticism.
- The long-standing debate between the hands-on school that endorses quality design and process engineering and the proponents of formal analysis (i.e. the probabilistic risk assessors) is unlikely to abate.

Paradoxically, given the means to overcome these problems, risk analyses could serve to demystify and open up the risk management process, particularly as assumptions and calculations necessarily become very precise, the importance of particular failures and accident sequences clearer, and the approaches and expertise of management more transparent. This was very apparent in the experience in Canada and the

UK. What is required is that the interested party acquires enough capability, knowledge and experience to overcome the entry prices or hurdles of involvement.

Beyond these impacts, a number of other potential broad societal and institutional implications merit continuing attention. An increase in the diffusion and use of probabilistic risk assessment carries with it the potential for undue overconfidence in the results of the particular assessment and an erosion in sensitivity to the inherent limitations of the models and techniques. As risk assessment becomes routine, so there may be a loss in quality as implementation becomes standardized and passes to those who are users rather than architects of the methodology. Also lurking is a danger that assessment results may come to exert undue influence in licensing and regulation, leading to inadequate emphasis upon more traditional (and potentially broader) means of safety assurance. This is precisely the issue that produced caution in the UK concerning the uses of probabilistic risk assessment.

The rate of transfer of these techniques to management systems may also be a concern, for incorporation in institutional processes may outstrip the relatively small accumulation of experience. Then the dominance of a small number of firms or groups conducting such assessments in any country may unduly concentrate too much authority, whether in public or private hands. Finally, in the utilities and regulatory agencies responsible for nuclear power, organizational change must be anticipated. Influence and authority are likely to pass to a small group of analysts who have had limited experience with traditional management approaches and who lack hands-on experience with the technology.

It is apparent, then, that large-scale risk analysis has made, and is evolving, important contributions to the understanding of nuclear risks and means for their minimization. The contributions have occurred, and will likely continue to occur, in the private dimension of nuclear power – with the utilities and with regulators charged by society with primary management responsibilities. As vehicles for formulating public policy, for resolving social conflict and for educating the public, their role appears sharply limited. Increased use of these techniques appears to bear strong ties to political culture, so that their role in broad societal management of technology will vary from country to country. Yet, in all cases it is evident that large-scale risk studies can involve broad societal impacts. Continuing vigilance is needed to ensure that such studies realize, rather than drive, societal values for responding to risk.

NOTE

1   The US Nuclear Regulatory Commission distinguishes among the studies according to scope: level 1 includes systems analyses for scenarios leading to core-melt frequency; level 2 also includes containment analyses leading to assessment

of releases; level 3 provides a full assessment of public risks by including off-site consequence analysis (USNUREG, 1984, p13).

Part 2

Corporations and Risk

4 Corporate Management of Health and Safety Hazards: Current Practice and Needed Research

Roger E. Kasperson, Jeanne X. Kasperson,
Christoph Hohenemser and Robert W. Kates

INTRODUCTION

Corporate management of health and safety hazards is *terra incognita*. As with all unknown lands, myths and stereotypes abound that are shaped, in this case, by the extensive chronicling of management failures, vivid risk events and images of corporate behaviour. No road map currently exists to the varied topography of widely differing industrial sectors, long-established and new technologies, large and small firms, and profitable and marginal companies, and we do not provide one here. Rather, we enquire into the characteristics of hazard management programmes in large industrial corporations. The corporations tend to lie at one end of the industrial continuum in terms of size and resources available for hazard management. Several are unquestionably among the leaders of industrial health and safety; one experienced one of the major failures of the 20th century. So the landscape of industrial hazard management as it exists during the 1980s here is complex, featuring both industrial success and failure, opportunity and limitation, and emerging trends in safety as well as past errors.

Note: Reprinted from *Corporate Management of Health and Safety Standards: A Comparison of Current Practice*, Kasperson, R. E., Kasperson, J. X., Hohenemsen, C. and Kates, R. W. with Svenson, O. (eds), 'Corporate management of health and safety hazards: Current practice and needed research', Kasperson, R. E., Kasperson, J. X., Hohenemser, C. and Kates, R. W., pp119–132, © (1988), authors, Perseus Books Group.

CHARACTERISTICS OF CURRENT PRACTICE

A larger hazard burden

Prior to the 1970s, managing technological hazards was never a major undertaking within industry. In a society that was much less environmentally and health conscious, where scientific knowledge of hazards was underdeveloped and with regulatory structures embryonic, employers were primarily geared to the obligations entailed by common law – to communicate to the employee the hazards involved in a work activity and to provide to employees and publics reasonable protection from harm. Acute hazards defined much of what constituted an employer's knowledge of dangers to employees and to the environment. And all too frequently in the marketplace the adage was 'let the buyer beware!'

Not so with today's large corporation. A single modern corporation confronts an oft-bewildering multitude of hazards – to employees, to consumers, to plant neighbours, to the environment – which it must assess and manage. This is no longer a side activity to be run out of the plant manager's back pocket or a minor office in the corporate division of public affairs. PHARMACHEM, a pharmaceuticals company that we have analysed (Kasperson et al, 1988d), for example, combines the occupational hazards of a small-parts assembly plant with the complex hazards involved in chemical storage, synthesis and blending. The production of one PHARMACHEM product involves some 30 chemicals of varying hazard and poses, in a single product, a substantial burden in assessing and managing hazards, including potential accidental release, combustion, absorption, inhalation or ingestion.

Meanwhile, PETROCHEM, a large petrochemical company that we have also studied (Kasperson et al, 1988b) must deal not only with the long-standing risks of accidents at drilling rigs and with fires and explosions involving petroleum products, but with the chronic hazards associated with the 1500 chemicals involved in its chemical divisions and newer less-understood chronic hazards of inhalation of gasoline vapours. Although Volvo (Svenson, 1988) in its early history could concentrate on the hazards posed by automobile collisions and industrial accidents, it must now deal with the more perplexing issues involved with automobile emissions, the use of new materials in the manufacturing of cars and trucks and increasing automation. The industrial process at Rocky Flats uses basic material so toxic that hazard management consumes most of the company's total effort (Kasperson et al, 1988c). Union Carbide found at Bhopal that even an extensive hazard-management programme proved inadequate to prevent disaster with a toxic substance such as methyl isocyanate (see Chapter 5 in this volume).

Yet, whereas hazard management commands increasing resources and a complex organization for many corporations, for many others in industry the picture is quite different. For them – as for the workers at the local gas station, for migrant pickers, for meat-packers or for construction workers – hazard management is still a peripheral and neglected activity. Probably all industries have experienced some upgrading of hazard responsibility over the last several decades; but for many of the large manufacturing corporations the changes have been dramatic. Here hazard management has become a major corporate activity, replete with a degree of professionalization, specialization and demand for expertise formerly confined to the governments of nations and states.

A growing recognition

As the burden of hazard management has escalated, so too has the view of safety, waste reduction and environmental protection. Society's values about environmental protection and health security have changed dramatically over the past several decades. The complex institutional structure of regulatory agencies and environmental groups reflects these changes. It is not surprising that, whether by persuasion or necessity, corporations have evolved a new commitment to health and safety. Since the 1970s, corporate statements and codes of social responsibility have become commonplace in industry. The Harvard Business School has accorded the study of ethical issues a place in the training of business executives. Cradle-to-grave programmes of protection have emerged in the chemical industry. Product stewardship and protection of workers have become more accountable and business managers have increasingly become liable, under both civil and criminal law, for the way in which they exercise these responsibilities. The Occupational Safety and Health Administration (OSHA) drove this point home to the Chrysler Corporation by fining the company US$1.5 million for a series of job safety violations (Holusha, 1987).

It is abundantly clear from the continuing presence of industrial hazards and notable cases of management failure that serious inadequacies persist. It is, for example, very uncertain what effect (if any) codes of social responsibility have had on corporate attitudes to hazard management. But it is also the case that the internal corporate expectation of its responsibilities is not the same in 1988 as it was in 1958 or 1968. Though stated goals generally preceded actual changes in behaviour – and, of course, some firms are innovators in safety, others followers and others highly resistant to change – a growing corporate recognition of its responsibilities for safety and environmental protection across the board is one of the realities of current trends in industrial hazard management.

A revolution in resources

It is scarcely surprising that industry possesses substantial resources for hazard management and that a major development in capabilities has occurred over the past decade. This has been extensive within some sectors of industry (particularly the nuclear power and chemical industries); but it is also highly variable within and among industries.

At PETROCHEM (Kasperson et al, 1988b), for example, the past 15 years have seen the health and safety staff at corporate headquarters grow from a handful to 125, the employment of some 30 industrial hygienists to work on location in the plants, the addition of a toxicology laboratory employing some 30 toxicologists, and the establishment of a formal risk-assessment unit within corporate headquarters. Such well-developed resources also appear to characterize other comparable large chemical and petrochemical companies, such as Dow, DuPont, Rohm and Haas, Union Carbide and Monsanto. The presence of 15 to 20 industrial toxicology laboratories of the size and resources of the US Environmental Protection Agency's most advanced laboratories speaks to an overall industry capability, particularly in light of the fact that a complete toxicology profile for one substance costs around US$2.5 million, takes six years to complete and consumes some 30 to 50 worker years of effort. Rohm and Haas, for example, logged in approximately 600 samples for toxicological evaluation in 1983 (*Chemecology*, 1983).

A 1983 survey of risk activities in chemical companies found that the mean firm surveyed had 84 health and environmental specialists and spent almost 4 per cent of its annual sales on toxicity testing and environmental pollution control (Peat et al, 1983, pp51–74), figures that have almost certainly increased over the past five years. Nor are extensive capabilities restricted to the chemical industry: Northeast Utilities, employing a staff of 15 professionals in probabilistic risk analysis alone, has one of the most advanced risk assessment units in industry.

But these gains are highly uneven, and many corporations lack the needed hazard management capabilities. Thus, MACHINECORP (Kasperson et al, 1988a, pp4–6, 122–123, 125), although a large multinational corporation and one of the Fortune 500, had a corporate headquarters health and safety staff of three, only one of whom was professionally trained. An internal survey of the chemical industry on environmental auditing in 1983 revealed fully 40 per cent of the respondents did not engage in this rather rudimentary (but critical) health and safety practice, suggesting that the auditing failures evident at Bhopal were not an isolated occurrence (*CMA News*, 1983). Other industries engaged in agriculture, mining and construction or industries dominated by service activities have very poorly developed health and safety functions in comparison with chemicals or nuclear power. And even in the chemical industry, cutbacks during the mid 1980s have rendered the hazard management functions to be among the first to go.

Finally, the increase in new capabilities at the industrial trade association level is an important component of the revolution in industry hazard management resources. The Electric Power Research Institute (EPRI) conducts an active programme in risk research, and the Institute of Nuclear

Power Operations (INPO) is a key institution in evaluating the safety programmes of individual utilities and in the training of operators. The Chemical Industry Institute of Toxicology (CIIT), with its US$10 million annual budget and a staff of 106, funded by 31 chemical manufacturers, has resources nearly 20 per cent that of the entire national Toxicology Research and Testing Program, and this is over and above all the toxicology programmes at individual corporations. The Chemical Manufacturers Association (CMA) with its staff of 165, augmented by approximately 2000 representatives from member companies who serve on its committees and panels, has emerged during the 1980s as both a vigorous political lobbyist for the chemical industry and as a major force in chemical risk assessment, as well. With its Community Awareness and Emergency Response (CAER) programme and its US$1 million per annum CHEMTREC programme (a hotline for information on chemicals in transportation accidents), the CMA extends corporate individual resources substantially. Although more pronounced in these industries, similar industry-wide resources in other parts of the private sector are also commonplace. With 15,000 industrial trade associations in existence, the possibility for industry-wide activities in risk assessment and management is substantial. Yet the actual effect of these programmes and resources on industrial hazard management as a whole has yet to be assessed and thus remains unknown.

Determinants of hazard management performance

Ultimately, it is essential to define the key variables that contribute to managerial success or failure. This chapter can but propose relevant categories and indicate possible hypotheses. Our studies suggest two major classes of determinants: exogenous and endogenous.

The *exogenous variables* are perhaps the more apparent. It is clear, for example, that the regulatory systems in health and safety that have been enacted since 1970 have extensively driven the revolution in industrial resources. Whereas all firms have greatly strengthened their hazard management functions in order to keep pace with federal and state regulation, external regulation is more important to some industries and corporations than to others. The nuclear industry, for example, is now a leader in assessing catastrophic risk, in emergency planning, in providing defence in depth and redundancy, and in implementing security measures largely because of regulations imposed on a reluctant industry. Those corporations with limited resources and internal organization, such as MACHINECORP, tend to have reactive management programmes largely determined by external regulations. Others have more complex and innovative systems. Although regulation has been essential for achieving the current plateau of industrial hazard management success, its limitations have also become very apparent.

Liability and insurance costs comprise another exogenous variable, increasingly important and effective over time, even prior to the Bhopal accident. Widespread corporate concern hovers over what industry characterizes as the 'deep-pocket' syndrome that persuades juries to make awards according to the ability to pay, rather than according to the responsibility for the harm that occurred. It has also been argued (Huber, 1986) that hazard information rather than hazard severity drives tort litigation. Experiences at Johns Manville, Three Mile Island and Bhopal have driven home the clear financial vulnerability of corporations to failure in hazard management. In the trucking industry, requisite coverage for a small firm has increased from the traditional US$150,000–300,000 cost of protection to US$5 million. Insurance rates, which have been greatly underestimated by insurance companies, have increased over recent years in a number of industries by several hundred to 1000 per cent. In some cases, insurance is difficult to find at any cost. The causes are multiple – courts are making large awards and adding on punitive damages, the basis of claims is being broadened, and environmental clean-up and health costs have skyrocketed. Little of this was predictable to earlier rate setters in corporate insurance. Small wonder that issues of product liability and insurance are stimulating upgraded hazard management in corporations. Liability, in short, has become a powerful force for improving corporate hazard management.

The effects of hazard liability extend not only to corporate insurance but to the reinsurance market, as well, where Lloyds Insurance Group, which has provided much of the reinsurance for the chemical industry, has virtually abandoned the field because of the 'unbelievably excessive litigation', high damage awards and disproportionate defence costs (Baram, 1985, p35). By 1987, concern was mounting that fear of litigation was becoming a significant impediment to industrial innovation in the US (Broad, 1987).

Public scrutiny may be another important exogenous factor, as general societal concern over 'toxic' chemicals and radioactive materials indicates. Continuing questions over possible harm and public reviews to detect any failures contribute to a social environment that corporations cannot ignore: 'When an industry or company is under intense scrutiny, when it is hot, every operation must be squeaky clean' (Pinsdorf, 1987, p105). Union Carbide officials learned this lesson, albeit too late, after blundering through the incident at Institute, West Virginia. Be it the continuing demonstrations and blue-ribbon citizen committees at Rocky Flats (Kasperson et al, 1988c), the 1750 news media queries to Johnson and Johnson following the first Tylenol scare (Kniffin, 1987, p22) or the 'chemophobia' with which the chemical industry believes it must deal (Cox et al, 1985), the climate of public opinion can be a powerful stimulus for forcing reluctant corporations to upgrade their hazard management programmes.

The *endogenous determinants*, although likely no less significant, are less clear. Profitability is a key condition that often correlates with strong hazard management performance, whereas the deteriorating maintenance

and operating situation at the unprofitable Union Carbide plant at Bhopal
suggests the reverse (see Chapter 5 in this volume). New plants and new
technologies, *ceteris paribus*, allow for the incorporation of the latest
standards and engineering safeguards. Thus, for example, the layout and
construction of Shell Moerdijk, a new plant on a new site, could easily take
into account new developments in and prevailing views on health and
safety (Shell, 1987). The chemical industry has stood at or near the top of
safety performance of American industries; the long-established
industries, by comparison, tend to have weaker records. Caution is in order,
however: the poor record of waste disposal and occupational exposures in
the high-technology industries of Silicon Valley indicates how easily other
determinants (high venture capital/short-term profit horizons) can
compromise this generalization.

Commitment of high-level management to health and safety appears
to be an important, if somewhat elusive, contributor to effective hazard
management. Few corporations boast organizational structures that
accommodate the effective use of risk assessment techniques at higher
corporate levels (Rowe, 1982, p167). Health and safety issues must
penetrate the highest levels of corporate decision making if hazards are to
receive primary consideration. Nowhere is this clearer than in Volvo's long-
standing emphasis on the innovating and marketing of safety (Svenson,
1988). Lagadec (1987) notes the importance of top-level management's
ensuring that hazard problems are not covered up at lower levels and
insisting on an effective system of upward flow of information. It is worth
noting that such safeguards are difficult to implement in tightly coupled
or command-and-control management systems (Perrow, 1985).

Market diversity, organizational diversification and, perhaps, product
diversity are all relevant to different corporate responses. Again, the size
and scale of the corporation seemed important to risk management
programmes. Finally, the degree of hazard associated with the product
itself may drive internal hazard management. Had managers of the Bhopal
plant accorded to methyl isocyanate (MIC) the special handling
commensurate to the level of risk, the accident would have been less
disastrous.

In short, some evidence exists to support all seven of these exogenous
and endogenous variables. More research is needed, however, to delineate
which assume greater importance in what particular situations and
contexts.

The shadow government of corporate health and safety

Although not widely recognized, large corporations have an internal
regulatory system that 'shadows' that of the public sector. In the industrial
leaders in health and safety, these are elaborate systems with extensive
procedures for identifying and monitoring risks, formal procedures of
standard-setting, numerous committees and role specification,

mechanisms for conflict resolution, and elaborate means of seeking compliance. This 'shadow government' emerges in detail in our PETROCHEM study (Kasperson et al, 1988b), where the corporate regulatory system employs three tiers of standards: external (federal regulatory agencies such as the OSHA, the US Environment Protection Agency and the Consumer Product Safety Commission, or CPSC) and industrial consensus standards, internal (carrying legal implications) and 'targets' (more informal and specific objectives). PETROCHEM also allocates substantial emphasis in its regulatory system to maximizing information and minimizing surprises and error. The company shows the same conflict between benefit maximization (or minimization of regulatory burden) and risk minimization that occurs in the public sector, with similar means, both formal and informal, for reaching 'balanced' decisions. Moreover, personal 'clout' in such decisions appears to have parallels with influence in decision making in the public sector.

Internal industrial regulatory systems are not always well developed, of course, and may be prone to failure. MACHINECORP, despite its size, has a less articulated and formal system and operates with a single tier of externally defined standards (Kasperson et al, 1988a). Its audit system is limited and the corporation allocates relatively little effort to ensuring compliance. Similarly, the internal regulatory system at the Bhopal plant, which did have a formal audit programme, suffered widespread breakdowns in ensuring compliance. A strength of the Volvo Corporation, by comparison, is the degree of integration between a well-developed accident feedback programme (the accident investigation centre) and automobile design and manufacturing (Svenson, 1988).

Anticipatory behaviour

Another requisite of effective corporate hazard management is that it should be anticipatory with regard both to hazard occurrence and to external regulation, liability and insurance costs. Historically, the predominant norm for successful industrial hazard management was quick reaction – when problems occurred, the corporation should be able to act swiftly to identify and rectify the situation. The traditional hazard management system has not generally (drugs, agricultural chemicals and nuclear power are obvious exceptions) been geared to anticipating hazard problems.

The norms are changing. Speedy reaction is no longer the standard of a well-developed corporate hazard management programme. Except for Volvo Car Company, whose safety standards have been well ahead of national public regulations, corporations generally are only now beginning to engage in anticipatory hazard control. PETROCHEM has adopted a number of internal standards that are more stringent than those required by OSHA, seeks to anticipate external regulation by altering its productive processes and lowering (in a cost-effective manner) occupational exposures, and by the mid 1980s was conducting prospective epidemiological studies to identify potential hazards (Kasperson et al, 1988b). Monsanto has been a leader in actions to enhance community knowledge of chemical hazards and to increase local emergency-response programmes. The 3-M Corporation has pioneered innovative programmes in waste reduction, achieving 50 per cent waste reductions between 1975 and 1986 (Koenigsberger, 1986). Independent of regulatory requirements, liability and insurance are driving stronger anticipatory hazard management, a situation that poses new challenges to industrial organization but which may also reduce the role of governmental regulation.

Low-probability risks and defence in depth

Low-probability, catastrophic risks pose inherent difficulties for corporate hazard management. Where experience provides little guidance and actuarial indicators of risk are unavailable, formal probabilistic risk analysis must substitute for experience. Such assessment requires analytic resources not available in most corporations. When the authors visited a modem computer chip plant, programmes for handling routine risks appeared well developed, formally stated and enforced. Questioning revealed, however, that low-probability events were not in the consciousness of corporate health and safety managers and had not undergone formal (or even informal) assessment.

Defence in depth – multiple layers of safety systems incorporating redundancy – is well developed in certain industrial sectors in which issues of catastrophic risk have been the focus of scientific and public controversy. Rocky Flats has 11 levels of defence in depth; nuclear plants have similar well-developed layering and redundancy (Kasperson et al, 1988c). But Bhopal, Three Mile Island, Rocky Flats, the space shuttle *Challenger* and Chernobyl all speak persuasively to the vulnerability of such systems to compromise by shoddy maintenance, faulty design, management failure and human error. The case studies deliver a somewhat ambiguous message – defence in depth is a sound safety philosophy, but its success requires formal assessment (beyond much current corporate capability) for its emplacement, a willingness to invest heavily in safety, and vigilant monitoring and implementation. Paul Shrivastava (1987a, p132) cautions that defence in depth includes redundancy in personnel as well as in engineered safety systems.

An agenda of vital research

The results of our studies point in three major directions for future research on the corporate management of hazards. The first is a need to broaden current understandings by moving from individual cases to comparative studies. This could treat the overall structure and determinants of industrial hazard management, including analyses of entire industrial sectors, as well as individual firms, and extending the comparative perspective to other industrialized countries and the developing world. The second direction recognizes that exploratory ventures cannot achieve the depth requisite for a full understanding of the inner workings of corporate hazard management. Accordingly, it is essential to initiate projects that probe the internal dynamics of management systems and decision-making processes. The third direction is to follow up in detailed study some of the opportunities for improving hazard management by more constructively integrating the private resources of corporations into overall efforts of the public sector.

The determinants of hazard management

Wide variation, or substantial unevenness, exists in how corporations manage hazards. In general, in our research on corporations, we have had access to large, wealthy corporations that took pride in their programmes for managing hazardous products. Building upon this experience in learning how to characterize the hazard management commitment, function, organization and resources of corporations, further research should undertake a more systematic exploration of how these characteristics vary by hazard and by corporation size, age, profitability, planning horizon, market and product diversity, and organizational structure. Does depth in hazard management increase with hazardous products and processes, large size and plentiful resources, profitability and experience? What role does tight and loose coupling play in the response of corporate management systems to hazard surprises? Is it true, as is commonly suggested, that much of the failure in industrial hazard management is dominated by small, marginal firms that cannot, will not or do not invest in safety and health protection? Systematic study is needed to analyse the behaviour of a spectrum of firms within a given industry.

Management convergence

Modern hazard management employs a wide variety of technological and behavioural controls designed to prevent or reduce hazard. In contrast with the stereotypes, it is apparent that safety is not always in conflict with other corporate goals. The experience with nuclear power, and at Volvo, indicates that a well-developed quality assurance programme enhances *both* reliability and safety. Effective waste-reduction programmes can result in actual cost savings and materials recovery while protecting the corporation from uncertain environmental or health liability claims. Thus, opportunities exist for structuring hazard management to converge with production or product management goals and to afford long-term income protection for the corporation. Such hazard controls reduce the threat of accidents that interrupt production or that contaminate products even as they protect the health and safety of workers and consumers. Other controls, however, do not serve any inherent dual purpose and are employed only under the demands of external regulation, or the threat of a lawsuit or an insurance bill. Can the convergences and divergences between profit and health and safety be more clearly identified? Can ways be found to encourage the convergences and reduce the conflicts?

Trade unions and hazard management

Traditionally, trade unions have not been a strong force for protecting the health and safety of workers in the US. Despite participation in corporate health and safety committees and advocacy in particular hazard cases, the union presence has been anaemic overall and other priorities have generally dominated union actions. Over the past decade, however, some unions, such as the Oil, Chemical and Atomic Workers International Union and the United Steelworkers of America, have become decidedly more aggressive on health and safety questions. Some union contracts, for example, are now stipulating corporate health and safety commitments and programmes. But, US trade unions have very limited resources and only modest programmes. Except at Volvo in Sweden, our studies of corporations have found no substantial union role in hazard management. Meanwhile, the trade union movement in the US has faltered during the 1980s. A more searching assessment needs to address the changing role of unions and the means of increasing their effectiveness in occupational hazard management.

National comparisons

Comparative studies of hazard management in industrialized countries have suggested a greater reliance on corporations and trade unions in hazard management and more confidence in regulatory discretion outside the US. Some evidence indicates that other countries are heavily reliant upon US risk assessments but are quite efficient, with fewer resources, in setting standards and informally negotiating compliance in a flexible manner with industry. Whereas some studies (Kelman, 1981; Pijawka, 1983; Brickman et al, 1985) conclude that the results are quite similar despite the varied approaches, other studies (e.g. Kasperson, 1983b) suggest that the outcomes may be significantly different.

A systematic set of national comparisons of industrial hazard management is needed to address such questions as: do Europeans and Japanese trust their corporations more than Americans do? If so, is this trust justified? Do they make better public use of corporate resources? Does the consensual, negotiated approach to standard-setting and implementation produce higher or lower levels of safety and compliance? What is the role of the trade unions in industrial hazard management? What novel mechanisms have been developed as alternatives or complements to regulation? How transferable is hazard management success across political cultures and economies? In short, how well does successful hazard management travel?

Corporate social responsibility programmes

In the wake of the environmental and civil rights movements, US corporations have widely adopted social responsibility programmes. Such programmes generally provide the philosophical underpinnings for and include stipulations affecting corporate management. Thus, the product stewardship programme at Dow Chemical has for years recognized an obligation for 'cradle-to-grave' care of chemicals (Buchholz et al, 1985). A social responsibility programme explicitly commits PETROCHEM to a goal of leadership in health and safety protection in the petrochemical industry (Kasperson et al, 1988b). A similar programme was credited by Johnson and Johnson with playing an important role in the Tylenol scare (Kniffin, 1987); but limited knowledge exists as to how such programmes really function, the extent to which they make a difference in the corporation, the methods for verifying compliance, and their long-term impacts on managerial behaviour. At a visit to Dow during the late 1970s, company responses to our queries indicated that despite a well-articulated programme and extensive investments, independent validation of the implementation and compliance achieved had not occurred. Post-mortems on the Bhopal catastrophe attest to similar defects in verifying implementation (see Chapter 8 in this volume). A careful study of whether these corporate ethical principles have actually affected corporate objectives and the behaviour of company managers, or whether they are simply window dressing, would substantially add to current knowledge.

Auditing the audits

Formal auditing programmes have been widely adopted to assume compliance with corporate objectives in hazard management (as in the stated goal for lowering the occupational injury rate). Such audits are mounted internally in most large corporations; smaller firms often turn to consulting firms (e.g. Arthur D. Little) that specialize in such activity or depend upon the review of insurance companies. The ambition, scope and rigour of audits vary enormously, as do the resources allocated on their behalf, the frequency of application, the degree of integration with incentives to assume compliance by plant managers, and the corporation's determination to verify compliance. At Bhopal, it was apparent that the auditing procedure lacked teeth and follow-through (see Chapter 5 in this volume). The Chemical Manufacturers Association (CMA) has established auditing guidelines to improve on the performance of member firms, 40 per cent of which in 1983 neither had such programmes in place nor claimed any intention to develop them (*CMA News*, 1983, p12). A rigorous analysis of the scope and depth of such audits, their respective content, measures of success and failure, and effectiveness in correcting weaknesses and problems would be extremely valuable.

Hazard communication and worker participation

Worker exposure to risk, it is sometimes argued, is voluntary or semi-voluntary, based upon workers' knowledge of the hazards to which they are exposed and the degree to which they are compensated to take risks. Whereas it is apparent that the growth in industry programmes designed to communicate risks to workers (Melville, 1981; Viscusi, 1983, 1987) and the OSHA rule for hazard communication (OSHA, 1983) are major developments, evidence suggests that worker understanding of risk is often very inadequate (Nelkin and Brown, 1984). Whereas Rocky Flats speaks to a relatively high level of understanding of radiation hazards, Indian workers at Bhopal had a totally inadequate level of knowledge (see Chapter 5). Meanwhile, Sweden provides a telling case of much higher levels of ambition and programme development (also not, however, rigorously evaluated as to impact). Although the communication of risk is fashionable in the Washington regulatory community, as well as among scholars, relatively little is known about risk communication in the workplace. A searching examination of worker knowledge of hazards, the effectiveness of the hazard communication rule as it is implemented, the attributes of successful programmes, and how such knowledge can be utilized to improve worker participation in identifying and reducing risks would greatly enhance industrial hazard management efforts.

The culture of safety

Corporate officials involved in hazard management speak persuasively of the 'culture of safety' that accompanies upgraded safety, an issue we take up in depth in Chapter 7 of this volume. A representative of an oil company described how it happened at his corporation: as a result of an industrial accident, the chief executive officer (CEO) proclaimed a corporate commitment to substantially increased levels of safety. The enactment of new programmes required the breakdown of traditional ways of thinking and the encouragement of new attitudes. The effort consumed much of the corporation's management energy over a three-year period but did make major progress, in the CEO's view, for creating this new culture.

The corporation most frequently cited for having such a culture is DuPont, where a mature corporate culture has emerged and aggressive hazard management programmes have spanned a 15-year period (see Chapter 7). A different picture emerges in the Bhopal case, where understanding of toxic materials and the need for preventive maintenance was sadly lacking (see Chapter 5). How does a 'culture of safety' arise and how is it maintained? What is its content? Does it transcend changes in executive officers or crises in profitability or product line? Is it transferable to new locations, does it survive divestment and is it teachable in business schools?

For much of the past decade, the low-level chronic health hazards have dominated thinking and attention in industrial hazard management. The past disasters at Three Mile Island (an economic disaster), Seveso (an ecological disaster), Bhopal (a human disaster), Chernobyl (a human and ecological disaster), Mexico City (a human disaster) and the Rhine pesticide contamination (an ecological disaster) have fundamentally changed the corporate hazard agenda. The management of catastrophic risk is very demanding, requiring extensive analytic resources, meticulous corporate attention and well-developed databases. It is undoubtedly most advanced in the defence sector and in nuclear power plants. The use of probabilistic risk analysis has increased dramatically in the nuclear industry over the past decade (Kasperson and Kasperson, 1987b) – some 20 nuclear power plants now have, or are in the process of preparing, site-specific studies (Kasperson et al, 1987). Bhopal made it clear that the chemical industry faces low-probability/catastrophic risks, as well, and must incorporate relevant policies for assessing and managing low-probability risks. Other industries, such as biotechnology, may yet experience these problems. What is the state of catastrophic risk assessment in industry? To what extent are generic means for preventing catastrophic risk – 'defence in depth', equipment redundancy, advanced accident analysis, emergency training, remote siting and the use of simulators – routinely employed in other industries? How serious are the gaps?

Public use of private resources

Ideally, the hazard makers are potentially the best hazard managers. But inherent conflicts of interest and frequent failures undermine the potential. Are there societal mechanisms for reducing these conflicts of interests? Are these mechanisms adaptable to corporate hazard management? Are there analogues in decentralized review mechanisms (e.g. biohazard review committees with public members and professional certification in engineering)?

A National Research Council (NRC) report, based on a sample of 53,500 distinct chemical entities in commercial products, found that minimal toxicity data in publicly accessible sources existed for only one-third of drugs and pesticides, one-quarter of cosmetics and one-fifth of chemicals in commerce (USNRC, 1984c). Yet large amounts of high-quality toxicity data not reportable under the Toxic Substances Control Act (TSCA) and not available for public use reside in corporate files. Corporate officers characteristically raise numerous objections and obstacles to sharing such data: the data may not be useful or comparable; they reveal research and development strategies of the firm; toxicity testing is a major cost of product development and corporations compete in this activity; firms might be liable for the information provided; and so forth. Can ways be found to meet the reasonable concerns while affording greater use of this major resource?

Hybrid institutions

A growing number of novel hazard-management institutions combine industry, public interest groups and government. The Chemical Manufacturers Association has joined in negotiated rule making (as with inadvertent releases of polychlorinated biphenyls, or PCBs) in joint suits with environmentalists (as in hazardous waste health effects), in co-sponsoring independent studies (of hazardous waste) and in broad-based efforts for non-regulatory solutions to hazardous waste clean-up (Clean Sites, Inc.). The 1980s have also witnessed efforts to find new ways of involving the public in hazard control – the Ruckelshaus efforts to involve the public in risk assessment at the Tacoma shelter and the current US Environmental Protection Agency efforts to communicate the risks of indoor radon to homeowners are examples. Environmental mediation has been previously used with effectiveness by the Keystone Center in dealing with radioactive wastes.

Another class of hybrid is mandated by regulation, such as the so-called 'squeal law' aspects of TSCA that mandate the inclusion of private information into the public sector. A systematic effort to conceptualize such opportunities, to evaluate outstanding examples and to assess their future promise might answer such questions as what is the potential of hybrid public–private hazard management to provide cost-effective hazard reduction? Is voluntary corporate action a viable alternative to regulatory mandate? What are the dangers of collaborative industry–environmental initiatives that substitute for the normal regulatory process? Are there situations where mediation is preferable to regulation? Where is it not viable?

THE FUTURE OF CORPORATE HAZARD MANAGEMENT

Hazard management in corporations simultaneously functions on three time horizons: legacies, usually grim, from the past; ongoing management of current production processes and products; and the development of future processes and products. Much of the corporate hazard management effort directed at the past is defensive or compensatory, trying to correct, cover up or forget hazards created in the past. Current activity is both complex and variable among corporations. Corporations, at present, are defending products, reducing hazards where feasible, complying with regulations while resisting future ones, seeking to prevent disasters, identifying hidden hazards, and avoiding liability wherever possible. At the same time, some corporations are seeking safer products and improved engineered safety in new or rebuilt plants, whereas others cling to past practices that endanger workers and publics.

Because all three levels of activity are ongoing and involve very different kinds of responses and resources, the public image and private practice of corporate hazard management are confounded. The cross-fires over past decisions and newly proposed regulations should not, in our view, consume all of society's attention to corporate hazard management. The most pressing existing need is to decrease the great disparities in corporate hazard management, to bring all corporations in a particular industrial sector to the standard of the best – to spread, in short, the best of current corporate practice throughout the private sector. A focus on the future is also needed to examine how new products and new plant designs can better reduce hazards; how cultures of safety germinate and are nourished; how industry can be made more socially responsible; and how industrial capabilities can be more effectively used on behalf of the larger societal good. These, in turn, call for embedding the lessons of the past – in new technology development, in corporate organizational structures, in venture capital, and in the new constantly arising hazards whose impacts are imperfectly understood but whose benefits continue to be needed.

5 Avoiding Future Bhopals

B. Bowonder, Jeanne X. Kasperson and
Roger E. Kasperson

INTRODUCTION

Runaway chemical reactions are rare events, particularly in this heyday of the redundant and 'defence-in-depth' safety design for complex, high-risk technologies. Yet, during the chill of night between 2–3 December 1984, a statistically improbable worst-case scenario moved from the computer simulations of the risk assessors and played itself out on the unsuspecting citizens of Bhopal, India. A parade of failures – in design, in maintenance, in operation, in emergency response and in management – conspired with a southerly wind and a temperature inversion to push a lethal cloud of methyl isocyanate (MIC) out to kill and injure thousands of people, animals and plants in the area. By sunrise, the unprecedented horror had catapulted Bhopal to the head of history's roll of industrial disasters (see Table 5.1).

The inevitable spate of articles and conferences on the perils of technology transfer is in full force. Post-mortems on the accident are likely to proliferate for some time as the courts and the risk analysts puzzle over the catastrophic chain of collapses, each trivial in its own right, that sent MIC on its destructive path.

Indeed, much is at stake in the responses to the accident, for the post-mortems may select the 'wrong' lessons and thus fail to avert future calamities, place unwarranted crippling restraints on the chemical industry, or impede the flow of needed and generally beneficial technology to developing countries. The chemical industry, with a job-related lost-workday incidence of 2.43 per 100 full-time workers in 1983 (compared with an all-industry incidence of 6.84), is an undisputed leader in industrial safety (National Safety Council, 1984, p32). Union Carbide Corporation, the parent company involved in the disaster at Bhopal, has more than 20 years' experience in the safe manufacture, use, transport and

Note: Reprinted from *Environment*, vol 27, Bowonder, B., Kasperson, J. X. and Kasperson, R. E., 'Avoiding future Bhopals', pp6–13, 31–37, © (1985), with permission from Heldref Publications

Table 5.1 *Major industrial disasters during the 20th century*

Year	Accident	Site	Number of fatalities
1921	Explosion in chemical plant	Oppau, Germany	561
1942	Coal-dust explosion	Honkaiko Colliery, China	1572
1947	Fertilizer ship explosion	Texas City, US	562
1956	Dynamite truck explosion	Cali, Colombia	1100
1974	Explosion in chemical plant	Flixborough, UK	28[a]
1975	Mine explosion	Chasnala, India	431
1976	Chemical leak	Seveso, Italy	0(?)[b]
1979	Biological/chemical warfare plant accident	Novosibirsk, USSR	300
1984	Natural gas explosion	Mexico City	452+[c]
1984	Poison gas leak	Bhopal, India	2500[d]

Notes:
a 3000 evacuated
b 700 evacuated, hundreds of animals killed, 200 cases of skin disease
c 4258 injured, 31,000 evacuated
d 100,000 evacuated, 50,000 severely impaired
Source: data from Lagadec (1982); *Financial Times* (1984); Kottary (1985)

storage of MIC (to say nothing of a host of other hazardous products). With a cadre of scientists and technicians and an institutional structure for environmental protection, India is better equipped than other developing countries to manage hazardous technologies. Given this framework, other industries in other places are more likely candidates for catastrophic disasters. Thus, it is essential to understand how and why this particular surprise occurred at Bhopal if we are to ward off future similar tragedies.

JOBS AND SELF-SUFFICIENCY

Union Carbide was scarcely an unwelcome intruder in Bhopal. The Indian government promoted the siting of industries in less developed states such as Madhya Pradesh, where Bhopal is located. Eager to attract major industries, Madhya Pradesh leaders offered incentives to companies that would bring jobs and indigenous manufacturing to its unindustrialized cities; Union Carbide, for example, built on government land for an annual rent of less than US$40 an acre (0.4 hectares). A plant that would manufacture the carbaryl pesticides to fuel India's ongoing Green Revolution was particularly welcome as another step toward self-sufficient food production. Hence, the 1970 decision of Union Carbide of India Limited (UCIL) to manufacture the pesticide Sevin in an advanced facility in central India was met with great fanfare.

Sevin, manufactured from MIC, had received the endorsement of the Indian Council of Agricultural Research. Use of the pesticide decreases

Table **5.2** *Workplace limits to chemical exposures*

Chemical	Parts per million[a]
Carbon monoxide	50.00
Chloroform	25.00
Methylamine	10.00
Benzene	10.00
Acetic acid	10.00
Cyanogen	10.00
Phosgene	0.10
Methyl isocyanate	0.02

Note: a Time-weighted averages for eight-hour exposure
Source: ACGIH (1984)

insect damage of cotton, lentils and other vegetables by as much as 50 per cent. Even in the wake of the accident, few serious observers have suggested that India do away with Sevin and other carbaryl pesticides, which, ironically, are substitutes for 'more dangerous' DDT and organophosphates. Given the high toxicity (Kimmerle and Iben, 1964; Union Carbide Corporation, 1976) of MIC (see Table 5.2), however, it is clear that the chemical requires, at all stages, special handling commensurate to the risk.

It is easy to contend that high-risk facilities have no place in densely populated urban areas. Yet, such a facility is apt to attract squatter settlements to its gates, whether it be a liquefied natural-gas facility in Mexico City, a petrochemical complex in Cubatao, Brazil, or a pesticides factory in Bhopal. The showpiece UCIL factory and other industries that set up shop in Bhopal surely contributed to the staggering rise in population – from 350,000 in 1969, to 700,000 in 1981, to over 800,000 in 1984. As a Union Carbide official put it:

> In India, land is scarce and the population often gravitates towards areas that contain manufacturing facilities. That's how so many people came to be living near the fences surrounding our property (Browning, 1984, p3).

It is also, of course, how risks come to fall so disproportionately on the poor (Susman et al, 1983; Wijkman and Timberlake, 1985).

The showpiece factory never lived up to its promise of production and jobs for the area. A drought in 1977 forced many farmers to take out government loans, many of which began to fall due in 1980. The farmers then exchanged the expensive Union Carbide pesticides for others less costly and less effective. Meanwhile, the Indian government was providing incentives for small-scale manufacturers to produce pesticides that they could afford to sell at half the price of Union Carbide products. In addition, inexpensive, non-toxic synthetic pyrethroids made their debut, and sales of traditional pesticides began to drop throughout the industry. The Bhopal operation, never very profitable, broke even in 1981, but thereafter began to lose money. By 1984 the plant produced less than 1000 of a projected 5000 tonnes and lost close to US$4 million. UCIL, contemplating selling the operation, began to issue incentives for early retirement and cut back on its workforce. Many of the skilled workers left for securer pastures. Things were not going well.

EARLY WARNINGS

Whether cost-cutting measures and the departure of skilled personnel caused lapses in safety is difficult to ascertain. Nevertheless, the Bhopal plant experienced six accidents – at least three of which involved the release of MIC or phosgene, another poisonous gas – between 1981 and 1984. These accidents scarcely presaged the catastrophic release; but taken together they surely could have pointed to safety problems at the plant. Indeed, a phosgene leak that killed one worker on 26 December 1981 generated an official inquiry; but the findings (filed three years later) gathered dust in the Madhya Pradesh labour department until after the Bhopal accident, when two officials lost their jobs for having failed to act upon the report's safety recommendations (*Times of India*, 1985).

Meanwhile, a local journalist warned that the plant's proximity to Bhopal's most densely populated areas was inviting disaster. In 1982, Rajkumar Keswani took on UCIL in a series of articles in the Hindi press. 'Sage, please save this city', 'Bhopal on the mouth of a volcano' and 'If you don't understand, you will be wiped out', the headlines warned (Keswani, 1982a, 1982b, 1982c). On 16 June 1984, he tried again, this time with what he calls 'an exhaustive report on the Union Carbide threat'. 'The alarm fell on deaf ears', he wrote one week after the Bhopal accident (Keswani, 1984).

A 1982 safety audit by an inspection team from the parent company cited a number of safety problems, including the danger posed by a manual control on the MIC feed tank, the unreliability of certain gauges and valves, and insufficient training of operators (Kail et al, 1982). UCIL claims to have corrected the deficiencies; but auditors have never confirmed the corrections.

Just before the accident in Bhopal, Union Carbide Corporation's safety and health survey of its MIC Unit II plant in Institute, West Virginia, cited 34 less serious and two major concerns, the first of which was the 'potential for runaway reaction in unit storage tanks due to a combination of contamination possibilities and reduced surveillance during block operation' (Union Carbide Corporation, 1984; US Congress, 1984). Why the parent company, which owns 50.9 per cent of the Bhopal plant, failed to share with its subsidiary its two major concerns (the second was the serious potential for overexposure to chloroform) is unclear. Some Union Carbide officials contend that the different cooling systems – brine at

Institute and Freon at Bhopal – made the hazard communication unnecessary; but this is difficult to square with the recommendation:

> The fact that past incidences of water contamination may be warnings, rather than examples of successfully dealing with problems, should be emphasized to all operating personnel (US Congress, 1984, p462).

Equally puzzling is the parent company's earlier overriding of an alleged UCIL protest against the installation of such large storage tanks – 15,000 gallons – at Bhopal (Bhushan and Subramaniam, 1985).

In any event, MIC sat in storage at the Bhopal plant for at least three months prior to the accident (Delhi Science Forum, 1985). Such storage invites disaster, for the tiniest ingress of water, caustic soda, or even MIC itself is sufficient to set in motion an exothermic (heat-producing) chemical reaction (Worthy, 1985). One Indian scientist has even hypothesized that the reaction at Bhopal began slowly and imperceptibly at least two weeks before the fateful night in which it reached violent runaway proportions (Muralidharan, 1985).

ACCIDENT ANALYSIS

Some time shortly before the 10:45 pm shift change at the Bhopal plant on 2 December 1984, water and/or another contaminant entered MIC storage tank 610, thereby triggering a violent chemical reaction and a dramatic rise in temperature and pressure. It is not known whether the incoming control room operator was aware that the 10:20 pm tank pressure read 2 pounds per square inch (psi); but the 11:00 pm reading of 10 psi does not seem to have struck anyone as unusual. Nor should it have, since normal operations ran at pressures between 2 and 25 psi.

By the time the operator did take notice of the rising pressure – from 10 psi at 11:00 pm to 30 psi at 12:15 am – the reading was racing to the top of the scale (55 psi). Escaping MIC vapour ruptured a safety disc and popped the safety valve. On the heels of this initial release came a series of compromises and failures of virtually all the safety systems designed to prevent release (see Box 5.1). The deadly gas spewed out over the slums of Bhopal.

Some of the details surrounding the release remain sketchy; yet it is possible to construct a reasonably plausible analysis of the accident. Figure 5.1 depicts the accident in terms of a general model that views hazards as threats to humans and to things they value. Hazards, which arise out of human wants, develop in stages and end in unintended and undesired consequences. The stages flow into each other much like a succession of reservoirs, each leading to the next. Each linkage in this causal chain presents an opportunity for intervention via control measures designed to arrest the evolution of the hazard.

Box 5.1 A leak at Institute, West Virginia

Is safer safe enough?

The leak on 11 August 1985 of methylene chloride and aldicarb oxime at Union Carbide's plant in Institute, West Virginia, attests to the vulnerability of back-up safety systems that fail even in the face of US$5 million worth of upgrading.

The Institute plant resumed operations in May 1985 after undergoing extensive retrofitting and modernization that plant officials had touted as a precautionary measure, a way to make a 'safe plant safer'. In place were a new chemical process that would lower intrinsic risk by reducing the quantities of methyl isocyanate (MIC) in transport and a spanking new computerized warning system that would pinpoint the speed and direction of any airborne chemical release. A review by the US Environmental Protection Agency essentially endorsed the company's confidence that the plant posed no major risk to the surrounding community.

But by the time the noxious gas mixture had dropped in unexpectedly on four communities and sent at least 135 people to the hospital, assurances that Bhopal 'couldn't happen here' rang hollow. Once again, an intruder – steam, this time – entered the jacket surrounding an unusual storage vessel, triggered a temperature and pressure rise that overwhelmed three gaskets on the tank, thereby producing a leak that eluded the multi-layered safety system and escaped directly into the atmosphere. The high pressure also ruptured a safety valve so that additional gas flowed to a neutralizer and a flare tower, two back-up safety devices that only partially destroyed the vapour. As at Bhopal, another safety system, the water spray, was insufficient to arrest off-site migration of the gas. That some gas bypassed the safety systems altogether and escaped directly into the atmosphere and some gas sneaked past both the scrubber and the flare has posed the potential need to retool valving and piping systems throughout the chemical industry.

Just as MIC had surprised Bhopal, the gas mixture announced its own arrival – before the public siren sounded, before anyone notified local authorities, before the emergency response teams got wind of it. The newly installed cloud-dispersion computer modelling was not programmed to track an aldicarb oxime leak. In short, exposure preceded the activation of any plans to prevent it.

These multiple failures in the face of the intense corporate and public scrutiny lavished on this particular plant – perched, as it is, in a setting conducive to high safety performance – will confound the management of catastrophic risk in developing countries. No technological or managerial quick fix is equal to overcoming the vulnerability, pervasive even in well-conceived safety systems; the flaws in design; the inadequacies in computer modelling (or failure to input relevant data); the human reluctance to acknowledge failure when it does occur (no one wants to sound the siren); and the intrinsic limitations in the depth of regulatory inspection, review and enforcement.

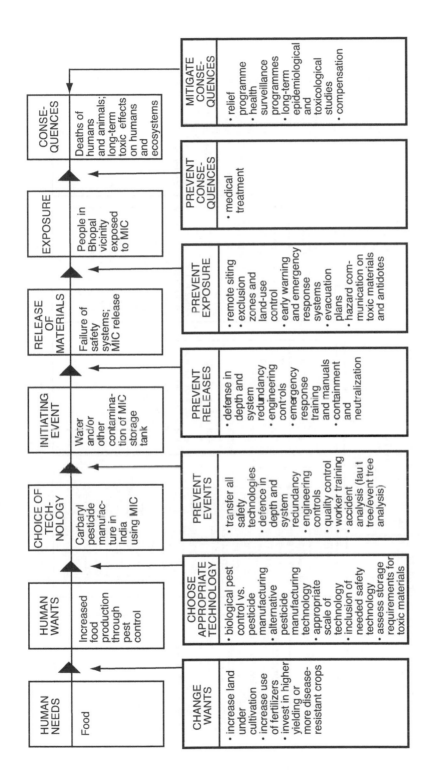

Figure 5.1 *The causal structure of hazard: Application to the Bhopal accident*

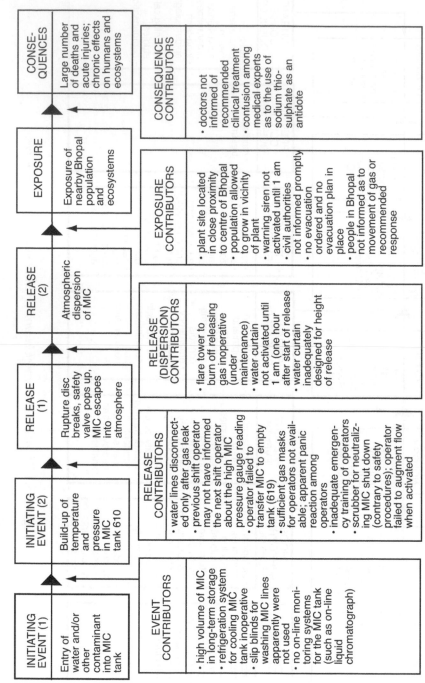

Figure 5.2 *Detailed model of contributors to the Bhopal accident*

In the case of the Bhopal accident, the basic human need for food generated a *human want* of increased food supply through pest control (i.e. the manufacture and application or pesticides). The particular *choice of technology* at Bhopal was the indigenous manufacture, using MIC, of carbamate pesticides. As Figure 5.1 indicates, this choice of technology entailed, at least implicitly, a series of important choices, ranging from the basic selection of chemical over biological pest control to a series of considerations relevant to the storage of toxic materials and the scale of technology. These decisions fundamentally shaped the inherent hazard that was set off by the initiating event: the contamination of MIC.

All the while, standard means to *prevent exposure* – remote siting, exclusion zones or so-called greenbelts, early-warning and emergency-response systems, evacuation plans and hazard communication – never materialized at Bhopal. The surprise release of poison gas thwarted any concerted effort to *prevent consequences*. For want of instructions to breathe through a wet towel, scores of people died ghastly deaths. For lack of information on the toxicity of MIC, Bhopal's medical community was hard put to *mitigate consequences*. Figure 5.2 shows how these contributors at each stage of the accident in effect hurried a low-probability hazard along to catastrophic consequences.

CONSEQUENCES OF THE ACCIDENT

More than six months after the accident, even the early (acute) consequences of the accident are not well defined and the longer-term (chronic) effects are very uncertain. Estimates of human fatalities range from 1400 to 10,000 (see Table 5.3). The Indian government's official count as of June 1985 of 1762 fatalities – based on death certificates – undoubtedly underestimates the toll because many people fled from the city and died in outlying regions; the deaths occurred disproportionately among people living in the nearby squatter settlements, about whom little information exists; and the upcoming elections provided incentive for official minimization of the number of fatalities.

More recently, Asoke K. Sen, the Indian law minister, put the fatalities at more than 2000 (Diamond, 1985b), closer to the more widely accepted estimate of 2500. This figure, based on body counts conducted by members of the press, is subject to errors in observation and tabulation and also misses the deaths that occurred outside the city. Because the exodus from the city was extensive within 24 hours of the release, a significant number of fatalities may have escaped count.

The Medico Friends Circle estimated at least 4000 deaths, based on the sale of death shrouds in the week following the accident. Other estimates have reached 10,000; but they are based on impressionistic information. Estimation is confounded further because over 80 per cent of the deaths occurred outside the hospitals. Only 438 deaths are recorded in

Table **5.3** *Major human and environmental consequences of the Bhopal accident*

Consequence	Source	Estimated number affected	Base of estimate
Mortality	Indian government	1762	Death certificates
	Independent Indian agency	1400	Residential survey
	Delhi Science Forum	5000	Survey of hospitals
	Newspapers and magazines	2500–10,000	Body counts, unsystematic interviews of doctors, victims, and local officials
	Medico Friends Circle	4000	Sale of death shrouds
	Asoke Sen (Indian law minister)	2000	Unspecified
	Indian, officials and embassy personnel[a]	5000	Unspecified
Morbidity	Newspaper reports	200,000 exposed; 50,000–60,000 suffering ill effects	Survey of hospitals and interviews with medical doctors.
	Delhi Science Forum	20,000 severely affected in lungs and eyes	Survey of hospitals, medical doctors, affected areas, and victims
	Newspaper reports	Effects on women's health: excessive vaginal discharge, excessive menstrual bleeding, retroverted position of the uterus, cervical erosion and inflammation; unusually high numbers of premature or underweight births, physical deformation among 20–30 daily births	Reports from survey of two Indian camps
	Indian Council of Medical Research	No indication of adverse foetal effects	Survey of 500 babies born since the accident
	Asoke Sen	17,000 persons in Bhopal permanently disabled, largely from lung ailments. Doctors have treated 200,000 (25 per cent of Bhopal's population)	Unspecified
Mental Health Effects	King George Medical College	168 mental cases, especially of neurotic depression or anxiety neurosis	Study of one clinic in affected area
	BusinessIndia	193 suffering from mental disorders (22.6 per cent of survey; majority are women under 45 years of age)	Survey of 855 patients, in 10 government hospitals

Evacuation	Newspaper reports	70,000 people left Bhopal before 14 December 1984	400 buses evacuated people for two days; interviews with officials, residents, and evacuees
Non-human Mortality	*BusinessIndia* and government sources	1047 animal fatalities; therapeutic treatment to 7000; delayed poisoning in poultry; breakage and deformation in various phytoplankton cells; change in pigmentations of chlorophyceal algae; fish suffering anaemic conditions; 1600 animals and several fishes found dead (790 buffaloes, 270 cows, 483 goats, 90 dogs and 23 horses); some insect species killed; certain trees lost all leaves; other vegetation in vicinity of plant turned black	Body counts and field observations
Economic Loss	Ishwar Das (Indian relief commissioner)	800 small-scale manufacturing units and about 20,000 shops and establishments suffered business loss; jobs lost from closing of plant, death or incapacitation of wage earners, short-term evacuation costs, uncompensated medical expenses (no quantitative estimates available)	Government survey

Note: a Reported by *Chemical and Engineering News*, 1985
Source: Clark University Center for Technology, Environment and Development

the various hospitals on 3 and 4 December (*Sunday*, 1985). Although a precise breakdown of age among the fatalities is unavailable, the deaths were disproportionately concentrated among children and, especially, infants.

Information on the number of people exposed to MIC, as well as on the long-term health effects from exposure, is even less available (vegetation analyses are under way and may improve understanding of MIC's effects.) The most widely accepted estimates indicate that 200,000 persons were exposed, and that 50,000 to 60,000 received substantial exposure (Waldholz, 1985). Law Minister Sen recently indicated that doctors have treated 200,000 persons for exposure and 17,000 persons in Bhopal have been permanently disabled, largely from lung ailments (Diamond, 1985b). Evidence of continuing physiological effects – including abnormally high blood levels of carboxy haemoglobin and methaemoglobin, low vital lung capacity, neurological abnormalities, widespread gastritis, and vomiting – is accumulating. At the time of the accident, the German toxicologist Max Daunderer warned about the stages of MIC's effects: irritated eyes, skin and lungs during the first four to seven days; serious central nervous system effects developing after three to four weeks; and then delayed central nervous system disorders, including paralysis (Rout, 1985, p27). The degree of reversibility or irreversibility of these effects, however, is not known.

The Indian Council of Medical Research is coordinating a massive data-gathering effort on long-term morbidity conducted by the All-India Institute of Medical Science, the V. P. Patel Chest Institute, the Industrial Toxicology Research Centre, and the KEM Hospital in Bombay. Meanwhile, widespread distribution (often without medical supervision) of antibiotics and corticosteroids, as well as widespread malnutrition and chronic diseases among many victims, have complicated the assessment of MIC's effects on morbidity. Studies aimed at defining potential genetic and carcinogenic effects are only beginning, and results will require another three to five years. Although MIC passed the Ames test (a short-term test for mutagenicity), the US National Toxicology Program is planning an ambitious series of animal studies to elucidate long-term effects on respiratory, reproductive and immune systems (Dagani, 1985).

Special concern exists over possible damage to women's health. Amid accounts of abnormally high levels of uncommon vaginal discharge, excessive menstrual bleeding, retroverted position of the uterus with severe restricted mobility and other disorders, the press has alleged that government and medical teams are avoiding women's health complaints. Junior gynaecologists and midwives in hospital maternity wards have reported unusually high numbers of premature or underweight babies and physical deformities among the 20 to 30 infants born daily in Bhopal (*Sunday*, 1985). Despite the sketchy evidence, some local doctors had advised a number of the 1000 women who were pregnant at the time of the accident to undergo abortions (*Sunday*, 1985), advice which a government services department has contested. The Indian Council of Medical Research recently reported that a study of the effects of MIC exposure on foetal growth in some 500 babies does not indicate foetal damage.

Effects of the accident on mental health are acknowledged, but are the least studied. Reports of mental trauma and other psychiatric effects persist; yet there is no systematic programme for monitoring or treating mental health problems. A distinguished group of psychiatrists who visited Bhopal several weeks after the accident acknowledged weakly that people were, indeed, suffering from anxiety and depression; but it was difficult to attribute these symptoms to the accident (*Sunday*, 1985). More recently, the King George Medical College has found widespread mental disorders among the Bhopal population.

Other damages and burdens add to the human health problems. Some 70,000 to 100,000 individuals left the city for distances up to 50km, with resulting disruption and economic loss. An estimated 1600 animals died on the first and second days after the accident, posing a serious disposal problem – eventually solved by digging a 0.4-hectare burial pit 5km from the city. Ecological effects – among the least understood of the accident's long-term effects – include apparent damage to certain vegetation, animal and fish species, but not to others, and are under study through the Indian Council of Agricultural Research.

What of the relief efforts to avert further health and ecological effects? Despite the remarkable emergency performance of the Indian medical system, many victims of Bhopal are suffering further harm. As often happens in disasters, the non-affected residents of Bhopal show a general lack of interest for the victims, many of whom are poor immigrants, not from Bhopal.

Numerous private relief organizations appeared quickly on the scene, administered their aid and departed. The Indian and Madhya Pradesh governments preside over an uneven relief programme, consisting of small doles of money, recently scheduled to be revised to US$180 per affected family (from US$833 for each death and US$12–240 for each person suffering injury or illness, depending upon its severity), and a free ration package of 12kg of cereals, 0.5kg of oil and 0.5kg of sugar per adult each month, plus 200 millilitres of milk daily per child. But of the 18,000 families expected to receive compensation, only half have been identified by June 1985 and of these only 4000 had actually received payment (Khandekar, 1985). The Indian government announced in June 1985 that 1500 housing units and a 100-bed hospital would be built to accommodate the most seriously ill Bhopal survivors (Diamond, 1985b). But despite the substantial efforts by the Indian government and a reported US$27 million expenditure by the Madhya Pradesh state government, the relief effort falls short of what is needed.

HAZARD MANAGEMENT

The Bhopal accident raises a number of basic issues for the management of industrial hazards, in general, and for disaster prevention in developing countries, in particular. The more prominent of these are characterized below.

Choice of technology

Few basic decisions affect hazard potential more than the initial choice of the technology. In the case of Bhopal, the choice involves the long-term storage of MIC, a chemical so extremely hazardous that some countries expressly prohibit long-term storage. Bayer AG, a large multinational corporation, manufactures MIC in West Germany and in Belgium; but the process uses the non-toxic intermediates dimethyl urea and diphenyl carbonate and involves no dangerous phosgene or chlorine. Moreover, Bayer promptly converts MIC into end products that are safe to store (Worthy, 1985), and temperature and pressure gauges on the tanks automatically control inconsistencies and can immediately trigger an alarm system (Ramaseshan, 1984). France prohibits domestic manufacture of MIC and requires that special MIC storage drums be maintained in separate sheds, equipped with automatic water sprinklers and sensitive gas detectors (Delhi Science Forum, 1985). England allows only one company to handle MIC and the gas must be stored at a site 3.2km out of the town of Grimsby (Delhi Science Forum, 1985). Such alternative technologies, replete with added automated safeguards, pose a lower inherent risk of catastrophic releases than the dangerous process chosen for the Bhopal plant.

Until 1978, UCIL did not store MIC at Bhopal. At that time, US corporate headquarters decided, apparently to promote efficiency, to utilize a technology that favoured large inventories of MIC, despite its high toxicity and in the face of reservations from the Indian subsidiary. The decision to store MIC at Bhopal was taken without apparent considerations of the particular safety issues posed by a location in India, compared with one in North America.

In addition, the Bhopal plant relied on manual, labour-intensive controls, while Union Carbide's plant in Institute, West Virginia, used a computerized monitoring system. These basic choices about the technology to be employed and the extent of needed safeguards at the Bhopal plant ultimately contributed to the disaster that occurred.

Siting and land-use control

Bhopal is situated in central India, and the plant site is astride main rail lines leading north to New Delhi, west to Bombay, east to Calcutta and south to the cotton-growing areas near Madras. For Union Carbide, the area was one of the few regions without an electricity shortage; as well, Bhopal's 16km-long Upper Lake provided an ample water supply for manufacturing chemicals, and a ready pool of labour existed in the city.

Since India has no policy on the siting of hazardous industries, no one contested the location of the plant a scant 3.2km from the centre of the city. Although the state department of town and country planning was aware of the plan to manufacture and store MIC, it approved the location and, in preparing the Bhopal Master Plan in 1975, subsequently classified the plant as 'general', rather than 'obnoxious', and did not require it to relocate to a more remote site (although 16 other smaller industries were relocated) (Qureshy, 1985). Indeed, in the aftermath of the accident, the Indian press engaged in controversy over whether an official who may or may not have issued an eviction order in 1975 was consequently transferred (Qureshy, 1985).

The absence of an exclusion zone or land-use control around the plant exacerbated the high risk of the location. Bhopal is one of the fastest-growing cities in the world: between 1961 and 1981 its population increased by nearly 75 per cent per decade, crowding people living in primitive huts onto government lands. The state government, having failed to prevent the population growth adjacent to the plant, compounded the problem by conferring legal status on those living in the squatter settlements in April 1984 so that they could vote in the December elections. This action was taken with knowledge of the accidents that had occurred at the plant and of the highly critical results of the 1982 safety investigation conducted by Madhya Pradesh.

Risk management

The Bhopal plant employed defence in depth: layers of safety systems intended to prevent major releases even in the face of individual system failures. Its five major safety systems were:

1   *Refrigeration system*: the MIC storage tanks were connected to a refrigeration system that circulates the liquid MIC and keeps it cool. In the event that MIC becomes contaminated, the refrigeration slows the reaction that may occur, thereby increasing the time available for safety response.

2   *Spare tank*: one of the three 60-tonne tanks at the plant is always left empty so that, in the event of an accident, MIC from a leaking tank can be diverted to the spare.

3   *Flare tower*: a 30m-high pipe located a short distance from the MIC unit is used to burn toxic gases high in the air, thereby rendering them harmless.

4   *Vent gas scrubber*: a tall, rocket-shaped unit is intended to detoxify any releasing gas by spraying it with caustic soda solution and converting it into a harmless vapour.

5   *Water curtain*: the plant was equipped with a network of water spouts that, in the event of an accident, shoots jets of water 12m to 15m in the air, forming a water curtain around the gas leak. The water neutralizes the MIC vapour to form dimethyl urea or trimethylbiuret, both comparatively safe substances.

Even if one or several of these systems were to fail, the others should successfully protect against any massive off-site release.

The Bhopal accident testifies to the vulnerability of even well-founded safety philosophy to diverse implementation failures and human error. At the time of the accident, with or without Union Carbide's authorization, the refrigeration system was not working. The use of a requisite spare tank – which may or may not have been empty – required that the operator manually open the valves connecting the two tanks, an operation taking no more than three minutes. In the confusion of the accident, however, the valves were not opened. The vent gas scrubber, on standby mode since 23 October 1984, failed – possibly because the operators neglected to augment the flow of caustic soda required to neutralize MIC. The flare tower designed to burn off escaping gas was under maintenance (because of pipe corrosion) and, thus, was inoperative. And the spouts designed to shoot jets of water into the air to quench a gas leak could not cope with the gusher of MIC some 35m high. In short, design errors, sloppy maintenance, poor safety practices, inadequate operator training and human error hopelessly compromised a many-layered safety system that should have worked (see Box 5.2).

Emergency preparedness

Bhopal was ill-prepared for the disaster. No emergency manuals or evacuation plans were available to local officials who, along with nearby hospital officials, were not aware of the toxic substances at the plant, their degree of toxicity, potential health effects or the recommended medical treatment. The exposure had occurred before the warning siren sounded (some hours after the beginning of the release), and no public warnings were issued. No information on the movement of the gas cloud – broadcast to the factory workers – alerted people in the vicinity about the best direction in which to flee. Tragically, people poured out of their homes and ran toward the factory. Although Union Carbide Corporation headquarters sent a telex on 5 December 1984, indicating that victims might be given amyl nitrite, sodium nitrite or sodium thiosulphate if cyanide poisoning (a possible sequel to MIC exposure) was suspected, UCIL did not divulge it to the public, arguing that administration of an antidote for cyanide would create widespread panic (Ramaseshan, 1985). Even basic instructions to breathe through a wet towel, which could have saved the lives of hundreds, were never forthcoming.

Box 5.2 Major findings of the Union Carbide Corporation investigation team (March 1985)

1 The incident was the result of a unique combination of unusual events.
2 The team believes that the safety valve remained open for approximately two hours before it reseated. During that period, more than 50,000 pounds of MIC vapour and liquid escaped through the safety valve.
3 The team's hypothesis is that the reaction in the tank occurred when a substantial amount of water was introduced into tank 610.
4 The exact source of water is not known; but laboratory work demonstrated that 1000 to 2000 pounds of water would have accounted for the chemistry.
5 The refrigeration system provided to cool the MIC in the storage tank had been non-operational since June 1984.
6 It is not known whether the 2 pounds per square inch (psi) pressure reading observed 40 minutes earlier by the operator on the previous shift had been communicated to the new operator.
7 The vent gas scrubber had been on standby mode since 23 October 1984. The return to an operating mode depended upon the operator being alerted to a problem and taking prompt action to activate the circulating pump.
8 The flow meter did not register a circulation of caustic solution in the vent gas scrubber.
9 The increase in temperature (of the MIC tank) was not signalled by the tank high-temperature alarm since it had not been reset to a temperature above the storage temperature.
10 Extensive experimentation suggests that residues in tank 610 are attributable to the reaction at high temperature of MIC with large amounts of water, higher than normal amounts of chloroform and an iron catalyst.

Source: Union Carbide Corporation (1985)

Given the lack of emergency preparations and critical information during the crisis, the response by India's medical system was nothing short of remarkable. Granted, much of the treatment was strictly symptomatic; but even timely eyewashes warded off some serious injury. It happened that the accident occurred 3.2km from the region's biggest, best staffed and best stocked hospital: Hamidia. Although its capacity was 760 beds, the hospital admitted 1900 seriously ill patients the first day after the accident and eventually treated more than 70,000 victims. By the second and third day, 35 mobile medical teams, each consisting of two doctors, a nurse, a pharmacist and paramedics, were operating outside the hospitals. Quickly the medical resources of Bhopal – 500 doctors, 5 major hospitals and 22 clinics – were supplemented by hundreds of doctors, nurses and paramedics flown in from the outside.

Amazingly, doctors never ran out of needed medicine and drugs in the aftermath of the accident. The location of the accident and the 'fast, intelligent, comprehensive marshalling of manpower, supplies and equipment' (Boffey, 1984) carry a sobering message. Had the spontaneous Indian medical response been more limited, had the medicines not been available, or had the accident occurred in other parts of the developing world lacking India's medical capabilities, the fatalities could easily have reached as high as 70,000 (Lepkowski, 1985).

Institutional issues

More dramatically than Three Mile Island, Seveso or Mexico City, the Bhopal accident has highlighted a host of far-reaching institutional problems. Although India is more advanced technologically and institutionally than most developing countries, the key legislation covering industrial hazards (the 1948 Factories Act) is geared toward mechanical rather than chemical hazards. Enforcement of environmental occupational health standards has been left to the state governments and is weak or non-existent. It is noteworthy, for example, that among 250,000 textile workers in India, not a single case of byssinosis has ever been reported (Lepkowski, 1985). The Madhya Pradesh labour department employs 15 factory inspectors for more than 8000 factories, and those employed at the Bhopal office are mechanical engineers with little knowledge of chemical hazards (Ember, 1985). Although the Madhya Pradesh government commissioned an investigation of the serious 1982 accident at the Bhopal plant, no corrective actions resulted. Trade union action on behalf of worker safety in India is also rare. Finally, although India has some active environmentalists, developing societies generally lack a well-developed network of environmental groups to contest and watch over policy and law enforcement – a key resource in developed countries. The accident also points up the absence of export controls over the transfer of hazardous industries to developing countries. Heretofore, much attention has focused on the export of hazardous technological products. The extent to which US regulatory systems function to review safety issues involved in the transfer of technology is currently unclear. In addition, the Reagan administration rescinded the Carter administration's Executive Order 12264, which placed restrictions on the export of hazardous substances. Similarly, at the international level, there is a general absence of codes and legal mechanisms to oversee the operation of multinational corporations.

All told, as Thomas Gladwin (1985) so aptly put it: 'The firm's technological reach had exceeded its managerial, cultural and institutional grasp.'

LEARNING FROM BHOPAL

Definite conclusions concerning the long-term lessons of the Bhopal accident must necessarily await fuller information and the analyses that are in process. Such efforts will certainly diminish many of the problems that thwarted effective hazard management at Bhopal; but some resolution of the deeper problems will require a longer time frame.

Values in conflict

The Bhopal accident spotlights an inherent tension between values in the *policies* surrounding the transfer of technology to developing countries and in the very nature of complex technologies. The Union Carbide plant came to Bhopal as part of an Indian plan for self-sufficiency in agricultural production and for industrialization of its underdeveloped regions through a labour-intensive technology transfer. But ensuring the prevention of catastrophe will probably require tightening controls by parent companies, which possess the overwhelming capability and knowledge for controlling industrial risks. Regulatory systems, whether in developed or developing countries, will always find it difficult to replicate such resources. Thus, the management of catastrophic hazards presses for centralized control by the parent corporation and extensive automation; at the same time, participatory economic development presses for decentralized control, indigenous resource development and labour-intensive production.

Transfer of technology

Probably no decision will more profoundly affect the balance of benefits and risks than the choice of technology to be transferred to developing countries. Whether in bioengineering, chemicals, energy or transportation, the technology in question will enhance development and economic gain only at some price in risk. But in response to the Bhopal accident, the flow of needed technology to developing societies may be impeded. Indeed, capital spending for developing countries by US chemical companies has been falling since 1981 (Webber, 1985).

The choice-of-technology mistake at Bhopal was not the transfer of the modern formulation plant for bulk pesticides needed for Indian agriculture. Rather, it was the construction of such a large MIC-based plant of this particular design (the most advanced pesticide plant in the developing world); the choice of a particular production process that carried higher inherent risk for a catastrophic accident; the choice of a scale of plant and equipment (designed to produce 5000 tonnes a year) that involved large inventories of highly toxic substances; and the failure to incorporate technological designs and practices for defence-in-depth protection against major accidents.

In the future, economic considerations alone should not drive the choice of the imported technology. Rather, they should be meticulously balanced with opportunities for risk minimization. In order to determine a safety goal, the realities of the developing country will have to be confronted. At Bhopal these included a shortage of experienced operating, maintenance and management personnel, and major structural obstacles to effective emergency response. The transfer of an *equivalent* safety goal is likely to involve *different* and *higher*, not identical, supporting safety systems.

Siting hazardous facilities

Past industrial disasters, including the Mexico City explosion and the Three Mile Island accident, have taught that remote siting provides a major protection against potential catastrophes. Particularly where engineered safeguards may be compromised through poor maintenance and equipment failure or operator error, physical distance – by affording greater dilution of the release and increased time for responses – offers an overall redundancy in safety. The tragic loss of life at Bhopal was concentrated in the densely populated squatter settlements located within 3km of the plant. A remote site for the plant or an exclusion zone of 3km at the Union Carbide plant might have averted most of the fatalities.

One apparent message from Bhopal would call for remote siting of facilities that carry the potential for rare but catastrophic events. Such a policy must be augmented by control over population growth around the facility. Short of truly remote sites, well-considered locations and land-use controls can provide important means of disaster minimization. Exclusion zones similar to those surrounding nuclear plants (in India, as well as elsewhere) should be provided, with the size of the zone determined by comprehensive, site-specific risk analyses. Use of company buses can readily overcome the problem of access to needed labour pools.

Risk assessment

The past ten years have witnessed a major development in the methods of probabilistic risk analysis (PRA), particularly for the nuclear industry. It is high time that other industries, including the chemical industry, begin to catch up. The US Nuclear Regulatory Commission (NRC) has recently summarized the current status of these techniques and their potential contributions for plant design and maintenance (USNUREG, 1984). These methods are likely to be more extensively used in Europe as chemical plants prepare the various safety studies required by the Seveso Directive, the European Economic Community's response to the 1976 accidental dioxin release at Seveso, Italy.

Although the absolute numerical values produced by such analyses should be viewed cautiously, the PRAs now being widely performed for nuclear power plants (and some other industrial facilities) have considerable power for identifying potential accident sequences that could lead to catastrophic consequences. These analyses should be more widely used in the licensing process for technologies with catastrophic risk potential. They should be comprehensive in scope – covering manufacturing operations, intermediate steps, storage and transportation of materials – and should more effectively integrate behavioural and cultural considerations, going beyond current practice (even in the nuclear case, human behaviour databases remain underdeveloped).

A comprehensive risk analysis, such as that for the Canvey Island petrochemical complex in England (UK Health and Safety Executive, 1978, 1981) or the Rijnmond Public Authority in Holland (Rijnmond Public Authority, 1982), should not only assess major risks but also identify cost-effective opportunities for risk reduction. Leading candidates for such analyses would be the petrochemical complexes at Baroda in India (Davidar, 1985) and in the Cubatao Valley in Brazil (Simons, 1985), where the major ingredients for industrial disaster exist.

Hazard communication and response

Some of the clearest failures in the Bhopal accident involved inadequate information flow and emergency preparedness. These failures occurred at various levels: inadequate worker understanding of MIC's toxicity and health threat; lack of knowledge by local government and medical officials of the plant's chemicals and their hazards; poor information during the accident to guide nearby residents; and lack of advice to local medical personnel as to recommended treatments. There was also no evacuation plan, and some workers may have escaped danger only because the door they chose to flee through happened, fortuitously, to open north rather than south, although some reports say that an announcement warned workers of the wind direction (Khandekar and Dubey, 1984).

An important lesson of the accident at Bhopal is how formidable the social, cultural and institutional impediments are to effective hazard communication and emergency response programmes. Such impediments include slow bureaucratic response, cultural differences in chemical experience and underdeveloped social communication networks. That the state government only convened at its normal meeting time, ten hours after the accident, to handle the emergency indicates the need to change the basic societal structure for responding to hazards. Bhopal has only one telephone per 1000 people, running water for only a few hours per day, few street signs or traffic lights, and crowded 3.7m-wide thoroughfares in which cows, goats, water buffaloes, taxis and horse-drawn carriages travel simultaneously in both directions (Diamond, 1985a). Local residents, with little experience with chemicals as part of everyday life and with no direct information provided by the plant, viewed the plant as producing 'medicine for the crops' and not substances harmful to people (*New York Times*, 1985).

Preventive maintenance to avoid accidents is a concept largely foreign to Indian culture. Although most workers at Bhopal had seen industry information on the hazards of MIC (Union Carbide Corporation, 1976), few understood it; nearly all underestimated the toxicity. And the lack of any local- and state-level emergency response organization made coordinated response to the accident impossible. In short, safe here is likely not to be safe there.

Institutional changes

The accident at Bhopal clearly reveals that at various levels institutional safeguards are inadequate. Multinational corporations will need to do more than conduct the initial reviews of toxic substance storage and handling as they did in the several months following the accident. They will need to re-examine, and undoubtedly increase, their capabilities in formal risk assessment for safe plant siting, design and management. Most will require new resources to appraise both catastrophic risk and the relevant societal and cultural factors. This upgrading should be modelled after the Institute for Nuclear Power Operations, which was established after the Three Mile Island incident, and should consider the use of simulators in operator training (Phung, 1984). The corporate codes of social responsibility will also need to address explicitly the lessons from Bhopal and then act on the more demanding obligations by improving auditing, monitoring, and compliance programmes.

Host countries will need to institute licensing and regulatory structures that are sensitive to their own 'risk-carrying capacities'. They will need to stipulate more stringent requirements for highly toxic substances or for industries possessing potential for other catastrophic accidents. Such controls should emphasize the development of hazard databases (Ashford, 1984), which will need to be linked for easy communication and exchange of information. The Bhopal accident also highlights the necessity for more openness by both the host government and corporations in siting decisions, environmental impact assessments, accident inquiry reports, and basic information on hazards. India has no provision for public participation or public hearings, and most governmental and official documents are confidential. The 'right to know' principle, if applied in India and other developing countries, would afford additional protection to workers and the public.

THE NATURE OF RESPONSE

As the Bhopal tragedy nears its first anniversary, broad-based efforts to avoid such future accidents have begun in earnest. Nearly all the large chemical corporations have examined catastrophic risk potential and safety practices at their plants in North America and abroad and have discovered problems, particularly with the size of chemical inventories. This has led to changes in corporate policies governing storage and emergency-response planning (Sapulkas, 1984). A task force of the US Environmental Protection Agency reviewed the issue of airborne toxic substances, recommended reinforcement of the national contingency plan for emergency response, and underscored the need to strengthen hazard communication. At the same time, the task force decided against special restrictions on the chemical industry (USEPA, 1985).

In addition, the World Bank has reviewed all of its major projects that have potential environmental impacts and is developing hazard prevention guidelines, due this month, based on the Seveso Directive. The Chemical Manufacturers Association has established two new programmes designed to increase member company and local community emergency response and the flow of hazard information to local officials and publics (*Chemical and Engineering News*, 1985). In 1985 a number of multinational corporations joined with the US Agency for International Development and the World Environment Centre in Project Aftermath, an assessment of information and technical assistance needs for industrial hazard management in some 40 countries (World Environment Centre, 1985).

These undertakings are a start; but it is essential to recognize that no quick fix exists to erase the problems that led to the Bhopal accident. Advances in formal risk analysis have yet to permit definitive quantification of accident probabilities and sequences for complex technologies. The cultural and social impediments to well-intentioned plans remain poorly understood and are only weakly incorporated within safety programmes. Corporations and regulatory agencies alike change slowly.

Previous experience in disease prevention attests to the long time frames and the degree of determination necessary for fundamental gains. The European effort to implement the corrective measures set forth in the Seveso Directive will not reach fruition until 1989, some 13 years after the accident. Despite the massive response by industry and government to the Three Mile Island accident, the extent of safety upgrading achieved and the degree to which attitudes towards safety have been altered remain uncertain six years after the event. So, too, one should expect the depth of the problems posed by Bhopal to require a decade of determined response, the adequacy of which will require testing over the long term. Moreover, it should surprise no one, as the accident recedes from societal attention and urgency, that conflicting values will tend to erode early progress in resolving the issues raised by the Bhopal accident, in developed and developing countries alike.

Of critical importance is the creation of a culture of safety among those who transfer technology to developing countries, those who woo it for the development of their national economies and localities, and those who manage it when it arrives. The continuing agony of the Bhopal victims argues poignantly that the magnitude of response be commensurate with the tragedy that necessitated it.

6 Emergency Planning for Industrial Crises: An Overview

Roger E. Kasperson and Jeanne X. Kasperson

INTRODUCTION

Serious industrial accidents come as surprises. As the 20th century winds down its final dozen years, the likes of Seveso, Bhopal, Mexico City and Chernobyl will catch society unaware and trigger, again, by now predictable responses. Whereas catastrophic events are invariably unanticipated, accidents that threaten the public are commonplace. In the US alone, a warning notifies the public of an industrial accident about once a day and several evacuations occur each week (Sorensen et al, 1987, p1). Moreover, some of the thousands of shipments of hazardous materials that move each day by truck, rail, air, barge and ship inevitably spawn transportation accidents that require emergency response.

For all the experience, however, the major industrial disasters of the past decade have revealed that emergency preparedness all too often falls short of societal expectations, if not regulatory requirements and professed readiness. Furthermore, each accident highlights distinctive shortcomings or differences in readiness for coping with the emergency at hand. So the accident at Three Mile Island revealed large-scale institutional disorganization and confusion in diagnosis of the incident; Bhopal pinpointed the lack of exclusion zones and emergency communication systems; the explosion of the space shuttle *Challenger* laid bare organizational imperatives that override risk warnings; and Chernobyl drew attention to the lack of transnational coordination, the fragmentary nature of existing standards for environmental protection, and the inadequacy of planning for distant accidents.

The aftermath of a major accident always produces some welcome improvements. Indeed, much has occurred to upgrade our understanding of industrial accidents and to improve emergency preparedness. Notable developments include advances in the use of probabilistic risk analysis

Note: Reprinted from Kasperson, R. E. and Kasperson, J. X., *Organization and Environment* (formerly *Industrial Crisis Quarterly*), vol 2, pp81–87, © (1988) by Roger Kasperson and Jeanne Kasperson. Reprinted by permission of Sage Publications, Inc

(PRA) that will permit greater precision in defining accident sequences and the nature of low-probability/high-consequence events; the tapping of accumulated experience with natural disasters to enhance understanding concerning the effectiveness of warning systems and alternative organizational arrangements; the regulatory regime and extensive efforts around the 102 nuclear power plants in the US; and the chemical industry's CHEMTREC system for transportation accidents and its more recent Chemical Awareness and Emergency Response (CAER) programme. The recently enacted Superfund Amendments and Reauthorization Act (SARA) Title III amendments, with stipulations for extensive dissemination to communities of information concerning chemical releases and the provisions for emergency planning, promise a new and quite remarkable societal venture into participatory emergency planning.

The US is not alone, of course, in these preparations. Under the Seveso Directive, European countries are, in a number of respects, more advanced in the formulation of emergency plans for potentially hazardous industrial establishments, although the scope of public involvement appears more limited than that unfolding in the SARA amendments. Meanwhile, the Bhopal accident has stimulated an international effort to develop guidelines for the transfer of technology to developing countries and to foster greater attention to emergency planning. Chernobyl has also engineered various national actions to improve communication with the public and international efforts at improved coordination in information flow and response measures.

Yet, for all this, significant uncertainties plague emergency planning and preparedness systems. This is perhaps inevitable, and to some extent the uncertainties remain irreducible, because:

- The causes and particular form of major accidents exist in astonishing variety.
- Human contributions and responses are often unpredictable.
- The low-probability nature of catastrophic accidents means that the adequacy and appropriateness of emergency plans for worst cases are rarely tested.

And then, of course, there is the inevitable tendency to correct for the last mistake rather than for the one not yet experienced.

In the following section, we note a number of the issues that lie in wait for the various national and community initiatives in emergency planning.

ANTICIPATING INDUSTRIAL ACCIDENTS

For most of history, it has been enough to deal with accidents by trial and error. As experience mounted with our interactions with nature and technology, initiatives sought to make events more unlikely (accident prevention in industry) or to decrease vulnerability to events (remote siting of hazardous facilities). During the 1960s, quick response to newly appearing hazards was typically the accepted standard for ensuring safety.

That standard has not been sufficient for some time. The array of problems associated with chronic hazards, the increasing complexity associated with anticipating major accidents and the greater risk aversion of the public have demanded that industrial hazard management become anticipatory in nature. In essence, this has meant that analysis must attempt to estimate rare events and that simulation must substitute for experience – that industry and government alike, in short, must develop new capabilities in identifying, assessing and planning for major risks, both acute and chronic.

Such a capability is mature only in several industries – in the aircraft industry where growth has depended upon the steady improvements in air safety over the decades and in nuclear power where the military origins of the technology and the spectre of the rare but catastrophic accident have forced investments in risk analysis and defence-in-depth approaches to safety unparalleled in industry, more generally. The development of large-scale, plant-specific computer models employing increasingly sophisticated PRA techniques has become an important tool in improving safety designs and procedures in the nuclear industry (Dooley et al, 1987, pp10–11). The potential of PRA is yet to be exploited fully in emergency planning; but the enhanced ability to define accident spectra, accident evaluation and release timing, and potential health and environmental consequences is a critical consideration in shaping appropriate decision structures, geographical zones of preparedness, warning systems, provisions for medical mobilization and efforts at public education.

Risk analysis techniques and capability need to be developed in various other industries, including the transportation sector. This will involve new expertise and an investment in analytical capabilities unfamiliar throughout much of industry and state governments. It seems unlikely that the implementation of the SARA amendments can go forth without significant embedding of risk analysis capabilities in both industry and state regulatory agencies.

Distributed decision systems

Emergency response to industrial accidents necessarily involves integrating information and action across a wide variety of expertise, public and private sectors, differing levels and agencies of government, responsible officials, and the public. Major accidents cry out for what Lagadec (1987, p29) terms 'organizational defence in depth'. It is estimated that the Three Mile Island accident involved up to 1000 different decision makers in responding to the accident. Ensuring timely flow of information and coordinating efforts in such a widely dispersed decision structure can be a Herculean problem, with countless opportunities for failure. Another key problem is that emergencies frequently necessitate that organizations assume roles and responsibilities that are unfamiliar and that they interact with other organizations with whom they are unfamiliar. Indeed, the long experience of members of the Disaster Research Center has prompted the identification of organizational coordination and performance as a central factor in emergency response success and failure.

Distributed decision structures raise a number of intrinsic problems that are not easily overcome. The agencies involved have competing missions and priorities – and, not infrequently, rivalries. Crises are not only substantive problems requiring solutions, but opportunities for settling old scores, enlarging funding sources, and changing organizational mandates and programmes. Problems tend to be defined in different ways and information often carries conflicting symbolic and substantial meaning. Some agencies gain by acting in pre-emptive fashion, others by deferring. Such decision systems are, in short, a Pirandellian delight.

It is not uncommon that emergency planners seek answers to institutional fragmentation in command-and-control structures, often modelled on the military (Quarantelli, 1988). Indeed, the extent to which ex-military officers populate emergency planning offices is striking. The command-and-control structures that have evolved do have a number of attributes responsive to emergency situations. Ordinarily, such structures provide clear delineation and enumeration of tasks. A strong vertical organization with well-developed channels of communication ensures the rapid dissemination of information. To a varying degree, the inclusion of back-up systems permits accommodation should particular links in the system fail. Overall, such structures attempt to deal with the authority dispersion problem by imposing detailed articulation of roles and tasks and by invoking predefined categories of accidents and emergency response. At their best, such structures are capable of rapid diagnosis, complex task accomplishment and timely response.

But they are also prone to a number of failures. The intelligence system depends heavily upon the accuracy of assumptions and diagnosis at the apex of the system, for the upward flow of communication is characteristically weak. Even when signals come in from outlying areas, the command function often has difficulty interpreting or assigning priority to them. Command-and-control structures also tend to develop rigor mortis as guidelines become regulations, assumptions reality and broad categories substitute for rich variance. As a consequence, when unexpected situations occur that do not fit preconceived structures, response can easily become delayed or maladaptive. Meanwhile, most participants know only their specific role and have little sense of the response system as a whole.

So, decision structures for emergency planning confront a central problem: how to ensure timely and effective response from multiple and dispersed decision makers, on the one hand, and how to maximize localized knowledge and adaptiveness, on the other. A central problem is how to make decision processes resilient to surprise and failure.

Public perceptions and behaviour

The success of any emergency response system ultimately depends upon its sensitivity to the perceptions of risk by and the behaviour of various publics during an accident. Much knowledge exists about how these publics view different types of risk, about the elements of a warning system that may shape public responses, and about likely public behaviour under high stress.

Yet stereotypes and misconceptions abound. Many emergency planners fret about the panic they deem as almost certain to erupt during an emergency despite the rarity of such instances. Others seek plans to overcome the likelihood that emergency workers will fail to perform their specified duties, although cases of emergency role abandonment are almost totally absent in disaster experience. Still others worry about overwhelming the public with information, although experience suggests that the public seeks and processes an extraordinary amount of information during serious emergencies. Even the nature of human nature comes into question, with fears over antisocial behaviour at odds with the observed tendency for widespread altruistic actions. At root, the problem is that everyone is an expert on the public, so that the need to be informed by science is left to other domains. Major improvements in emergency planning are possible by narrowing the gap between what is known, what is practised and what is assumed by the professionals.

Beyond that, we do, of course, need to know more. The social amplification of risk (see Chapter 6, Volume I) that trails in the wake of accidents is a major driving force in defining the requirements for emergency planning. How prepared should society be for rare but catastrophic industrial accidents? To what extent can emergency planning ensure safety for any given facility? How much of society's scarce resources should be allocated to emergency preparedness? And how will the adequacy of emergency plans affect future siting of industrial establishments and the transport routes for hazardous materials? How much variation exists in the public responses to different hazards and to different hazard events?

RISK COMMUNICATION

Improved emergency preparedness places heavy burdens on effective communication of risk information prior to and during crises. The communication needs exist at multiple levels – the risk assessor must inform the industrial manager and the emergency planner, mass media representatives must be knowledgeable, information must flow effectively throughout the overall emergency organization, and the public needs some understanding of potential accidents, as well as available means for protecting itself. The accumulated experience with industrial emergencies increasingly points to an environment rich in timely and accurate information as, perhaps, the most precious of all emergency resources. At the same time, the likes of Bhopal, Mexico City, Chernobyl and Three Mile Island, as well as crises with chronic hazards (such as Love Canal), suggest how vulnerable risk communication systems are to breakdown and distortions when the event occurs.

Communication and educational systems for preparing communities and local publics for accidents at industrial facilities are still in their infancy. As Paul Shrivastava (1987b, p2) noted: 'Of all the tasks involved in coping with industrial crises, communications is one of the most important and least understood.' Industrial officials have understandably been reluctant to trot out all of the bad things that could potentially happen at plants that they want to present as 'good neighbours', while local government officials have been loathe to assume new burdens that they are ill-equipped and poorly supported to carry out. And lurking in the background is the ever-present fear that frank talk about potential accidents may spark overreaction by well-meaning publics and media representatives, and by mischief in the hands of the miscreants who are potential opponents with private agenda. The first telling obstacle, then, in effective preparatory risk communication is overcoming the reluctance of those who possess knowledge of the risk to share it openly with those who are at risk (see Chapter 1, Volume I).

Beyond that, of course, lurk the difficult questions regarding what should be communicated. The public information brochures that have been prepared for nuclear and chemical emergencies have concentrated primarily on what the public needs to do, but very little exposition on why. These brochures are directive and prescriptive, but certainly not pedagogic. Improved approaches will need to go well beyond prescriptive action guides to broad programmes aimed at enhancing local knowledge about risks and at enlarging the basic capabilities of the public to protect itself from industrial accidents. Such programmes will involve not only brochures and pamphlets, but also training, risk education, study circles and emergency-response simulations and exercises.

Communication during crises requires special handling, for time is characteristically short, stress high and confusion abundant. Much is known, particularly from experience with natural disasters, about the design and implementation of effective warning systems (Mileti, 1975; Nigg, 1987). Experience there speaks to the need for multiple channels of communication, the use of credible sources of information, the importance of care in designing messages, the frequency of message transmission, the need for message validation, consistency among message sources, prior training of mass media representatives, redundancy in communication mechanisms, and effective rumour control. It is also apparent that the characteristics of the risk event (e.g. rapid onset and the persistence of the hazard), as well as the level of public concern, are important considerations in the design of the communication. Again, a substantial gap separates what is known from experience and what is assumed in warning system designs.

7 Corporate Culture and Technology Transfer

Jeanne X. Kasperson and Roger E. Kasperson

INTRODUCTION

Locating industrial facilities in developing countries confronts corporations with a stream of choices and potential conflicts among corporate objectives. How these choices and competing objectives are interpreted and reconciled depends heavily upon the meanings that are attached to them and the values that are brought to bear in corporate decision making. Accordingly, much depends upon what has come to be called *corporate culture* and the extent to which this culture actually guides the behaviour of its managers through the various stages of developments that result in the acquisition, construction or operation of a facility in a developing country. This chapter enquires into this issue in the case of E.I. DuPont de Nemours and Company, or DuPont, widely cited as having a particularly well-developed corporate culture.

Corporate culture is, of course, not a new concept; it has, in fact, several diverse roots of origin. Members of the 'Carnegie School' of organizational theory (Simon, 1951; Homans, 1958; March and Simon, 1958; Blau, 1964) conceived organizations as entities that make contributions in exchange for inducements and incentives. Social anthropologists, such as Kluckhohn (1951) and Levi-Strauss (1963), have included organizations in their more general writings on culture. As early as 1951, for example, Elliot Jaques (1951, p251) observed:

> The culture of a factory is its customary and traditional way of thinking and of doing things, which is shared to a greater or lesser degree by all its members, and which new members must learn, and at least partially accept, in order to be accepted into service in the firm. Culture in this sense covers a wide range of behaviour: the

Note: Reprinted from *Corporate Environmentalism in a Global Economy: Societal Values in International Technology Transfer*, Brown, H., Derr, P., Renn, O. and White, A. (eds), 'Corporate Culture and Technology Transfer', Kasperson, J. X. and Kasperson, R. E., pp149–177, © (1993), by Brown, H., Derr, P., Renn, O. and White, A. Reproduced with permission of Greenwood Publishing Group, Inc., Westport, CT.

methods of production; job skills and technical knowledge; attitudes toward discipline and punishment; the customs and habits of managerial behaviour; the objectives of the concern; its way of doing business; the methods of payment; the values placed on different kinds of work; beliefs in democratic living and joint consultation; and the less conscious conventions and taboos.

Attention to corporate culture as a way of understanding the behaviour of corporations has ebbed and flowed since these early writers first proposed this approach. The focus of scholarly writings on corporations has shifted from scientific management, organizational structure and rational or strategic planning during the 1970s and early 1980s. The 'surprise' of the two Organization of Petroleum Exporting Countries (OPEC) oil crises during the 1970s, the growing competition for American industry from foreign competitors, and particularly the stunning competitive success of Japanese firms in both smokestack and high-technology industries have rekindled interest in corporate culture as a framework for analysing corporate performance and behaviour.

This chapter examines the ways in which corporate culture enters into, and affects, the process by which a multinational corporation locates facilities in developing countries. In particular, it seeks to clarify the concept of corporate culture, assess the extent to which such a culture exists at DuPont, and analyse the role that this culture played in the location of a DuPont facility in Thailand. The discussion begins by examining the concept of corporate culture and its relevance to the issues of the technology transfer treated in this book. Recent research into high-reliability organizations and their management of hazards and safety adds another dimension to the analysis. Using these conceptual approaches, the chapter then examines in detail the location of DuPont's Bangpoo factory in Thailand. DuPont has achieved global recognition as having a distinctive and well-developed corporate culture, and Thailand itself is a culture very different from settings in the US; thus, the case provides a particularly rich context in which to explore these issues.

CORPORATE CULTURE: Functions and processes

Given the several theoretical origins of research on corporate culture and the flourishing interest in it during the 1980s, it is not surprising that a variety of frameworks and theoretical perspectives has emerged. It is useful to characterize these major approaches before examining how corporate culture entered into technology transfer in the case of DuPont's Bangpoo facility. Distinguishing major approaches to the analysis of corporate culture is unavoidably somewhat arbitrary given the great diversity of available perspectives. Particularly relevant for this study are four approaches: culture as *social functions, systems of common beliefs, systems of shared knowledge or patterns of symbolic discourse*.

The *functionalist approach*, in search of patterns that may have implications for organizational effectiveness, examines variations in managerial and worker perceptions, attitudes and behaviour across organizations and countries (e.g. Ronen and Shenkar, 1985). In this view, corporate culture is an internal attribute of the corporation that can enhance the effectiveness of the firm and provide a potential competitive advantage over other corporations. Research, therefore, is a quest to understand how such cultures differ and the means by which they might enhance corporate performance. This approach led sometimes to what might be termed corporate culture engineering, in which managers emphasize the manipulation of corporate culture as a powerful lever to achieve some other end, such as safety performance or some specific corporate goal. Managers, proponents argue, must find ways of using stories, legends and symbols to achieve these corporate goals (Peters, 1978).

This culture–engineering approach is apparent in the nuclear industry's calls for a safety culture. In the Nuclear Regulatory Commission's (NRC's) view, a *nuclear safety culture* is:

> ... a prevailing condition in which each employee is always focused on improving safety, is aware of what can go wrong, feels personally accountable for safe operation, and takes pride and ownership in the plant. Safety culture is a disciplined, crisp approach to operations by a highly trained staff who are confident but not complacent, follow good procedures, and practice good team work and effective communications. Safety culture is an insistence on a sound technical basis for actions and a rigorous self-assessment of problems (USGAO, 1990, p28).

In analysing safety problems at the Savannah River nuclear plant, the General Accounting Office (GAO) found that this safety culture had not kept pace with changes in the commercial nuclear power industry and that the old culture (whatever that was) continued. Accordingly, Westinghouse, the operator of the plant, was required to produce a comprehensive implementation plan involving specific changes needed in the safety culture, milestones for their accomplishment and measurement indicators of cultural change (USGAO, 1990, pp31–32).

A second view of corporate culture, one that overlaps with the functionalist approach, sees corporate culture as a *fabric of shared beliefs and values*. For many analysts (e.g. Deal and Kennedy, 1982; Liedtka, 1988), values form the heart of corporate culture. Such values define in concrete terms success and standards of achievement for employees. In corporations with highly developed, or strong, corporate cultures, which will be discussed later in this chapter, members share a rich and complex system of values and talk about them openly and without embarrassment.

A third approach conceives corporate culture as *shared knowledge*, the conceptual and ideational systems and the language and metaphors that are enlisted to make sense out of the corporate world. The semantics reflect the worldview of corporate members, whereas language and communication offer a rich means of interpreting the thought patterns that sustain particular forms of behaviour. Social interaction occurs through the exchange of symbols that have a shared meaning for members of the set or organization. The methodology for this approach is akin to that of ethnomethodologists or the post-modernists in their embrace of qualitative interpretation of specialized meanings attached to various aspects of routine life, the rules and scripts that guide action, and a generally subjectivist view of social reality (Smircich and Calás, 1987). Gregory's (1983) study of how Silicon Valley technical professionals understand their own careers exemplifies this approach.

Finally, some see corporate culture primarily as *patterns of symbolic discourse* (Pondy et al, 1983). These patterns, or themes, require deciphering and interpretation, following the work of the anthropologist Clifford Geertz (1973). Ingersoll and Adams (1983), for example, seek to identify the meta-myths that lie behind organizational behaviour and managerial practices.

Analysing business literature and executive biographies, they find rationality celebrated continuously as a managerial meta-myth. Others (e.g. Rosen, 1985) use a social drama metaphor to examine how the symbolic rites and ceremonies of corporations contribute to social order and patterns of political control. Stories also may provide clues to culturally based values and assumptions (Martin, 1982).

It is important to appreciate that these approaches proceed from different and sometimes incompatible assumptions and concepts. They do each offer important insights into or illuminate faces of corporate culture. Rather than debate the relative merits or correctness of each, we draw upon these several approaches to define the concept of corporate culture that prevails in this chapter.

Corporate culture, in the authors' views, refers to the basic values, beliefs, norms and fundamental assumptions that members have about the corporation, its environment and human nature; the organizational structures and rules of behaviour that emerge from these values and shared meanings; the discourse, relationships and symbols that provide coherence to the organization; and the policies, programmes and procedures that implement the goals and norms.

Following Mitroff and colleagues (1989), the authors envision corporate culture as having an onion-like quality (see Figure 7.1). At the core of the culture is the root mission and identity of the corporation and its members' most basic beliefs about the world, the corporation's place in that world and the individual's place in the corporation. Although the factors that compose the core are usually the most decisive for shaping the behaviour of the corporation, they are often the most difficult to articulate or discern. The next layer involves key organizational assumptions and values that give particular direction to organizational functioning and

Organizational discourse, symbols, policies, and behaviour

Organizational structures and rules

Organizational assumptions/values

Core organizational identity, deep beliefs, shared meanings

Source: adapted from Mitroff et al (1989, p272)

Figure 7.1 *An onion model of corporate culture*

actions. These assumptions and values are typically expressed in particular organizational structures and systems of rules. The interactions, however, are two way throughout the onion layers, so changes in the outer layers may also alter inner layers. As Giddens (1979) has noted, social structures cannot be said to exist without reference to the meaning that such structures have for the affected individuals. Thus, core identity/beliefs and concrete behaviour/policies are mutually expressive. And the outermost layer of the onion composes the policies, behaviours, discourse and symbols that express and maintain the structures and rules of the inner layers (Schein, 1980, 1985).

Corporate culture, then, provides the core beliefs and assumptions that set the base for the rules and operating principles in which members of the corporation root their behaviour. It also establishes the frames of reference and perspectives from which to interpret events and respond to surprises or crises. In essence, it is a shorthand code that pervasively and subtly renders reality understandable and provides order and security to its members.

Not all corporate cultures are the same, of course; they differ in key attributes. Here we note, in the form of a matrix (see Figure 7.2), several variables that are central to shaping the effectiveness and capacity of the culture to respond to surprises in its external environment.

Some corporations have a clear sense of mission; others do not. *Mission*, in the authors' usage, refers to *the shared definition of corporate goals and*

MISSION

A	B
Clear Mission, High Consistency	Unclear or Divided Mission, High Consistency
C	D
Clear Mission, Low Consistency	Unclear or Divided Mission, Low Consistency

CONSISTENCY

Figure 7.2 *Mission and consistency in corporate culture*

functions vis-à-vis the larger world. A strong sense of mission provides clarity and direction to the behaviour of managers and workers and a capacity to take action in novel, ambiguous or even emergency situations. It also provides an intrinsic sense of meaning and worth to the contributions of the corporation. A highly internalized sense of mission allows members to shape their behaviour to distant ends that transcend short-term planning horizons or corporate targets. Denison (1990) cites the example of Medtronic, a corporation best known as a producer of cardiac pacemakers. A very distinctive part of Medtronic's unique culture was a powerful sense of mission 'to contribute to human welfare by application of biomedical engineering in the research, design, manufacture and sale of instruments that alleviate pain, restore health and life of man' (Denison, 1990, p98). This strong mission of caring and commitment served as a magnet for building the corporation's workforce, for developing an explicit ideology and behavioural norms, and for managing the corporation effectively with little internal control or accountability.

    Consistency refers to the extent to which managers and workers agree on corporate mission, values and rules. Highly consistent corporate cultures provide a strong basis for effective communication since values and meanings are widely shared. Agreed-upon expectations and norms shape behaviour in ways that formal rules and organizational structures cannot. Two types of consistency may be envisioned. *Horizontal consistency* occurs when members in very different parts of the corporation can easily communicate and coordinate efforts. Such congruence and common knowledge contributes to effective working relationships, tacit agreements and conflict resolution. *Vertical consistency*, by contrast, refers to a high understanding among workers regarding the expectations of their superior and, conversely, to the understanding, on the part of management, of workers' assumptions and norms. High vertical consistency typically involves a sense that a member will be rewarded over the long term for short-term contributions and sacrifices (Camerer and Vepsalainen, 1988, p121). Both types of consistency breed significant economic efficiencies, as well as a capacity to respond to challenges and surprises.

Corporations possessing clear missions and high degrees of consistency are often described as having *strong cultures*. In such corporations, employees have a well-developed sense of purpose and an attachment to the organization. Clear mission and informal norms provide strong guidance for behaviour – little time is wasted in seeking direction or interpreting events. Moreover, corporations with strong cultures are resilient to surprise and adversity. When Delta Airlines Flight 191 crashed in August 1985, for example, volunteers from Delta's management ranks immediately came forward to assist the families of victims and survivors. Although the corporation did not solicit or recruit them, many of its executives flew to Dallas where they gave their time and emotional energy, sometimes at great personal cost, to help in victim identification and burial arrangements. This response appeared to emerge largely from their identification with the 'Delta family' and their sense that Delta was in a time of need (Isabella, 1986, p175). Procter and Gamble is frequently cited as another corporation with a strong and highly consistent corporate culture, with a powerful tradition of listening and responding to its customers and a socialization process that builds values in key areas of the corporation's operations (Deal and Kennedy, 1982; Denison, 1990, pp147–160). In their qualitative analysis of differential performance of some 80 companies, Deal and Kennedy (1982, p7) found that the only common denominator among the highly successful companies was that all were consistently 'strong culture' companies.

To the extent that a culture's consistency is undeveloped or that other strong cultural identities exist and compete for loyalty among its members, however, a corporation is apt to harbour various *sub-cultures*. Such sub-cultures may be quite well developed and the systems of common beliefs and assumptions may depart substantially from those in the dominant corporate culture. This observation is akin to the argument of Thompson et al (1990) that cultures are plural and not singular. Members of the health and safety departments of a corporation, for example, may embrace rather different beliefs and expectations about the corporation than do members of the accounting or marketing divisions. Blue-collar workers recruited from a particular ethnic group may differ substantially from white-collar managers. And, of course, in cases of corporations locating plants in developing countries, locally recruited employees may bring cultures profoundly different from those existing in the parent corporation.

Although much discussion and writing have centred on how strong cultures arise and are maintained, in fact, these processes are not well understood. The notion that corporate cultures, like an administrative flow chart, can be easily engineered is particularly naive. It may well be, as Camerer and Vepsalainen (1988) argue, that strong corporate cultures simply arise without benefit of purposeful design. Some of the factors that other studies have found to be important in such emergence include:

- *Early history*: the early history of a company can provide a strong stamp for mission, values and assumptions. And since precedents, myths and stories can be powerful forces, such cultures may change slowly.
- *Heroes*: charismatic leaders can be pivotal figures in establishing and embodying corporate values. They are symbolic figures who are both motivators and role models for members of the corporation.
- *Recruitment and socialization*: some companies consciously recruit particular types of individuals whom they feel personify or will fit and reinforce the corporation's culture. A variety of training and socialization processes may be used to create and maintain cultural identity, norms and learning (Hayes and Allinson, 1988).
- *Incentives and sanctions*: reward and penalty systems are often employed to reinforce corporate objectives, norms and rule. Members who integrate well with the culture are rewarded and promoted; others languish or are let go.
- *Rites and rituals*: how things are done in the company, even when highly routine, can develop and express corporate culture. Ceremonies place the corporate culture on display and provide memorable experiences; rituals express myths that sustain values, and scripts that create meaning.

Later in this chapter, each of these factors is discussed in the context of DuPont's establishment of the Bangpoo plant.

Some see a hierarchical structure for management in the dynamics of corporate values. In this view, top managers function as the cultural *value formulators*, those who by their actions and priorities signal key values and mission to the members of the corporation. The president and chief operating officer, in particular, 'provide the map that ultimately directs the behaviour of employees' (Isabella, 1986, p188). The middle managers are the *value translators*, who convey key values and assumptions into the structures, policies and everyday procedures of the corporation. As translators, these managers reinterpret events and situations in a manner that communicates actual norms and rules to subordinates. Finally, the *value maintainers* are the lower managers who implement corporate values by making decisions and engaging in behaviour that is consistent with the corporate culture. Although it is doubtful that any corporate culture functions in quite this idealized, top-down way, the distinctions among roles do relate to the onion conception of corporate culture and the cultural processes discussed here.

This brief overview of corporate culture provides a backdrop for a second stream of research. Work on high-reliability organizations is particularly relevant to the issues involved in the location of potentially hazardous facilities in developing countries.

HIGH-RELIABILITY ORGANIZATIONS AND RISK

Most studies of corporate culture give short shrift to the ability of high-technology firms to prevent catastrophic accidents; but this issue has emerged as a major societal concern. The Bhopal, Mexico City, Three Mile Island and Chernobyl accidents all focused attention on questions of safety and reliability. Although still very exploratory, research during the 1980s addressed hazard management issues that warrant integration with the more established, traditional approaches to studying corporate culture.

In 1984, Perrow published *Normal Accidents*, a book concerned with the increasingly complex technologies needed by society and the inherent problems in ensuring safety, especially what is necessary to prevent catastrophic accidents. Perrow devotes particular attention to two problem areas – systems complexity and tight coupling – that confront high-risk organizations. Complexity refers to such issues as potential for unexpected sequences in accident development, technological complexity, potential for interactions among apparently separate or incompatible functions, and the indirectness of needed information sources. Tight coupling refers to the lack of slack in systems, the time-dependent nature of many processes, invariant sequences of operations and single means for reaching particular goals. In Perrow's view, serious accidents are normal in high-risk organizations because they are destined to fail. Because these organizations deal with highly sophisticated technologies that involve unanticipated interdependencies and interactions, they face a constant struggle to ensure high reliability. Indeed, reliability and safe operations, rather than productivity, become the chief goal (or, in the authors' usage, 'mission'). In such systems, problems propagate in unexpected ways, and even high reliability in each and every part fails to ensure high system reliability. The coincidence of tight coupling and technological complexity, in short, creates conditions of interactive complexity, which, in turn, pose unique problems and generate a new family of system-level failures that end up as catastrophic accidents.

Although Perrow sketched this argument in vivid hypothetical terms, the empirical basis and confirming evidence that he offered were still rather meagre and debatable. Over the past seven years, a group of scholars at the University of California at Berkeley has further developed and modified the initial hypotheses. High-reliability organizations, in their view, are those in which managers pay extraordinary attention to operational reliability because of the inherent dangers of a situation, and because achieving desired outcomes (e.g. safety) that are related to reliability is otherwise impossible. Highly reliable organizations are considered to be those that repeatedly avoid catastrophic events even though the potential for such accidents is high (Roberts, 1989, pp112–113). Put another way, the research centres on those cases that fall in quadrant 1 of Figure 7.3.

TECHNOLOGICAL RISK

	HIGH	LOW
HIGH	1 Air Traffic Control	2 Urban Water Supply
LOW	3 Bhopal, Three Mile Island	4 Handicrafts

RELIABILITY

Figure 7.3 *The relationship between reliability and technological risk*

La Porte (1987), arguing that organizational research has generally shed few insights into the nature of high-reliability organizations, has called for detailed studies of the structures and behaviour of such systems. He defines five pairs of knowledge/behaviour patterns that make for high reliability:

1   Nearly complete causal knowledge of the functioning of the technical and organizational system helps to ensure expected outcomes. Based on this knowledge, both personnel and machines exhibit nearly error-free performance.
2   Error regimes specify the small deviations from routines for both machines and operations and signal potential onset of failures in critical components. Associated with this are well-developed alerting systems and personnel with strong error-identification capabilities.
3   Improved knowledge permits characterization of the consequences of given failures, particularly their potential for spawning damage throughout the system. Associated with this are strong error-absorbing capabilities in people responsible for appraising the organization of technological malfunction or lapses in human performance.
4   Credible, exact knowledge of the environmental and social effects of technical operations provides the basis for comprehensive cost–benefit analyses. Associated with this knowledge is a continuous monitoring capability to detect external effects as the technology spreads and ages.
5   System error specification regimes are sensitive to public concern and interests. Alerting strategies address system errors and mitigate their consequences.

The Berkeley team has, over the past several years, examined four high-reliability organizations. The first is the Federal Aviation Administration's Air Traffic Control System in which controllers handle aircraft some 7.5 million times a year. Approximately 100 'near misses' (or, more accurately, 'near collisions') annually require quick action by controllers and pilots; but mid-air collisions are rare (Rousseau, 1989). The second is Pacific Gas and Electric Company, which provides 10,000 hours of electrical service to its 4 million customers, achieving 99.965 per cent reliability in terms of avoiding outages. The third is nuclear aircraft carriers, incredibly complex technological systems that have achieved remarkable safety records in their flight operations and handle up to 300 cycles of aircraft preparation, positioning, launchings and arrested landings, often at about 55-second intervals per day (Rousseau, 1989, p291). Finally, Roberts et al (1994) report on the culture of PAVEPAWS, an early-warning-system unit staffed by US Air Force personnel.

These studies yield important insights to supplement, extend or sometimes contradict research on corporate culture. Several are particularly germane to corporate culture and technology transfer as considered in this chapter. First, Rochlin (1989) argues that a key strength of high-reliability systems is their ability to be self-designing and to be able to translate organizational learning into standard operating procedures that members can be trained to follow. It is the organization that learns, Rochlin argues, and not its individual operators. System or organizational learning, in turn, depends strongly upon high accuracy in communication and quality of information. Hence the need for corporate cultures that make heavy investments in training and in developing detailed rules for making decisions. At the same time, these cultures need to be elastic in order, on the one hand, to promote conventional, routine behaviour under normal operating conditions, while on the other providing high discretion and information-scanning proficiency under crisis conditions.

Second, researchers have found that high-risk organizations face unusually uncertain external environments of hazard and yet have high interdependence associated with high technological complexity. The environmental uncertainty gives particular value to organizational decentralization; but the interdependence produces problems of coordination and control that managers often resolve through the use of extreme hierarchy (e.g. the ranking system used aboard aircraft carriers). Organizations appear to solve this conflict in different ways, such as through highly flexible hierarchy or special units to manage interdependence. Others fail to solve the problem and suffer severe failures.

An interesting finding from the aircraft carrier studies is that external forces continuously pressure for new technologies and technological fixes, whereas the operators of the carriers strongly resist their imposition. These operators view the new technologies as inherently less reliable than the old ones and as creating much more interdependent systems in which problems propagate. Accordingly, carriers turn out to be a fascinating mix of old and new technologies. Ship personnel manage state-of-the-art weapons systems, relying on 20-year-old computers and grease pencils. Carrier operators repeatedly told Berkeley researchers that the old technologies did the job when crises occurred or the electricity went off (Roberts, 1989, p122).

The findings also point to a distinctive role for corporate culture in contributing to high reliability. A basic challenge posed by new technologies is that more and more of the work and operations have disappeared into the machines (Weick, 1989). Trial and error are less available as a learning strategy because errors may produce catastrophes. As a result, managers and operators must increasingly rely on inference, imagination, intuition, mental models, symbolic representation of technology and problem solving to understand what is happening and to avoid accidents. Thus, a corporate culture that values and makes use of stories, storytelling and storytellers may develop greater reliability because members know more about their system and the potential errors that might occur and have greater confidence in their ability to handle errors (Weick, 1987, p113).

Finally, the research pinpoints a series of problems that all organizations with needs for high reliability (such as DuPont's facilities worldwide) must confront:

- High reliability requires both hierarchy and decentralization. Technological complexity makes centralization and integration absolutely indispensable; yet many decisions are optimal when pushed to the lowest level commensurate with the skill.
- Asymmetrical reciprocal interdependence leads to very tight coupling, whereas flexible reciprocal interdependence requires loose coupling.
- Demands on high-risk organizations typically stimulate more introduction of technology, whereas budgetary constraints reduce the workforce. What results is more tightly coupled systems and reduced slack.
- Training tends to be narrow and specialized, whereas uncertain events pose complex and interactive contingencies.
- High-reliability organizations are very expensive to operate; yet pressures almost always exist for efficiency and cost cutting (Roberts and Gargano, 1990).

Thus, when corporations locate facilities in developing countries, they not only confront issues of culture transfer but must also find solutions to these problems. Several of the problems were, in fact, important issues in the Bangpoo case.

TRANSFERRING CORPORATE CULTURES AND ENSURING HIGH RELIABILITY

A multinational corporation contemplating the location of a production unit in a foreign country, particularly if it is a developing country, faces

Table 7.1 *Motivations to invest in Thailand: The first three choices of importance*

	Number of firms by rank of importance		
Motivation factors	First	Second	Third
Diversification of business	–	–	–
Access to local materials	1	6	4
Low wage costs	13	3	1
Disciplined labour force	–	–	–
Fewer language problems	–	–	–
Investment incentives	2	5	8
Good and efficient government	–	–	–
Political stability of host country	–	–	–
Risk-free environment	–	–	–
Strategic location	–	–	1
Adequacy of infrastructure	–	1	1
Adequacy of communication facilities	–	–	–
Availability of land	–	–	–
Existence of industrial estates	–	–	–
Other factors	–	–	–

Source: Patarasuk (1991, p97)

issues that have the potential to affect substantially the corporate culture that emerges in the unit and the eventual reliability, and profitability, of operations achieved. Generally, these issues include the technological environment and capabilities prevailing in the host country, the market advantages to be gained, labour costs, incentives offered by the host country, the stability of the political environment, the business arrangement and, of course, cultural factors. Indeed, a recent study (Patarasuk, 1991) of 16 firms investing in Thailand suggests the importance of low wage costs, investment incentives and access to local materials as motivation for corporations to open facilities in Thailand (see Table 7.1).

Although the issues associated with cross-cultural transfer did not play a prominent role in these survey results, they do represent pervasive issues that corporations must address. For example, the host country culture may differ substantially from the corporate culture in attitudes toward authority and equality. Hofstede (1980), in a survey of some 88,000 employees of one multinational firm in 67 countries, found that respondents in developing countries were generally more receptive to hierarchical, authoritarian and paternalistic relationships, and that collectivism, rather than individualism, was a common norm. In many developing countries, kinship networks and family ties are powerful forces in the organization of the business environment and often serve as channels of communication among businesses and between business and government. If corporations adapt to these networks, significant advantages can accrue to the corporation through employee loyalty or superior access to information; conversely, failure to take account of such differences can leave corporations at a serious disadvantage (Austin, 1990).

Cultural attitudes and values carry far-reaching implications for organizational decision making and, thus, corporate culture. Scandinavian workers expect superiors to encourage their participation, whereas Latin American and Asian workers expect managers to be paternalistic. Japanese firms encourage decision making in groups with input from all organizational levels.

French firms generally have hierarchical structures in which decisions are made at the top and passed down through the organization (Ronen, 1986, p349). Italian managers see a need to provide precise answers to workers' questions, whereas many US, Swedish, Dutch, and British managers emphasize helping subordinates to find their own solutions (Laurent, 1983).

Even basic assumptions about and conceptions of incentives vary widely among cultures. In group-oriented cultures, singling out individuals for praise of their high performance – a practice of individually oriented cultures – sometimes produces a loss of face and threatens group harmony. Economic incentives work in some cultures but not others. Adler (1986, p132) cites the case of an American company's raising the hourly rate of Mexican employees to encourage them to work longer hours; instead, they worked less, observing that they now had enough money to enjoy life. Some cultures stress heavily the degree of trust that exists in superior–subordinate relationships; others emphasize productivity and performance (Choy, 1987).

Obviously, subtle cultural factors lie in wait for corporations involved in technology transfer. Table 7.2 suggests the rich diversity of such factors, which corporations need to consider and assess. It is not surprising, in view of this list, that transferring a corporation's culture and reliability system is much more complex and difficult than transferring its technology or management procedures. In Thailand, for example, Japanese firms have dominated the economy since the 1960s, and Japanese management systems apparently have become quite popular with Thai companies. Nonetheless, transferring the Japanese management system takes longer than transferring the production technology. American-style corporate management systems, for their part, have not become commonplace in Thai corporations, despite the fact that they are more decentralized and individualistic and, thus, superficially more adapted to Thai cultural values. The reasons for these difficulties are unclear, although public concern about conspicuous consumption is evident in Thailand, as is concern about a loss of cultural identity by Thais working in transnational corporations and about the possibility that Thais are becoming pseudo-Westerners (Patarasuk, 1991). Indeed, where Thai managers have been trained in Western management concepts, the result appears to be not the transfer of Western

Table **7.2** *Potential impact areas of cultural factors*

Management area	Cultural value and attitude parameters
Organizational structures	**Social structures:** Hierarchical ⟷ Egalitarian Vertical ⟷ Horizontal Familial ⟷ Institutional High-status differentiation ⟷ Low-status differentiation
Manager–employee relations	**Societal relationships:** Authoritarian ⟷ Democratic Paternalistic ⟷ Self-reliant Personal ⟷ Functional
Decision making	**Decision processes:** Autocratic ⟷ Participative Unilateral ⟷ Consultative
Group behaviour	**Interpersonal orientation:** Collectivism ⟷ Individualism Group welfare ⟷ Self-interest Other oriented ⟷ Task oriented
Communication	**Personal interactions:** Personalized ⟷ Impersonalized Closed ⟷ Open Formal ⟷ Informal Indirect ⟷ Direct Passive ⟷ Aggressive
Incentives and evaluation	**Motivation:** Economic ⟷ Non-economic Work ⟷ Pleasure Loyalty ⟷ Competency Role ascription ⟷ Merit achievement
Control	**Human nature perceptions:** Intrinsic good ⟷ Intrinsic evil Trust ⟷ Distrust Cooperative ⟷ Conflictive Malleable ⟷ Unchangeable
Time management and planning	**Time perceptions:** Infinite resource ⟷ Finite resource Present oriented ⟷ Future oriented Imprecise ⟷ Precise Controllable ⟷ Uncontrollable
Spatial relationships	**Space perceptions:** Private ⟷ Public Nearness ⟷ Farness Aesthetics ⟷ Functionality

Source: Austin (1990, p355). Reprinted with the permission of the Free Press, a Division of Simon and Schuster Adult Publishing Group, from *Managing in Developing Countries: Strategic Analysis and Operating Techniques* by James E. Austin. © 1990 by James E. Austin. All rights reserved.

corporate culture and management, but a hybrid of Western and Thai norms and cultural behaviour patterns (Runglertkrengkrai and Engkaninan, 1987).

Gladwin and Terpstra (1978), viewing the wide array of host country cultural variables that impinge upon technology transfer, suggest five cross-cultural factors that all corporations need to address:

1 *Cultural variability*: the degree to which conditions within a culture are stable or unstable, and the rate at which they are changing. Variability creates uncertainty and requires a corporation to be highly flexible and adaptive.
2 *Cultural complexity*: the extent to which the rules of human behaviour are explicit or implicit. In many Western countries, the tendency is for communication to be direct and explicit, whereas in many Asian countries, non-verbal cues and symbolic uses of language are important.
3 *Cultural hostility*: the degree of negative reaction in the host country to the incoming corporation. Such hostility, of course, has deep implications for building corporate values and trust.
4 *Cultural heterogeneity*: the degree of sharing of cultural values and meanings between the host country and the country of origin of the parent corporation. Congruence or lack of congruence may set limits as to what is possible in transferring corporate cultures and may indicate necessary adaptations.
5 *Cultural interdependence*: the extent to which developments in other cultures are related and affect the host country culture. This variable suggests cultural openness to events or change to external events or development, an important source of uncertainty or ambiguity.

These five factors may then be related to the within-culture and between-culture dimensions shown in Table 7.3.

The plant manager occupies a key position in dealing with the cultural issues attendant on technology transfer. The manager acts as the translator and advocate of corporate culture, the analyst of the host country environment, and the interpreter of host country culture and sub-cultures.

Table 7.3 *Dimensions of the cultural environment of international business*

Dimension		Continuum	
		(Low)	(High)
Within cultures	Cultural variability	Low and stable change rate	High and unstable change rate
	Cultural complexity	Simple	Complex
	Cultural hostility	Benevolent	Malevolent
Between cultures	Cultural heterogeneity	Homogeneous	Heterogeneous
	Cultural independence	Independent	Interdependent

Source: Gladwin and Terpstra (1978, pxvii). Reprinted with the permission of South-Western, from *Cultural Environment of International Business* by Terpstra, V. (ed). copyright © 1978 by V. Terpstra. All rights reserved

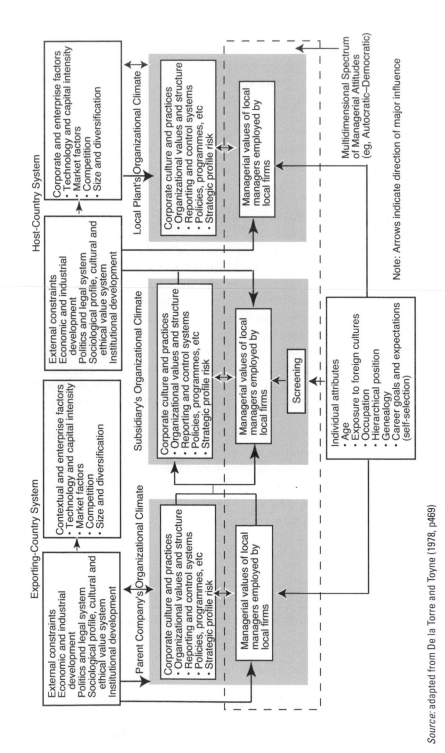

Source: adapted from De la Torre and Toyne (1978, p469)

Figure 7.4 *Interactions among corporate culture, organizational climate and national setting in technology transfer*

Because of this role, multinational corporations (including DuPont) often pursue a strategy of management indigenization in which they appoint host country nationals to key management positions in the local plant. Although this strategy enhances the corporation's knowledge of local circumstances and may help to identify opportunities for cultural congruence and adaptation, it also carries the risk that subsidiaries or local plants will substantially modify elements of the onion structure of corporate culture in order to satisfy host country goals and assumptions.

Corporate and host country culture operate, then, in a multi-tiered system in which the parent company, the subsidiary and the local plant all interact (see Figure 7.4). Managers at each level play particularly crucial roles as culture formulators, translators and interpreters. At each level, the corporation exists within the environment of a home country with its own culture, goals, expectations and constraints. Typically, individuals are recruited from these home country systems into the corporation or move between levels (and cultures). How these various transactions are managed extensively shapes the performance and outcomes of technology transfer.

From the foregoing conceptual discussion of corporate culture, high-reliability organizations and cultural issues in technology transfer, the following section considers the location of the DuPont Bangpoo plant in Thailand. This case provides a rich context for exploring the various ways in which corporate culture enters into technology transfer. In addition, it affords opportunity for interrelating corporate culture and value questions since DuPont is widely regarded as a multinational corporation with a strong corporate culture.

THE CORPORATE CULTURE OF DUPONT

In early 1990, *Fortune* carried an article assessing retrospectively the corporate culture movement of the 1980s, suggesting that, in the midst of all the trendy notions, there was reason to 'take heart'. 'An increasing number of enterprises are at last figuring out how to alter their cultures and more than ever are doing it' (Dumaine, 1990, p127). The models and illustrations used in this article accorded DuPont a particularly conspicuous place. This prominence fits closely the self-image of DuPont's top managers, who energetically embrace the company's corporate culture as a primary vehicle in the success of the corporation and its role in 'corporate environmentalism' and industrial health and safety. As DuPont's chairman and chief operating officer, E. S. Woolard, recently put it, the primary challenge DuPont faces is to ensure 'that we excel in environmental performance and that we enjoy the non-objection – indeed, even the support – of the people and governments in the societies where we operate' (Woolard, 1989a, p6). This is a tall order in the US and elsewhere, where social trust in corporations has eroded to very low levels.

What are the key components and characteristics of corporate culture at DuPont, how did this culture arise, how is it maintained, and how does the culture affect DuPont's operations in the US and abroad?

Origins and early history can put an enduring stamp on a corporation's culture. DuPont had its birth in 1802 in explosives. The founding father, Pierre S. du Pont de Nemours, had been a prominent economist and politician in France prior to taking refuge in America during the so-called Reign of Terror attendant on the French Revolution. Originally, he planned to establish a colony of balanced agricultural and industrial ventures. When this scheme collapsed, Pierre's son, Eleuthère Irénée, discovered that gunpowder produced in the US was expensive but poor in quality. With financial backing from France and the US, he established one of the largest explosives plants in the US outside of Wilmington, Delaware, on the Brandywine Creek (Taylor and Sudnik, 1984).

During these early years, the company operated with a strongly paternal bent that reflected both the need to recruit and retain highly skilled workers and Pierre du Pont's goal to create a harmonious society, as well as a productive industrial enterprise. All du Pont family members shared equally in management and profits, and property was held in common and apportioned according to need. Eleuthère du Pont viewed the family as personally responsible for the well-being of its workers; accordingly, the earliest company rules (1811) made it clear that the company was responsible for maintaining safe workplaces. Employees were forbidden to enter a mill before management had first surveyed the mill for hazards and operated the equipment. Although the manufacture of gunpowder at that time typically proceeded in vertical stages on floors of a wooden building, thereby increasing the risk of explosions, the du Ponts set up their works in horizontal stages along the river bank in permanent stone structures (Blanchard, 1990). The du Ponts themselves lived on the site of the mills and worked alongside the employees. They also established pensions for families of employees killed on the job and provided company housing for employees. These paternal and safety-oriented traditions were effectively wedded with profits, and by the end of the century DuPont was the largest explosives producer in the US.

The early history of family idealism, a productive process associated with high hazards and a strong paternalism provided a base for a succession of health and safety innovations at DuPont. During the early 1900s, the first company physician was retained to deal with work-related injuries. In 1915, the company instituted annual physical examinations and hired a full-time medical director. In 1925, DuPont included X-rays as a routine part of physical exams. A year later, it created a corporate safety and fire-protection division and began to conduct safety audits of its various plants. In 1935, DuPont established the Haskell Laboratory for Toxicology and Industrial Medicine, one of the first such laboratories in American industry. In 1956, a formal epidemiology programme, including a cancer registry, was initiated.

In 1979, DuPont became one of the first corporations to establish a computerized exposure tracking system to document and monitor employee medical histories. The 1980s saw occupational medical programme surveys and an alcoholism remediation programme for retirees and employee families and survivors (Karrh, 1984). Recently, the creation of a health improvement programme for employees has addressed a broad range of health issues, including back injuries, nutrition, smoking cessation, blood pressure control and off-the-job safety. Over its history, in short, DuPont has been a clear leader in initiating a wide variety of health and safety innovations in the US (Kasperson et al, 1988a).

Even its top managers in health and safety will pointedly tell you that DuPont 'is in business to make money; we do not make decisions that are not sound business decisions' (DuPont, 1990). But it is also clear that this historical record of health and safety innovations has something to do with its early history, its founding values and its sense of corporate mission.

When DuPont describes itself as 'a company of principles', it does, indeed, specifically state its intent to adapt well 'to rapidly changing conditions in a highly competitive market' (DuPont, 1989, p14). But in that highly competitive world, it seeks to maintain a distinctive reputation:

> ... as a company that is committed to the safety of its employees, its neighbours and its customers; as a company of unquestionable integrity and social responsibility; as a company whose operations are environmentally sound; and as a company whose products and services are consistently of the highest quality (DuPont, 1989, p14).

DuPont officials consistently cite several principles relating to the environment and safety as central to DuPont's sense of mission:

* There is no privilege without duty. Selling products and employing workers carry key obligations.
* All injuries, occupational illnesses and, more recently, environmental releases are preventable.
* DuPont will not sell any product that cannot be manufactured, distributed, used and eventually disposed of safely.
* All makers and users of DuPont products must be informed of the attendant risks.
* Management is fundamentally responsible for safety.

So, although DuPont is surely a business that aims to be profitable, it also sees itself as occupying a particular niche in industry. Indeed, these goals are seen as interrelated and mutually reinforcing, not in conflict. DuPont aims to be an industry leader, an innovator in health and safety programmes, and to have one of the safest places to work in the industrial world (Woolard, 1989b, p2). That means that DuPont cannot settle for even comparatively favourable safety rates or good grades on its environmental protection performance. Thus, when internal statistics record an increase in spills and accidents, DuPont increases the number of safety-conscious and disciplinarian types in its managerial ranks. The worker who gets to be plant manager needs to know instinctively that if you leave a hose lying across a walkway, sooner or later someone is going to trip. DuPont must be willing, for example, to spend money for safety on the margins, where other corporations will not. In return, however, DuPont has emerged as a major marketer of health and safety programmes and points with pride to a study of 178 of its clients, which indicates that, as a result of DuPont expertise, they were able to improve their safety records by an average of 37 per cent in the first year and 89 per cent by the sixth year (DuPont, 1989, p14).

Over the past five years, DuPont has sought to extend its corporate mission and identity to be an environmental leader, as well as a safety innovator (Woolard, 1990). During the early 1980s, it became apparent that DuPont's environmental performance did not match its occupational health and safety performance. A preoccupation with economics and improved labour/management relations had compromised the company's environmental record, and the high-volume producer of chemicals found itself at or near the top of lists of major polluters (Dean et al, 1989; USEPA, 1989; Citizens Fund, 1990). In 1986, it committed to a 33 per cent reduction in all environmental emissions by the end of the 1990s. It has also adopted an assumption, that of the National Safety Council, that all environmental releases are preventable (which, of course, they are not). Thus, DuPont has very consciously embarked on a leadership role in *corporate environmentalism*, which it defines as an attitude and a performance commitment that places corporate environmental stewardship fully in line with public desires and expectations (Woolard, 1989a, p6). This means that DuPont sees itself as assuming a leadership role in developing an industry-wide environmental agenda over the next decade and will seek opportunities to align itself with the environmental community and to demonstrate that environmental and industrial goals are compatible (Woolard, 1989a, p6) and even 'sustainable' in the long run (Woolard, 1992).

Critics label such pronouncements as mere 'environmental etiquette of the 1990s' (Doyle, 1992, p90). Indeed, this is not a matter of altruism; rather, DuPont sees such leadership as directly in its material self-interest: 'The ultimate competitive advantage is to remain in business when your competitors are driven out. In the decades ahead, many companies that have not responded to the environmental imperative will be denied the privilege to operate in important trading nations in key markets' (Woolard, 1989c, p4). Or, again in Woolard's words:

> Industrial companies will ignore the environment only at their peril. Corporations that think they can drag their heels indefinitely on

genuine environmental problems should be advised: society won't tolerate it and DuPont and other companies with real sensitivity and environmental commitment will be there to supply your customers after you're gone (Holcomb, 1990, p20).

If a strong corporate culture does, indeed, require both a clear mission and a high degree of consistency, culture building must be omnipresent and effective. Indeed, DuPont officials will tell you that safety and (now) environmental protection are a condition of employment at DuPont. 'If we have a site manager who does not buy into the DuPont philosophy, he or she will not be a plant manager long' (DuPont, 1990).

But, interestingly, DuPont does not see recruitment as a key, nor does it have a training programme radically different from that of most of its competitors. Instead, it works from the top down in instilling attitudes, motivations and behaviour norms through an all-embracing corporate socialization programme. 'We expect all DuPont employees to have the right attitude, irrespective of job assignment. We especially expect all managers to constantly convey that attitude. It's from us that the organization learns whether we're serious about safety or whether we're just talking a good line' (Woolard, 1989b, p2). The first-line supervisor plays a particularly critical role with new employees and is expected 'to turn around their attitudes' (DuPont, 1990). Individual workers, meanwhile, are instilled with the notion that they are responsible not only for their own safety, but for that of their co-workers as well, and peer pressure occupies an important socialization and regulatory function.

The reward structure at DuPont is closely geared to these corporate cultural values and processes. Indeed, chairman Woolard has argued that the best way to create a more trusting corporate environment is to reward the right people: 'The first thing people watch is the kind of people you promote. Are you promoting team builders who spend time on relationships, or those who are autocratic?' (Dumaine, 1990, p131). Accordingly, all DuPont managers are evaluated on safety and environmental performance as key criteria for promotion and compensation increases.

The cultural values that support DuPont's mission and identity as a safety and environmental leader are seen as providing broad benefits to the corporation. 'Safety is a powerful lever to achieve a whole series of other corporate goals – we learned this long ago', a DuPont official noted (DuPont, 1990). Safety is viewed as a very effective locus for instilling discipline and building managerial leadership. It is also integrated with other values that support corporate performance. DuPont specifically espouses a 'total person' approach to safety. The company's survivor benefits are extraordinarily high. Fire hazard, seat-belt usage and other off-the-job safety information continually goes home with DuPont employees. Recently, DuPont instituted Health Horizons, an employee health improvement programme in which more than one half of its employees worldwide participate (Karrh, 1989).

Symbols, rituals and signals all play an important role in DuPont's corporate culture. Managers nurture pride in safety and environmental performance through a complex system of awards, flashing green lights, and accolades and awards from top management. The first item of business at all corporate executive board meetings is safety, and any lost workday cases are discussed specifically. 'Near misses' (by which is meant 'near mishaps') are major events and receive substantial visibility at top management and through the company. Safety is ubiquitous, almost a mania, in the DuPont work environment; signs and cues are visible everywhere. Safety is in front of people at all times and built into even the most routine procedures, such as the 'tool box' meeting in which safety issues are treated before a work crew undertakes even the most ordinary of maintenance tasks. Workers encounter a constant flow of admonitions to use seat belts even in the back seats of cars, to wear goggles when mowing their lawns, to avoid tilting their chairs while sitting, and always to hold the banister while ascending or descending stairs, at work or away from work.

For all this richness in cues, socialization, symbols and ritual, DuPont must struggle to maintain the culture upon which it heavily depends. In the midst of economic cutbacks and getting leaner during the mid 1980s, DuPont found that its safety performance had dropped. Preoccupied with the economic downturn, it discovered that its enunciation of safety principles in various corporate communications had slipped from the top to the bottom of the list. Meanwhile, cutbacks in management had reduced the culture building and personal relationship between supervisors and subordinates. And so corporate officials found the need to gear up the corporation for a reinvigorated effort if cultural values and behaviour norms supporting safety were not to be eroded. And so lessons were drawn: 'We know from experience that when we do stop pumping the safety commitment throughout the company, bad things start to happen' (Woolard, 1989b, p2).

Clearly, safety is a key ingredient in DuPont's corporate culture. This section has explored at length the nature of that culture, which permeates DuPont facilities worldwide. In the following section, corporate culture sets the context for DuPont's transfer of technology to its Bangpoo facility in Thailand.

DuPont corporate culture and technology transfer

DuPont is a strongly international enterprise. Its 140,000 employees work in 200 manufacturing and processing plants in some 40 countries throughout the world. It also markets in more than 150 countries, and international sales account for more than 40 per cent of its revenues, which were expected to increase to 50 per cent by the mid 1990s (DuPont, 1989, p2). One in four of its employees work in plants outside the US. During the 1980s, the international character of DuPont grew substantially, as the company invested more than US$10 billion to fund more than 50 major acquisitions and joint ventures with other companies.

DuPont has a long-standing policy of equivalency in its standards of safety and environmental protection wherever it operates in the world, with, as chairman Woolard bluntly puts it, no exceptions. Equivalency here means something more than general functional equivalency. Indeed, DuPont's central approach to dealing with the myriad health and safety challenges in locating facilities in developing countries is to transfer not only its safety standards, but its entire corporate culture. As one DuPont official described it:

> In DuPont I think there is only one environmental and safety rule –
> that is, the *E. I. DuPont Safety and Regulation Rules* apply anywhere
> that DuPont operates. If you can't have safety, then you don't
> acquire that plant. If you can't operate it safely, then you close it
> down (Chalat, 1989, p10).

So, during the mid 1980s, when the Occupational Safety and Health Administration (OSHA) required changes of all US companies in their accounting systems for occupational injuries, DuPont implemented the rule change worldwide, despite objections from its international facility managers about the need for such enlarged bookkeeping changes and dreary record-keeping. The motivation was a sound one – that DuPont could track and compare all accident statistics across all its facilities on the same database. In other words, once a change is adopted in corporate policies and procedures, all DuPont facilities must expect to live with it.

But formal rules and procedures are, it is worth recalling, only the outer layer of the onion of corporate culture. More critical is the development of a sense of corporate mission and patterns of attitudes and behaviour that realize safety goals. In locating plants in other countries, DuPont officials recognize that the host country culture necessitates some adaptations. Trophies and cheerleaders don't work in Asia, DuPont officials informed the authors, nor do financial awards for innovation. So, in Asia, DuPont emphasizes team building; in Germany the plant manager is a strict disciplinarian; and in Taiwan the plant manager is a father figure who seeks a close personal relationship with the workers.

DuPont attaches great importance to the selection of the plant manager. In Thailand, DuPont selected an Asian manager from its Singapore facility for that role. A key part of her responsibilities was to understand the local culture, to develop in the plant a specific cultural adaptation that made DuPont's core values and environmental philosophy work, and to recruit the plant's workers. Specifically, equivalence was sought not only in standards but in employee skills, knowledge and attitudes, commitment to safety philosophy, and sense of personal stake in the company. The corporation brought the plant manager and production supervisor, both of whom were also part of the DuPont task force that designed and started up the plant, to the US for six weeks of advanced training. By the time the first group of workers was hired, the local DuPont management had received extensive preparation as culture translators.

Much attention in the Bangpoo culture building centred on the selection and early training of workers. Preference was given to fresh technical school and high school graduates. Job interviews routinely involved a series of questions on personal attitudes toward risk-taking in personal life and a careful evaluation of the candidate's willingness to conform to group norms and practices. Workers received training three times longer than would have occurred in a comparable US facility. Meticulous observation and evaluation of safety habits at and outside of work occurred over the first six months, and workers' attitudes toward safety were elicited and monitored. DuPont also developed a system for recruiting and evaluating workers to determine which were most compatible with its corporate culture. Indeed, the entire Bangpoo operation carries the stamp of DuPont safety and environmental mission. The site itself could have been adequately developed for US$500,000 but was built at a cost of US$1.5 million in order to meet DuPont's own high standards for environmental and occupational health and safety. Safety hats and safety glasses are ubiquitous, doors have outlines on the floor to indicate the direction in which they open, and emergency equipment is in place for handling any chlorine release from an accident in the chlorine facility adjacent to the DuPont plant. Confronted by the lack of an adequate local sanitary landfill, DuPont constructed its own on-site waste incinerator. An engineered dust-control system was installed to avoid the need for uncomfortable face masks. DuPont also provided special training for the local fire department and hospital in order to upgrade the local emergency-response infrastructure.

CONCLUSIONS

These observations are not intended to suggest that the Bangpoo facility is extraordinarily safe or that DuPont's transfer of corporate culture to Thailand has been a resounding and unequivocal success. The authors have no independent verification, beyond a brief field visit by colleagues, to permit such findings. But several pertinent observations are possible:

- DuPont clearly has a well-developed corporate culture in which high safety and environmental performance are a key part of corporate mission and identity.
- In the Bangpoo case, the company undertook extraordinary efforts to transfer its corporate culture as intact as possible, making a variety of financial investments dictated primarily by safety and environmental goals and the need for culture building.

- The transfer did not simply reproduce a parcel of DuPont in Thailand. Rather, indigenization of management and some adaptations to local culture were used to realize basic corporate norms and to implement its organizational and rules structure.
- Corporate culture bore most of the burden for safety and environmental performance, as Thai environmental monitoring, regulation and inspection were weak and ineffective.

Much of this case is reassuring with regard to safety and environmental responsibility; but a downside also exists. DuPont is centrally committed to its corporate mission and culture wherever its plants are located. The centrality of that commitment reverberates in other value areas. Developing countries sometimes seek as a high priority the transfer of technological capability and the maximization of industrial development. A strong corporate culture such as DuPont's can – perhaps, even usually does – interfere with those objectives. If DuPont's corporate culture is to work, full control over operations is essential. Although DuPont does provide an unquestionably powerful model for a host country to emulate in building risk management and environmental protection regimes, the degree of indigenous rooting of the DuPont corporate culture is limited. DuPont, for all its emphasis upon environmental leadership and its worldwide marketing of safety systems, is not about to transfer its productive technology, marketing advantage or valued patents to other companies or host countries.

8 Industrial Risk Management in India Since Bhopal

B. Bowonder, Jeanne X. Kasperson and
Roger E. Kasperson

INTRODUCTION

The Bhopal accident marked a watershed in 20th-century industrial safety. It injured, displaced or killed record numbers of people, and caused untold delays in treating and compensating victims (Morehouse and Subramaniam, 1986; Shrivastava 1987a, 1987c, 1992; see also Chapter 2 in Volume I). The accident also had far-reaching effects on the chemical industry's management of hazards in both developed and developing countries. Many studies have addressed the responses in Europe and North America (Stover, 1985; Shrivastava, 1987a, 1992; Kletz, 1988; Hadden, 1994; van Eijndhoven, 1994). But India, of course, was the country most directly and profoundly affected by the accident. Accordingly, we enquire in this chapter into social learning and institutional responses to industrial risk in India after Bhopal, particularly in the area of legal and regulatory developments and the implementation of hazard assessment and management systems.

'Social learning' is the process by which a society or nation perceives, assesses and acts upon harmful experiences or past mistakes in purposeful ways. With respect to environmental risk, the learning process may incorporate decisions to bear or spread the risk, to prevent or reduce the risk, to mitigate the consequences of a particular incident, or to adapt to or accommodate the risk. The Bhopal experience offered a host of possible lessons to industrial societies. This is not to say, however, that the event led to radical change instead of mere incremental adjustment. Bhopal stimulated extensive soul-searching in India about technological threats and the Indian capacity to anticipate and manage them. Our objective is

Note: Reprinted from *Learning from Disaster: Risk Management Since Bhopal*, Jasanoff, S. (ed), 'Industrial risk management in India since Bhopal', Bowonder, B., Kasperson, J. X. and Kasperson, R. E., pp66–90, © (1994), with permission from University of Pennsylvania Press.

to assess the extent to which the Indian institutional responses reflected true social learning.

Social learning in its simplest form involves ongoing, incremental changes that render the institutional fabric of a society more responsive and efficient in dealing with future events and stresses. Learning from disaster, however, has the potential to be more purposive and may be more complex, radical and broad based. A crisis such as Bhopal may trigger signals or early warnings from which society can infer new causes for concern about the environment or discover that institutional structures are insufficient to appraise and respond to such stresses. P. W. Meyers (1990) distinguishes four types of learning: *maintenance learning* involves incremental improvements in organizational procedures and performance, all the while retaining the basic organizational fabric; *adaptive learning* aims at building the 'right system' and entails changes in roles, rules and procedures; *transitional learning* occurs when an organization shifts its major strategic emphases and often requires the 'unlearning' of methods and procedures that are no longer functional; *creative learning* involves a radical redefinition of both problems and solutions, along with more conflict and openness in assessment and decision-making processes.

Social learning, it is important to emphasize, is not the same as institutional change, which may or may not accompany increased learning and, indeed, may proceed without increased knowledge. Learning must include an element of generalizing that 'goes beyond simple replication to application, change, refinement' (Jelinek, 1979, p161). Crises such as Bhopal obviously are apt to stimulate both learning and change.

The Bhopal accident laid bare a number of significant shortcomings in India's legal and regulatory structure for managing risk. First, the state pollution control boards, India's primary agencies for enforcing environmental regulations, had no power to close down facilities, even those that posed imminent danger. Between 1978 and 1983, for example, the Bhopal plant experienced six major accidents, three of which involved spills of toxic materials. People notified the government that the facility had become intolerably risky (Bowonder and Miyake, 1988), and yet it continued to operate. Second, the Factories Act of 1948 failed to distinguish between non-hazardous and hazardous facilities, and the Factory Inspectorate lacked the power to halt a facility's operations even in the face of documented safety lapses. Third, the failure of Union Carbide Corporation to provide information on how to handle methyl isocyanate (MIC) exposures, apparent in the Bhopal accident, spotlighted the absence of any statutory requirement to disclose needed emergency response information. Fourth, plant siting and modification procedures failed to require hazard assessments for new facilities or for proposed changes in existing ones. Fifth, citizens lacked recourse to the courts even if they had been exposed to high levels of hazards or if firms had violated safety provisions. Finally, firms typically found it more economically advantageous to avoid compliance and pay the penalties than to meet statutory requirements. Legal proceedings against violators were hampered by a host of institutional factors: huge backlogs of cases, the difficulty of proving criminal liability, tortuous laws of evidence and the lack of technical sophistication among judges (Singh, 1984).

LEGISLATIVE INITIATIVES

During the years following Bhopal, India enacted new legislation and amended existing laws to upgrade the national system of industrial hazard assessment and control. What can be said about the efficacy and impact of these new initiatives? How much did India really learn from Bhopal?

The Environment Protection Act, 1986

In 1986, India enacted an umbrella law – the Environment Protection Act – in order to provide a holistic approach to risk management and to remedy shortcomings in existing environmental pollution laws (most notably, the Water Act and Air Act). The new statute, which came into force on 19 November 1986, sought both to protect and improve the environment and to prevent major hazards. It differed from earlier environmental acts in several notable respects.

The Environment Protection Act for the first time covers the control of hazardous substances and provides for inspection of facilities handling hazardous materials and for control of the storage, handling and transportation of such materials. It requires firms to disclose to the Ministry of Environment and Forests information on hazardous materials to which humans are likely to be exposed. It also establishes environmental laboratories for analysing water and soil samples.

Any person so authorized by the central government has the right to enter and inspect a plant at all reasonable times, may search any building for actual or potential offences under the act, and may seize any evidence necessary to prevent or mitigate pollution. Another significant provision, introduced for the first time, is the power to close down a facility. The central government may direct the closure, prohibition or regulation of any industry or operation, or regulate the supply of electricity or water or any other service. During the reporting year 1989–1990, the government of India issued closure directives to some 51 units for not implementing the required environmental protection measures (Central Pollution Control Board, 1991, pp27–28).

A more controversial provision allows citizens to approach the central government or the court directly to report violations by corporations or government agencies; but the complainant must first give 60 days' notice so that the authority concerned can pursue an action in court. During 1988–1991, citizen groups used this provision extensively to challenge corporations for alleged safety lapses arising from emissions, discharges and the storage of hazardous materials. Citing harassment, industry associations and representatives counter-petitioned the government to withdraw this provision (*Economic Times*, 1992b). Multinational corporations such as DuPont have made their coming to India (now that the government has embraced economic liberalization) contingent upon the withdrawal of the provision. Meanwhile, the 60-day notification period continues to draw fire from citizens' groups for essentially providing time for offenders to delay clean-ups (Chitnis, 1987, p155; Abraham and Abraham, 1991, p359). Recently, a proposed amendment to the Environment Protection Act seeks to cut the notification period to 30 days (*Business and Political Observer*, 1991a).

In a further direct response to Bhopal, the Environment Protection Act permits top corporate managers to be held personally responsible for violations of the law. Companies previously had the freedom to designate 'the occupier of the factory', who was responsible for any offence committed. The Environment Protection Act, however, requires the management to involve itself actively in preventing pollution or accidents. The law thus marks a substantial improvement over earlier legislation, though its effectiveness will depend upon the central government's ability to develop and implement its key provisions.

The Factories (Amendment) Act, 1987

Indian factories legislation was first enacted in 1881 and has passed through numerous subsequent modifications. Immediately after independence, the government of India enacted the Factories Act of 1948, revising and extending a 1934 law to include welfare, health, cleanliness and overtime payments. Modelled after UK legislation, the 1948 act first codified the principle that the health, safety and welfare of workers employed in hazardous manufacturing processes should be legally protected. Following Bhopal, the 1987 amendment introduced additional safeguards in the use and handling of hazardous substances in factories. It imposed a statutory obligation upon management for emergency response procedures and sought to control the siting process in order to minimize risk to host communities. It also included provisions for the participation of workers in safety management and upgraded the penalties for non-compliance.

Hazardous industries

A key section of the 1987 legislation identified a list of industries that government officials consider to be hazardous. The list, which encompasses 'all major chemical and other potentially hazardous operations and facilities', is not very significant in itself, but has far-reaching implications in conjunction with other sections of the act. Paralleling the Environment Protection Act, the statute also changed the definition of 'occupier' to 'the person who has ultimate control over the affairs of the factory'. This definition again sought to make the top management more personally committed to the safe operation of hazardous facilities and to compliance with existing safety regulations.

Duties of the occupier

The act specifies the occupier's *duties*, all of which aim at improving risk management. Thus, the occupier must undertake plant maintenance in a way that is safe and without risk to the health of workers; safeguard health and safety during the use, handling, storage and transport of hazardous substances; provide information, instruction, training and supervision to ensure the health and safety of all workers; and monitor the health and safety of workers. The occupier must also prepare a written statement of corporate policy with respect to the health and safety of workers and must give notice to workers regarding the hazards that they face. The onus of disclosing safety information, both to the Factory Inspectorate and to workers, is directly on management, a change from earlier regulations. The act also institutionalizes management responsibilities for reducing risks to people in the vicinity of the plant; earlier laws did not deal with risks to nearby residents.

Siting of hazardous facilities

To keep hazardous facilities away from major population centres, the act provides a new siting procedure. A Site Appraisal Committee, chaired by the state's chief inspector of factories, will examine applications for the establishment of factories that involve hazardous products or processes. It has the power to require a broad range of information from the applicant. Careful scrutiny and review prior to siting an industrial establishment has significant potential for minimizing exposure of the public to materials released during a breakdown or accident. In practice, however, the committees in many states frequently forgo stringency in favour of courting industry, as every state seeks lower hurdles to industrial development. Thus, the Maharashtra Pollution Control Board recently locked horns with the central government over the proposed expansion of a petrochemical complex in a densely populated area (*Business and Political Observer*, 1991b). And in Madhya Pradesh, the state that ought to have learned most from Bhopal, 'ad hocism' prevails in the shadow of an inconsistent siting policy (Joshi, 1991, p59). Factory owners with existing units complain of the extreme difficulty, given their inability to acquire and fence off a suitable buffer zone, of discouraging shantytown developments. Yet, only the government can legally acquire the large tracts of land that would be needed for this purpose. In addition, siting or relocation in remote locations is often costly in other terms, so that owners frequently try to avoid this outcome (Hadden, 1987, p716).

Disclosure of information

Disclosure obligations under the act take several forms:

- The occupier of every factory involved in a hazardous process must disclose all information involving dangers, including health hazards and mitigation measures, to the workers employed in the factory, the chief inspector of factories, the local community within whose jurisdiction the factory is situated, and the general public living in the vicinity of the facility.
- The occupier must establish a detailed policy regarding worker health and safety, communicate this policy to the chief inspector of factories, and draw up a site emergency plan and detailed disaster control measures.
- The occupier must inform the chief inspector of factories in detail of the nature of all hazardous processes.
- The occupier must establish measures, approved by the chief inspector of factories, for the handling, usage, transportation and storage of hazardous substances outside the factory, and the disposal of such substances. The occupier must also publicize these measures to the workers and the general public living in the vicinity.

Hazard information

The act specifies several responsibilities with respect to compiling and maintaining information. For example, the occupier must maintain medical records for workers exposed to any chemical, toxic or otherwise harmful, during manufacture, handling or transportation. Workers must have access to their medical records and to means for protection from exposure to hazardous substances. These requirements have tremendous implications for the empowerment of workers; but since levels of literacy are low, effective implementation will require determined education of workers and involvement of trade unions in areas that have few traditions and experiences upon which to build.

Permissible exposure

The act calls for permissible exposure limits for chemicals and other toxic substances in manufacturing processes and sets standards in terms of both eight-hour, time-weighted average concentrations and short-term (15-minute exposure) concentrations. The standard-setting process is not at all interactive. Indeed, the ministry usually sets exposure limits by simply copying standards from other countries. Concentration-based, as opposed to mass emission-based, standards have produced no real reductions in pollution. Even when individual factories in a highly industrialized area (e.g. Bombay) are in compliance, the combined emissions from several plants may reach deadly levels. Aware of the problem, the Ministry of Environment and Forests has begun 'to lay down mass-based standards which will set specific limits to encourage the minimization of waste, promote recycling and reuse of materials, as well as conservation of natural resources, particularly water' (MEF, 1991, p43). Meanwhile, the cost of implementation is high; many small facilities lack the necessary infrastructure and resources to comply with regulations.

Danger warnings
The amended Factories Act gives workers the right to warn the occupier or person in charge about an imminent danger. The warning can be given directly by the worker or through the worker safety committee and must be brought to the notice of the factory inspector. Upon receiving the warning, the occupier is required to take immediate action and to send a report to the factory inspector.

Penalties
Violation of any of the provisions of the Factories Act is punishable with a two-year term of imprisonment, a fine of 100,000 rupees or both. If the violation continues after conviction, the fine may be extended to 10,000 rupees for each additional day.

These new amendments make the Factories Act, in principle, a powerful regulatory instrument. In fact, however, state governments lack both the resources and the regulatory infrastructure to implement the upgraded provisions. In Madhya Pradesh State, an average factory inspector must inspect about 280 facilities annually. Inspectors, moreover, are mostly mechanical and electrical engineers, with precious little experience in hazard assessment, industrial medicine, toxicology, emergency management and occupational health issues. Worse yet, as one Indian journalist quips: 'The ball-point pen is the only "equipment" with a factory inspector to monitor hazard in industrial units and help check pollution in Madhya Pradesh' (Joshi, 1991, p59). The inadequacy of coordination among central, state and local government agencies exacerbates the problem of implementation. A series of industrial accidents since Bhopal – a chlorine leak in densely populated Bombay in 1985; a serious oleum leak in Delhi in 1985; a fire in a large semiconductor complex in Chandigarh in 1989; an accident in a gas cracker unit in Bombay in 1991; and a trucking accident in Mahul (near Bombay) in 1991 – all illustrate persistent inadequacies in the management of risks at hazardous plants.

The Air (Amendment) Act, 1987

The 1987 amendment to the 1981 Air Act also took note of problems observed at Bhopal. As discussed above, every industry must have the consent of the State Pollution Control Board to release pollutants in accordance with applicable emission standards. Under earlier laws, such consent was irrevocable, once granted. Under the 1987 amendments, however, the state board may cancel its consent at any time or refuse renewal if regulatory conditions have not been fulfilled. Prior to cancelling or refusing a further consent, however, the authorities must grant the person concerned a reasonable opportunity to be heard. State pollution control boards may also petition a court to restrain persons from causing air pollution, mainly as a means of handling individual release events.

The Hazardous Wastes (Management and Handling) Rules, 1989

On 28 July 1989, the government of India promulgated new rules concerning the management and handling of hazardous wastes. These rules specify what constitutes a hazardous waste and require that a person who owns or operates a facility for collection, reception, treatment, storage or disposal of hazardous wastes take all of the practical steps needed to ensure proper handling of such wastes. All generators and managers of hazardous wastes must apply to the state pollution control board for authorization to pursue any of these activities. The rules also prohibit the import of hazardous wastes from any country to India for dumping or disposal.

Though the rules have been on the books since 1989, implementation has been slow. The state pollution control boards lack adequate technical skills for handling the technologically complex problems posed by hazardous wastes. The Central Pollution Control Board, for its part, has failed to conduct the background work needed to determine the sources and quantities of hazardous wastes. As of early 1992, most of the states had not yet specified sites for the disposal of hazardous wastes. Developing the needed technical competence to implement upgraded regulatory procedures continues to be one of the weakest links in India's environmental management system.

The Public Liability Insurance Act, 1991

Bhopal highlighted the need for a procedure to provide immediate relief to the victims of industrial accidents and to address the now-legendary economic and psychological difficulties confronting the poor. Accordingly, on 7 January 1991, parliament passed the Public Liability Insurance Bill, which requires every owner to carry insurance to cover death, injury or property damage resulting from an accident and provides for compensation for permanent or partial disability. This act is designed mainly to protect people living in the vicinity who are not covered under any other compensation law (workers, for example, are currently covered under the Workmen's Compensation Act of 1923). The procedures for settlement of claims are very simple, with disbursement provided by the district collector.

Although this act may be seen as another striking example of social learning in the aftermath of Bhopal, full implementation again may prove elusive. For an existing industry, both the extent of liability and the basis upon which it will be estimated are unclear. The law fails to distinguish between highly hazardous and less hazardous facilities, and, more important, it does not address how the dynamics of settlement growth in the vicinity of a hazardous plant will be incorporated into the liability estimation process. Nevertheless, it is an important step in institutionalizing compensation arrangements for persons other than workers who are affected by accidents or releases from hazardous facilities.

Environment (Protection) Second Amendment Rules, 1992

A recent amendment to the Environment (Protection) Rules, 1986, requires all firms that operate under the Environment Protection Act to conduct environmental audits. Beginning with the period April 1992–March 1993, any firm that discharges effluents or emissions must submit to the relevant state pollution control board a report that details:

1   consumption of water and other raw materials;
2   air and water pollution;
3   quantities (by category) of solid and hazardous wastes generated;
4   waste-disposal practices; and
5   investments in environmental protection and pollution control.

The worthwhile objective of environmental audit is to encourage companies to work aggressively to reduce their pollution burden and to make optimal use of natural resources; but the regulation is fraught with weaknesses. It is difficult to verify or use the audit information submitted to the already overburdened pollution control boards. Even as the government of India has endorsed environmental audits (MEF, 1992, p7), it has failed to identify certified environmental auditors, and requisite expertise is virtually non-existent in most firms, which, incidentally, had no say in designing and formulating audit formats. A handbook might have helped. Indeed, the Indian government might have taken a page from the Environmental Protection Agency (USEPA, 1986) or United Nations (UN) agencies (UNEP/IEO 1990) and prepared a detailed audit manual prior to launching the new regulation. Such preparatory work would have eased the task for implementing agencies and audit-performing firms alike, and enhanced the prospects for achieving pollution reduction. As it is, the essentially volunteer procedure for disclosure offers no way to cross-check the information provided. Prior stringent regulations have not themselves yielded effective compliance; so it is unlikely that environmental audits will improve the environmental–pollution status of industries.

The impacts of legislative initiatives

Taken together, these six laws have developed a new and potentially far-reaching management structure for reducing or mitigating industrial risks in India. But two major realities restrict their potential. First, the legislation overemphasizes the very regulatory approaches and procedures that had proved ineffective in Bhopal, while neglecting economic instruments that might have helped to internalize damage costs. Numerous studies have demonstrated the effectiveness of economic incentives in enhancing environmental quality, particularly in developing countries (Project 88, 1988; Stavins, 1989). In countries as disparate as Malaysia and Germany, for example, high charges for pollution provide a continuing disincentive to discharge contaminants. Malaysia's introduction of discharge fees proportional to the pollution load has effectively induced corporations to reduce emissions. Meanwhile, the Malaysian government lowered the permissible standards for various pollutants slowly over a five-year period. Economic instruments also stimulate permanent efficiency improvements, so that the costs of pollution prevention and conservation are gradually internalized within the production process.

Yet in India the continuing reality is that the cost of compliance with standards often far exceeds the cost of non-compliance, and reliance on regulation only encourages industries to find ingenious methods for avoiding or delaying compliance. The recently enacted requirement for environmental audits speaks to an overuse of regulatory mechanisms or 'command-and-control' approaches to curbing pollution and reducing industrial hazards. India has still to learn that yet another regulation, devoid of built-in incentives for compliance, will not improve implementation of existing regulations. It will only weaken the working of existing institutions, since it will force them to spread meagre resources very thin over a variety of regulatory procedures rather than concentrate on critical environmental problems or major polluters.

Second, regulatory reform after Bhopal has run substantially ahead of broader institutional changes. Existing institutions often lack the structure, skills and resources to implement the new regulations. Hazard management requires diverse specialized skills not currently represented on India's pollution control boards, which are staffed mostly by civil engineers from public health departments with little expertise in hazardous substances control. Worse yet, recent years have witnessed a proliferation of political appointees, whose sometimes selective prosecution of violators has eroded public confidence (Lalvani, 1985). Similarly, factory inspectors, who are mostly mechanical or electrical engineers without appropriate training, have been assigned additional responsibility for hazard assessment, hazardous substances control and emergency management. Pollution control boards and factory inspectorates remain seriously understaffed. Recent piling on of duties attendant on enactment of environmental audit regulations will only further compromise ongoing monitoring and implementation of existing regulations. The shortfall between new responsibilities and needed expertise is likely to undermine the implementation of the environmental protection measures included in the six recent laws.

DEVELOPMENTS IN THE COURTS

The Indian judiciary's role in institutionalizing risk management and risk compensation extends beyond the specifics of the Bhopal litigation (see Rosencranz et al, 1994). Judicial actions have reinterpreted existing constitutional provisions to incorporate a new awareness of the deterioration of environmental quality and the potential for industrial accidents. Certainly, the Bhopal accident and the passage of the Environmental Protection Act stimulated a quick response from the courts. Several cases deserve special notice.

The public litigation cases

The constitution of India originally made no mention of environmental protection. In 1976, however, parliament added two new articles: Article 48(A), which stipulates that the state shall endeavour to protect and improve the environment and to safeguard forests and wildlife, and Article 51(A(g)), which establishes the fundamental duty of every citizen of India to protect and improve the natural environment, including the forests, lakes, rivers and wildlife, and to have compassion for all living creatures. In recent years the Supreme Court of India has agreed to hear public-interest lawsuits filed by voluntary organizations, by public-minded citizens and by judges (Shastri, 1988). Since the Bhopal accident, the supreme court has taken a particularly serious view of environmental offences, as illustrated by a recent mining controversy.

In the Doon Valley in the state of Uttar Pradesh, a number of mining companies quarry limestone in hilly areas covered with forests. A voluntary agency – the Rural Litigation and Entitlement Kendra – filed a public-interest suit alleging that mining was destroying the environment. The supreme court issued an interim judgment on 19 October 1987, ruling that stone quarrying in the valley should generally be stopped. In its final judgment of 30 August 1988, the court ruled that the ongoing mining lease should be terminated without provision for compensation. Observing that natural resources are permanent assets of mankind and are not intended to be exhausted in one generation, the court found that such termination was in the broad interests of the community.

The judiciary has also interpreted the basic articles of the Indian constitution as supporting environmental protection and minimization of industrial hazards. Prompted, in part, by the Bhopal catastrophe and the massive suffering it induced, the judicial arm of the government has taken a long-term view that mandates the curtailment of human activity resulting in serious or irreversible environmental damage. The judiciary's relatively forward-looking stance on environmental issues has counterbalanced to some degree the more conservative and incremental perspective of the administrative and regulatory wing of the government.

The absolute liability case

The oleum leak that occurred in Delhi in December 1985 at an industrial unit of the Shriram Food and Fertilizer Industries (SFFI) represented a milestone for Indian tort law. On 7 December 1985, the Supreme Court of India accepted for consideration a public-interest petition requesting that the SFFI unit be relocated away from Delhi and that a permit be required to restart it. A little more than two years later, the court issued a landmark judgment dealing with safety management, the liability of hazardous facilities, and compensation for workers and the public (*All India Reporter*, 1988a). The judgment made it clear that corporations in India in charge of hazardous facilities have an absolute and non-delegable responsibility to prevent hazards; this liability, moreover, was not subject to any of the standard common-law defences (see Rosencranz et al, 1994). Finally, the supreme court held that for any damage arising from a hazardous activity, the measure of compensation must be commensurate with the magnitude and capacity of the enterprise. In other words, the deeper the pocket, the larger the fine must be to achieve an adequate level of deterrence.

From the standpoint of risk management, however, the SFFI accident revealed that even one year after Bhopal, the chemical industry and the government had not really learned how to reduce the risks of serious industrial accidents or how to respond effectively to emergencies. SFFI's various units occupied a single complex in the Delhi metropolitan area and were surrounded by thickly populated settlements. Some 200,000 people resided within a radius of 3km. Nevertheless, at the time of the accident no emergency response system was in place and no coordinated effort was made to communicate to people in the vicinity of the plant. Instead, confusion reigned, with various agencies providing conflicting advice to local citizens. The events at SFFI clearly indicated that hazard assessment capability had not yet developed in major industrial facilities. On the positive side, the Delhi Municipal Corporation did take up the safety issue upon receiving a complaint about SFFI, and the supreme court, of course, did rule on a public-interest writ petition. Neither action would have occurred but for the heightened awareness of risk created by Bhopal.

The vicarious liability case

Another major case dealt with the issue of the vicarious liability of senior executives in cases of chemical damage. A distillery located at Modinagar discharged untreated effluents into a river, producing a fish kill. The Uttar

Pradesh State Pollution Control Board brought suit, arguing a violation of the Water Act. The chairman, the vice chairman, the managing director and the directors of Modi Industries filed an application with the state high court asking to have the charges set aside since the executives were not directly responsible. The state court accepted this contention and absolved the officers of vicarious liability; but the Supreme Court of India set aside the decision and ruled that top executives in manufacturing firms are, indeed, responsible for implementing the Water Act. They are liable for offences under the act unless they can prove that the violation was committed without their knowledge or that they exercised due diligence to prevent its commission (*All India Reporter,* 1988b).

The tannery pollution case

Many small-scale industrial enterprises in India are household units or cottage industries that, for the most part, do not adhere to hazardous waste or pollution control regulations. This smaller and generally unorganized industrial sector remained relatively unconcerned about its risk problems until the Supreme Court of India intervened. Leather tanneries are a major example. At one time, they used non-polluting natural products from trees and barks in a process known as East India tanning. The chrome tanning that came into India after 1950 caused water pollution and land degradation. Since most of the leather tanning units were in the cottage-industry sector and were operated with very little capital or technical expertise, it was difficult to secure their compliance with pollution control regulations. Pollution from tanneries continues to be a serious national problem, particularly since many untreated effluents make their way to the Ganges.

A public-interest lawsuit in the Supreme Court of India argued that tanneries should be prevented from discharging effluents until they had installed appropriate treatment systems. For their part, the tanneries pleaded that they were small and could not afford such remedial action. After considering all of the issues, the supreme court ruled that the financial capacities of industrial establishments should be considered irrelevant and ordered the tanneries to undertake primary treatment of the effluents. The court also held that, just as an industry that cannot pay minimum wages to its workers cannot be permitted to continue, so the adverse environmental effects caused by the discharge of effluents justified the closure of the tanneries, despite the inconvenience caused to management and labour (*All India Reporter*, 1988a).

INSPECTION AND ENFORCEMENT

Changes in inspection procedures and regulatory enforcement constitute the third set of Indian institutional responses to Bhopal. Two changes occurring during the past five years deserve special notice.

Hazard assessment

Following the Bhopal accident, all of the major industrial states in India appointed multidisciplinary task forces to analyse their existing hazardous facilities. These teams visited large industrial sites, reviewing emergency management plans, safety practices and storage facilities, and, based on preliminary hazard assessments, made recommendations for improving safety management and emergency planning. In certain cases in which facilities were located in close proximity to dense human settlements, the task forces recommended shifting industries to new sites. In Karnataka state, for example, the task force studied 19 hazardous facilities for which it identified safety problems and inadequacies that required correction. Similar exercises were also carried out in all states with major hazardous processing or storage facilities. The officials of the Factory Inspectorate also used a task-force approach for identifying and assessing hazards (Government of Karnataka, 1987; Bowonder and Arvind, 1989). These review exercises have helped state governments to develop local expertise in hazard assessment. They also produced widely publicized reports through which ordinary citizens in many states became aware for the first time of the hazards of industrial facilities.

Another new development is the requirement that manufacturing facilities must prepare detailed hazard assessment reports, which must be approved by the Factory Inspectorate before they may construct and operate new facilities. Firms may be asked in this process to make specific – and burdensome – changes in equipment and procedures, as well as in the location of facilities. In Bombay, a number of units were required to move outside the city limits when they planned further expansion, even though trade unions opposed such actions because of their employment implications. The major shortcoming of these activities was that the hazard-assessment task forces could only make recommendations since they lacked any statutory powers. The Factory Inspectorate, however, has enforced the safety-related recommendations as part of its safety inspections, which are statutory. Furthermore, the assessment by an external committee was, in most cases, a one-shot affair, with little or no follow-up. Nevertheless, the exercises established the principle that private as well as public units could be inspected for safety by an external agency; many units, in fact, improved their safety procedures and made safety investments as a direct result of the committee recommendations.

Environmental impact assessment

Since 1985, the government of India has required that every new project obtain environmental clearance. The operational framework for this requirement is a set of administrative procedures (see Figure 8.1) that, as yet, lack the backing of any law. All projects are first screened at the state

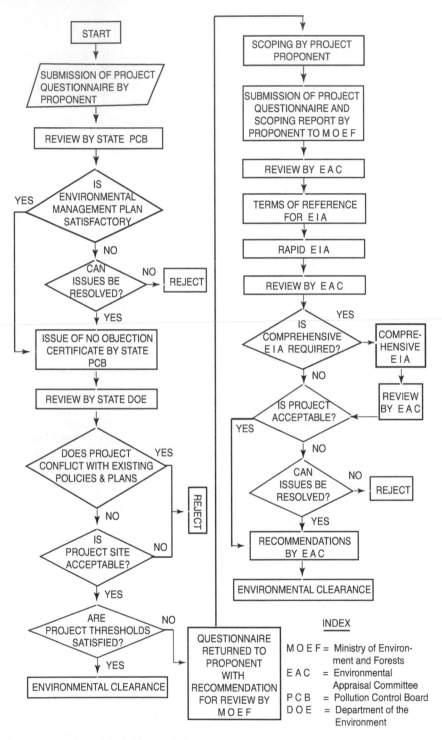

Source: Bowonder and Arvind (1989, p187)

Figure 8.1 *An operational framework for conducting an environmental review*

level to decide whether an environmental impact assessment (EIA) is required. The State Department of Environment then clears those projects that do not require an EIA and refers those that do to the Ministry of Environment and Forests.

EIA procedures require various agencies to examine the necessary safeguards at the project initiation stage. To initiate the review, the project developer completes and submits to the State Pollution Control Board a project questionnaire provided by the Ministry of Environment and Forests. The board reviews the environmental management plan and either rejects the proposal or, if the plan is satisfactory, issues a 'no objection certificate' and recommends the project to the State Department of Environment. No specific format has been proposed for the EIA as yet, however, and this has resulted in the exclusion of certain important impacts (for example, growth-inducing impacts) from review. The administrative apparatus for EIA at the state level is also technically very weak. If model EIAs or model environmental management plans were prepared, and project sponsors were properly trained and sensitized, the anticipatory value of the EIA process would be greatly enhanced.

The norms and procedures for environmental clearance were vague until February 1992, when the Ministry of Environment and Forests classified all projects and industries according to two schedules (*The Hindu*, 1992). Projects listed in Schedule 1 – which includes atomic power, thermal power, river-valley projects, refineries, chemical facilities, and so on – require clearance by the relevant state government as well as by the ministry. Schedule 2 lists projects that require only state clearance. Yet the ministry reserves the right to review state clearances if it receives challenges in writing or if the project lies within 10km of ecologically sensitive areas. For projects under Schedule 1, a committee will assess the environmental impacts prior to issuing a preliminary clearance. Although the procedures are clearly stated, the detailed format of the EIA and the specific aspects to be included are left to the developing agency. Moreover, since no formal law exists to support a given site clearance, subjective elements inevitably compromise the process and may well render an EIA ineffective.

Implementation of pollution control regulations

The implementation of pollution control regulations was generally ineffective in India prior to the enactment of the Environmental Protection Act of 1986. A study by Bowonder (1988; see also Bowonder et al, 1988), based on extensive interviews and a survey covering about 200 industries, provides insight into the major obstacles and areas of weakness (see Figure 8.2). Since 1986, however, the central government has taken a tougher stance on pollution. As noted earlier, the 1986 act empowers the Ministry of Environment and Forests to order plant closures without going through the state pollution control boards. In 1991, the ministry closed 51

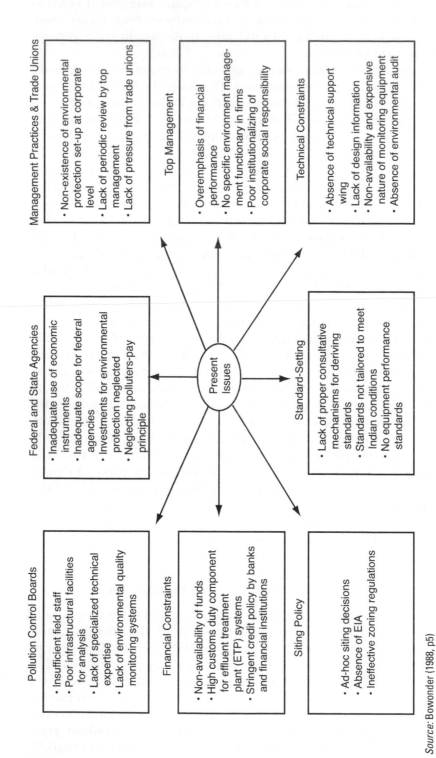

Figure 8.2 *Major obstacles to environmental management in India*

polluting units. In addition, it now has the authority to intervene directly in cases where a company or government agency has failed to comply with environmental standards.

Units that do not plan to install pollution control equipment within a specified time period receive closure notices, which are, in turn, a prelude to the closure directive that the ministry issues for continued failure to comply with regulations. The ministry may ask for a timetable for compliance, as well as proof of the firm's commitment to comply with the regulations.

Inspection and classification of industries

The Bhopal accident indicated the need to revamp completely the inspection procedures of the factory inspectorates and state pollution control boards. Beginning in 1988, the government of India divided industries into three major categories and proposed a tentative inspection schedule. All large-scale industries were placed in the 'red' category; for this class, the frequency of inspection was to range from once a month to once every six months, depending upon the size of the industry. Medium-scale industries in the 'orange' category were to be inspected once every two years, whereas small industries in the orange category were to be inspected once a year. Finally, those in the 'green' category were to be inspected once every two years.

This classification represents a laudable attempt to standardize and prioritize inspection procedure; but it has serious shortcomings. Since the nature of the inspection is not specified, considerable ambiguity remains about the depth and scope of review. The depth of inspection is not related to the age of the facility and its equipment, although older facilities frequently demand more in-depth inspection to uncover and correct hazardous situations. A system for comprehensive analysis of near-mishaps has yet to be developed. Procedures for the comprehensive logging and analysis of data have yet to be specified. More problematically, automatically classifying all large industries as hazardous appears to defeat the very purpose of ranking industries according to the degree of hazard. Finally, until the hazard-assessment expertise and skills of inspectors are seriously upgraded, the quality of inspection is unlikely to improve significantly.

CONCLUSIONS

It is clear that the Bhopal accident was a traumatic event that stimulated enormous attention to industrial hazards by Indian legislators and policy makers. The resulting actions were national in scope, unlike the activities of various groups whose advocacy of environmental conservation in India had concentrated on local problems and on rectifying particular harms, rather than on addressing more generic ones. Thus, in the Doon Valley case mentioned earlier, the concern was to preserve a specific mountain ecosystem in the north of India. The agony of the Bhopal gas victims greatly increased public awareness about the need for controlling pollution, protecting the environment, conserving natural resources and improving safety management practices in hazardous industrial facilities. The media devoted special attention to the consequences of Bhopal and, subsequently, at repeated and regular intervals, highlighted the hazards that other Indian industrial facilities posed to humans and environment. A wide array of institutional changes since Bhopal bears witness to the depth of concern in parliament and the central government over future accidents and threats to the environment.

A great deal of social learning, then, occurred in India as a result of the Bhopal accident, particularly in the formal regulatory arena. Bold legislative and regulatory reforms, bolstered by activist judicial decisions, put into place some much-needed institutional frameworks, but fell short of providing the infrastructure, resources and support needed to ensure implementation. Passage of the Public Liability Insurance Act of 1991 made India one of the first countries to introduce mandatory hazard insurance for people other than workers; but the formidable problems of implementation were not assessed before the law's enactment. Regulators continued, as before, to ignore affected parties in decision making, thereby forgoing opportunities for creative learning. Indeed, most of the institutional responses to Bhopal reflected maintenance or adaptive learning; transitional or creative learning appeared to be in much shorter supply. The goal of making the environment a public good, and of ensuring that environmental values would take precedence over private economic interests, remained distant as India moved incrementally to secure better compliance with environmental regulations. Indeed, an inclination toward economic liberalization, which facilitates the start-up of new industries, threatens to compromise India's environmental gains. A prime minister's assurance that 'no feasible power project will ever be held up for want of environmental clearance' heralds the accommodating 'green channel' that essentially waives licensing requirements (*Economic Times*, 1992a).

Non-governmental organizations (NGOs) and trade unions played, at best, a peripheral role in the process of social learning. Trade unions remained largely passive and reactive in their attitude to industrial hazards. An emphasis on economic benefits to workers, an indifferent trade union leadership and low worker awareness about safety and hazards account for this passivity. NGOs did somewhat better (see Reich, 1994; Rosencranz et al, 1994); but strong professional leadership and expertise are still very limited. The large increase in the number of NGOs and in their membership has not yet produced a commensurate increase in technical expertise and technical information support (Khator, 1988). India still needs professionally competent organizations with a sound and effective base of technical expertise, as well as political access.

Scant attention went into training and the development of skills: skills for carrying out hazard assessment and evaluation, for planning emergency responses, for undertaking emergency evaluation, for communicating risks to populations and workers, some of whom may be illiterate, and for improving industrial risk management generally. These shortcomings are particularly troubling in the wake of Bhopal, which starkly pointed up the need for better risk communication. As one analyst puts it: 'If we give dangerous plant or material to people who have not demonstrated their competence to handle it, we are responsible for the injuries they cause' (Kletz, 1988, p89).

In the long run, pollution prevention strategies and the polluter-pays principle, along with improved risk anticipation and prevention systems, may be the most effective way of controlling pollution in India. Long before Bhopal, Trevor Kletz (1978) championed *intrinsically* safe plants – namely, plant designs that use fewer and less hazardous materials (Kletz, 1985, 1993). 'Clean' new process technologies and 'environmentally friendly' products will have to be pursued progressively through the ecological modernization of the economy, as proposed by Simonis (1987). Indeed, the Ministry of Environment and Forests has recently initiated a system for certifying such products and encouraging consumer awareness (MEF, 1991, p43). The growth of the economy also needs to be reoriented in order to minimize environmental impacts through the promotion of cleaner, safer and less resource-intensive industries.

And so, even as the World Bank (1988) champions the development of national capacity, industrial risk management efforts move forward in India in a fragmented and haphazard way, involving overlapping agencies and institutions and a byzantine administrative process (Hadden, 1987; Bowonder and Arvind, 1989; Hadden, 1994; Reich, 1994). Industrial risk management has made important strides since Bhopal, but has yet to be institutionalized in a comprehensive and systematic way. A kind of collective organizational amnesia virtually ensures the recurrence of Bhopals (Kletz, 1993). The end of a 'decade of determined response' that we called for in Chapter 5 of this volume approaches, and much more must be done to achieve a holistic framework for managing industrial risks to humans and the natural environment. One promising sign is the embracing of 'sustainable development' by the Associated Chambers of Commerce and Industry of India (ASSOCHAM), which has presented a slate of suggestions for effective cooperative management of environmental pollution (Sankar, 1992). Meanwhile, the ideal of sustainable development of India's economy and technology continues to beckon.

Part 3

The Globalization of Risk

9 Hazards in Developing Countries: Cause for Global Concern

B. Bowonder and Jeanne X. Kasperson

INTRODUCTION

A flourishing literature on hazards, risks and disasters threatens to overwhelm the most assiduous of bibliographers. Emanating as it does from an impressively diverse array of disciplines, quasi disciplines and professions, this staggering body of research courts the likes of an 'annual review' that will take critical stock of prevailing strengths and deficiencies. One would expect an interdisciplinary state-of-the-art assessment to recognize and redress the relative dearth of attention to the developing countries, which, after all, bear a disproportionate brunt of the harmful consequences of hazardous events, activities and processes. Quarantelli (1988) suggests that 70 to 85 per cent of all disasters occur in the developing world. Even cursory looks at statistical compilations (OFDA, 1988; UNDRO, 1988) bear out his observation. A major report on natural hazards (USNRC, 1987) pinpoints the 'double blow' – of high death tolls and economic losses – that falls on developing countries.

Given the acknowledged magnitude of the problem, a comprehensive review of the literature on hazards in developing countries is very much in order. This chapter, however, by no means purports to tackle that onerous and long overdue task. Instead, the authors have settled for a more manageable effort to stimulate research and discussion on an important topic that has come up short in the overall hazards arena.

This relative inattention to hazards in the developing world has not escaped notice (Inhaber, 1985; Minor et al, 1986; Oliver-Smith, 1986; Texler, 1986; Kasperson and Kasperson, 1987a). Yet the unquestionable fragility of countries that must perforce juggle short-term basic needs with long-term goals of sustainable development demands immediate concern

Note: Reprinted from *Risk Abstracts*, vol 5, Bowonder, B. and Kasperson, J. X., 'Hazards in developing countries: Cause for global concern', pp103–109, © (1988), with permission from Cambridge Scientific Abstracts

and special handling on a global scale. Indeed, the handling of hazards in developing countries will reverberate not only on physical quality of life, but on the very survival of the Earth. The hazards that beset developing countries fall into three broad categories: *natural, technological* and *environmental*. One may well quibble about the artificial distinction between natural and technological hazards and contend that both types fit easily under the wide umbrella of environmental hazards; but we decline to enter that fray. Over all, we view hazards as threats to humans and the things that humans value (Kates et al, 1985). For purposes of this chapter, we find it useful to group such threats into three categories.

NATURAL HAZARDS

The likes of floods, droughts, earthquakes and volcanoes wreak excessive havoc in the less-developed countries of the Earth (White, 1974; Bowonder, 1981b). The North American drought of 1988 caused great damage and hardship, but it is no match for the Sahelian drought. An earthquake in Tokyo is unlikely to cause devastation on the scale of a comparable (on the Richter scale) earthquake in Mexico City.

Technological advances permit the developed world to make accurate predictions, to issue early warnings, and to order timely evacuations in the face of an impending natural or technological disaster. Meanwhile, the increase of poor populations in the most disaster-prone regions of the developing world has wrought an increase in the vulnerability of those regions to the catastrophic consequences of natural hazards (Oliver-Smith, 1986, p7). Witness the unprecedented toll exacted by the floodwaters that inundated three-quarters of Bangladesh in the summer of 1988.

The march of industrialization and urbanization has attracted more and more people to live in high-risk areas that are particularly prone to experience natural disasters (Sewell and Foster, 1976). A tendency to treat such calamities as 'acts of God' accounts for the dearth of systematic efforts to understand consequences or to develop long-term preventive strategies. People who are already barely eking out an existence will not avoid a risky floodplain or the shadow of a volcano any more than they will eschew the squatter settlements around a pesticides factory in Bhopal or a liquefied gas facility in Mexico City. In short, the poorest of the poor are probably likely to reside in the path of both natural and technological hazards. Small wonder that the impacts of natural hazards are on the rise in the developing countries (Oliver-Smith, 1986; USNRC, 1987).

TECHNOLOGICAL HAZARDS

Technological hazards often surface in the contamination of air, land, water and food, or in the failure of consumer products, energy and transportation systems, and medical devices. Indiscriminate deployment of technology or mismanagement of complex technologies accounts for numerous technological hazards (Bowonder, 1981a; Lanza, 1985; Perrow, 1985; and see Chapter 5 in this volume). The Bhopal accident demonstrates how readily the use of a complex technology in a poorly developed area that lacks a proper infrastructure for communications can spark the world's deadliest industrial accident (Lanza, 1985; Shrivastava, 1987a; see also Chapter 5 in this volume).

At Bhopal and elsewhere in the developing world, cultural factors, inadequate infrastructure, low levels of technical skills, blind transfer of technology, illiteracy, and the absence of local public-interest groups highlight the perilous mismatch between the social system and the technological system (see Chapter 5 for these issues in the Bhopal accident). Air pollution (Agrawal and Raju, 1980; Ghafourian, 1983; Smith et al, 1983; Quaraishi, 1985; Sani, 1985; Travis, 1985; Smith, 1986), water pollution (Sheppard, 1977; Walter and Ugelow, 1979; Middlebrooks et al, 1981; Davidar, 1982; Dissanayake, 1982; Bowonder, 1983b; Oluwande, 1983; Szekely, 1983; Wolfe, 1983; D'Monte, 1984; Bhargava, 1985; Ghinure, 1985; Moursy, 1986), and the dumping of hazardous waste (Suess and Huismans, 1983; Smith and George, 1988) attest to the interaction, if not collision, of technology with the numerous socioeconomic and institutional contexts of the developing countries. In the grip of cultural and socioeconomic constraints such as illiteracy, underdevelopment, corruption, weak political systems and the absence of infrastructure, such technological hazards wield an extraordinary hold on the developing country. To a disproportionate share of the consequences of natural hazards, the population of a developing country must add an excessive burden of the effects of technological hazards. And as if that combination were not lethal enough, enter the environmental threats.

ENVIRONMENTAL HAZARDS

Environmental hazards arise out of the interaction of fragile ecosystems with underdeveloped socioeconomic systems. Sometimes the interaction generates low-intensity hazards (Bowonder, 1980, 1981a, 1982, 1983a; Biswas and Biswas, 1984; Howard-Clinton, 1984) that are distributed over space and time. Slow-moving environmental hazards may disperse their impacts on humans directly or on the ecosystems upon which humans rely. Such impacts are apt to be particularly severe in tropical areas with poor soils; arid and semi-arid areas that are water deficient; regions characterized by low biomass productivity and high evapotranspiration; areas of high population density; and areas with high livestock populations.

In the many developing countries that harbour areas with one or more of these characteristics, ecosystem recovery from human disturbance is slow and sometimes impossible (Bowonder, 1986a; Texler, 1986; Blaikie and Brookfield, 1987). An overall tendency to downplay environmental hazards (Walter and Ugelow, 1979; Schaumberg, 1980; Bowonder, 1985a, 1985b) has thwarted the progress of sustainable development. Thus, deforestation (Aluma, 1979; Rajitsinh, 1979; Bunker, 1980; Bowonder, 1982, 1987; Dewalt, 1983; Kramer, 1983; Nations and Komer, 1983; O'Keefe, 1983b; Van Gelder and Hosier, 1983; Aiken and Leigh, 1985; Lundgren, 1985; Nautiyal and Babor, 1985; Bhagavan, 1986; Grainger, 1986; Tangley, 1986) has exacted an enormous environmental toll. One of the consequences of deforestation is soil erosion.

Soil erosion, whatever its causes, exacerbates environmental degradation (Sanchez and Buol, 1975; Biswas and Biswas, 1978; Brown, 1983, 1984; Finn, 1983; Lamb, 1983; O'Keefe, 1983a; Randrianarijaona, 1983; Singh, 1985; Rapp, 1986; Bowonder et al, 1987). The slow but deadly process of desertification (Le Houerou, 1977; Jodha, 1980; Kates, 1981; Darkoh, 1982; Walker, 1982; Fare, 1986; Luck, 1983) produces yet another round of degradation.

Environmental hazards also include an array of health hazards that are attributable to environmental factors (Obeng, 1978; Jalees and Vemuri, 1980; Rajendran and Reigh, 1981; Sridhar et al, 1981; Bowonder, 1983c, 1986b; Osore, 1983; Tucker, 1983; Bowonder and Chettri, 1984; Texler, 1986). Contaminated water supplies and unhygienic living conditions, particularly in overflowing squatter settlements of urban centres, are invitations to the spread of disease.

Another category of environmental hazards surfaces in the deterioration that befalls ancient monuments, sites of cultural value and tourist attractions as a direct or indirect consequence of human activities (Tam, 1981; Bowonder, 1983a; Hinrichsen, 1983; Kadry, 1983; Kramer, 1983; Lamotte, 1983; Singh, 1985). Such assaults are seldom intentional but, rather, the indirect outcome of some seemingly economically beneficial activity. Threats to the likes of the Taj Mahal (Bowonder, 1981b), Himalayan tourist spots (Nautiyal and Babor, 1985) and Caribbean islands (Tam, 1981) speak to a tendency to undervalue environmental benefits. In short, when it comes to a trade-off between tangible economic gain and intangible environmental protection, a built-in bias favours the tangible profit and the environment comes up short.

Similarly, the careless or inadvertent destruction of natural habitats spotlights yet another category of environmental threat. The attendant decline or disappearance of affected flora and fauna (Aluma, 1979; Ehrlich, 1980; Baidhya, 1982; Jackson, 1982; Ong, 1982; Glynn, 1983; Aiken and Leigh, 1985; Da Fonseca, 1985) may well be irreversible as areas lose their capacity to support life (Centre for Science and Environment, 1985; Texler, 1986).

THE NEED FOR GLOBAL ATTENTION

The survival of the planet is in some jeopardy. The hazards that threaten the developing world imperil the entire world and require global attention commensurate to the task. We suggest the resurrection of the notion of constructing national risk profiles (Whyte and Burton, 1980; Kasperson and Morrison, 1982; Kasperson and Kasperson, 1987a) that will permit comparative assessment of the natural, technological and environmental hazards that may overwhelm the developing world. The coordinated research programme of the International Atomic Energy Agency (IAEA), United Nations Environment Programme (UNEP) and World Health Organization (WHO) might well undertake the preparation of such profiles.

The IAEA–UNEP–WHO effort echoes the concerns of various other groups (WICEM, 1984; WCED, 1987). The director of the Disaster Research Center at the University of Delaware has proposed a set of criteria for disaster preparedness in developing countries (Quarantelli, 1988). The Center for Technology, Environment and Development (CENTED) at Clark University has launched a project, funded by the US National Science Foundation, on the ethical aspects of transferring potentially hazardous industries to developing countries (see Chapter 7 in this volume). Meanwhile, a committee of the US National Research Council proposes that 1990–2000 be an International Decade for Natural Hazard Reduction, during which society will cut by 50 per cent the current toll of natural disasters (USNRC, 1987).

And the 1988 Society for Risk Analysis (SRA) meeting included a special session entitled 'US Transfers of Technological Hazards to Developing Countries'. Perhaps some of the papers from this session will surface in the proceedings of the conference or the SRA's journal, *Risk Analysis*, which carries so few articles on developing countries (Inhaber, 1985; Minor et al, 1986). With all these noble efforts under way, the prospects are far from bleak. We hope that our brief chapter has driven home the merit of such endeavours. The hazards that must threaten the developing world are global and they require global concern.

10 Priorities in Profile: Managing Risks in Developing Countries

Jeanne X. Kasperson and Roger E. Kasperson

INTRODUCTION

The panoply of risk that envelops the everyday existence of those who dwell in the developing world threatens to eclipse the issues that command so much attention in more developed and more affluent societies. In the US and Canada, risk analysts engage in debates about whether an individual annual fatality rate of one per million marks a useful threshold for regulatory intervention, whether a *de minimis* standard might ward off over-response to insignificant hazards, whether mandatory air bags might produce further reductions in an already low rate of automobile fatalities, and whether one nation or another is accountable for the transborder consequences of acid rain. The recent 'Risk Assessment Issue' of *Science* (1987) runs the gamut from ranking potential human carcinogens to examining the safety goals of the US Nuclear Regulatory Commission (NRC); but six articles by expert risk assessors have little to say about risk in the developing world, which, in fact, may be bearing a disproportionate share of the burden. Meanwhile, the spiralling poverty-ridden populations of the developing world contend with the familiar scourges of human history: with natural disasters such as floods, droughts and earthquakes; with communicable diseases such as diarrhoea and malaria; with periodic famine and chronic undernutrition (UNEP, 1986).

Add to these traditional perils the array of threats that presides in the developed world – the degenerative cardiovascular and respiratory diseases; the unprecedented air pollution that chokes overcrowded urban areas; the prospect of contracting cancer from exposure to industrial chemicals; the contamination from pesticides, fertilizers and heavy metals of food, water and soils; and the potential for Bhopal- or Chernobyl-like calamities associated with the development of new technologies. The sum is staggering.

Note: Reprinted from *Risk Abstracts*, vol 4, Kasperson, J. X. and Kasperson, R. E., 'Priorities in profile: Managing risks in developing countries', pp113–118, © (1987), with permission from Cambridge Scientific Abstracts

The developing countries can muster only limited expertise, few financial resources and overtaxed institutions to tackle so bewildering an array of risks and hazards. Moreover, the context for coping differs significantly from that in more developed societies. Different hazards, vulnerabilities, managerial resources and timescales clamour for centre stage and highlight the need for special handling.

DIFFERENT HAZARDS

The majority of the world's people suffer most from familiar hazards, many of them rooted in nature and in poverty. Geophysical hazards claim an average of 250,000 deaths per year; 500 million to 730 million of the world's people are hungry (Hagman, 1984, p161; World Bank, 1986, p1); periodic famines such as that accompanying the Sahelian drought during the early 1980s can exact death tolls in the millions (Wijkman and Timberlake, 1985); communicable diseases still account for 10 to 25 per cent of human mortality; and vermin, pests and crop diseases destroy fully 50 per cent of the world's food crops (Harriss et al, 1985). At the same time, each year 11.3 million hectares of forests bow to destruction (WCED, 1987, p47), soil degradation depletes 15 million hectares of land required for food production, and desertification claims broad tracts of land, thereby threatening the long-run environmental sustainability requisite to support development (IIED and WRI, 1987).

Although such hazards predominate, industrialization and urbanization also exact their tolls: in air pollution in São Paulo (UNEP, 1986, p37), in hazardous wastes in Mexico's sewer systems and nearby agricultural lands, and in extensive pollution fouling India's rivers (D'Monte, 1984, p273). In the context of such formidable hazards, it is scarcely surprising that other hazards such as residues of pesticides and heavy metals in foods, soils and water, lead toxicity from automobile emissions, and indoor air pollution from domestic cooking fuels remain hidden (Bowonder, 1981a). As national governments court and harbour industrial development, it is incumbent upon them to look beyond the obvious pollution and confront the invisible, often extremely dangerous, hazards attendant upon particular industries (Leonard, 1985, p813). Ashford and Ayers (1985, p881) caution that an absence of long-range planning may bequeath a legacy of irreversible damage.

DIFFERENT VULNERABILITY

Because so many of the 350 million people who live in developing countries survive at the margins of human existence, they are highly vulnerable to the occurrence of hazards, whether from nature or technology. In inhospitable settings of chronic malnutrition and a lack of potable water and proper sanitation, even small perturbations in forest, soil and agricultural systems can wreak devastating consequences on human health. Overall increases in all major types of so-called natural disasters have resulted in a sixfold increase in the number of victims, most of them, disproportionately, in the poorest countries (Hagman, 1984, p44; Wijkman and Timberlake, 1985, p23), where the poorest of the poor live in the most dangerous places (Susman et al, 1983, p278).

Imported technology (with its own cache of hazards), far from coming to the rescue, often collides with an underdeveloped institutional infrastructure. Indigenous settings short on capabilities for management and emergency response cannot accommodate technological fixes that do not give due attention to the primary needs of the most vulnerable (Hagman, 1984, p86).

The most vulnerable include the children who die each year – 15 million from malnutrition, 4.6 million from diarrhoeal diseases and 4 million from respiratory diseases (UNEP, 1986; WHO and UNICEF, 1986, p8). The most vulnerable also include women, who gather firewood, encounter regularly a lack of sanitation and safe drinking water, and who breathe at close range the dangerous fumes from cooking fuels and from passing motor vehicles (D'Monte, 1984). In India, environmental degradation translates into long workdays for 44 million child labourers (Vallura, 1987) and still longer workdays (14 to 16 hours) for the landless women caught up in the relentless cycle of travelling farther and farther afield to collect scarce water, food, forage and fuel, and who bear more and more children to help with these chores that become inexorably more difficult and time consuming as resources become scarcer (Centre for Science and Environment, 1985, pp172–176). Given extreme poverty and high population density, India boasts few ecological niches that are not 'occupied by one human group or another for its sustenance'. Thus, any human activity that destroys, degrades or transforms such niches will inevitably harm those who were earlier dependent upon the space (Centre for Science and Environment, 1985, pp369–370).

Different managerial resources

Few resources are available to assess and manage the hazards that proliferate in the developing world. Databases for delineating the hazards remain woefully inadequate and expertise to analyse the sources, pathways, and human and ecosystem consequences is in short supply. External aid frequently lacks coordination and trains on industrialization 'at all costs'. Well-intentioned and even well-administered technical assistance rarely reaches beyond the particular agency or sector that enlists it or to which it is allocated. Meanwhile, the legislative, regulatory and institutional structures requisite for managing environmental and technological hazards are out of date, hopelessly overburdened, and fragmented in scope and jurisdiction.

DIFFERENT TIMESCALES

Because of the rapidity of change in developing countries, hazards occur in highly compressed timescales. Developing countries are experiencing, in decades, a population transformation in terms of death and birth rates, which took hundreds of years in Europe and North America. The United Nations projects that, by the year 2000, Mexico City will have a population of 30 million, and 82 other megacities, 61 of them in developing countries, will have populations over 4 million (UNEP, 1986, p50). Developed societies have 100 years' experience with chemicals – developing societies, often only a decade. Genetically engineered products are already finding their ways into societies with virtually no scientific and technological tradition, let alone any capacity to accommodate surprises.

The foregoing discussion points up the complexity of the risk domain in developing countries. The double burden of having to grapple simultaneously with old, traditional perils and new, often hidden or invisible hazards poses difficulties far more formidable than those in more advanced societies. How should developing countries cope with these seemingly overwhelming problems? We propose the resurrection of one approach – the national risk profile – as a potential tool for identifying and arraying potential assessment and management options.

THE NATIONAL RISK PROFILE

As originally proposed by Whyte and Burton (1980) and Kasperson and Morrison (1982), the national risk profile offers a means for taking stock of, and categorizing, the diversity of risk problems that confront a particular society. The profile allows for a systematic display of a range of risks, aids the clarification of deficiencies in databases, facilitates the establishment of priorities, defines and evaluates the context of current risk management activities, and enhances the prospects for formulating a coherent national policy for managing risks. These objectives, of course, ought to prevail in any society; but they have particular value for the developing country that is trying to order and manage the complex array of risk problems within its national borders.

The national risk profile, then, would include:

- identification and estimates of the major risks facing a given society, systematically treating those arising from nature, environmental degradation, infectious diseases, technology, and urbanization/ industrialization;
- actuarial evidence on the numbers and incidence of risk events and the types and magnitude of associated consequences;
- trend lines for risks, risk events and consequences;

- the distributions of risks among geographical regions, economic sectors and social groups;
- identification and estimates of the population at risk from particular hazards, including subgroups that are demonstrably or potentially particularly vulnerable; and
- an account of the coverage by existing databases of major risk areas and indicators.

Construction of such a national profile will be no easy task. The undertaking will have to contend with numerous gaps, omissions and unreliable data and a dearth of expert judgement. Beyond the tapping of official records and statistics, newspapers, local histories, records of various organizations and projects, and even folk records will require scrutiny and may yield valuable pieces to the puzzle. In many cases, gross estimates will be necessary for the short term, and the filling in and improvement of data and estimates may constitute longer-term objectives.

Indeed, the compilation of the risk profile will, undoubtedly, pinpoint outstanding shortcomings in the database. Both the possibilities and the constraints of estimating risk in various areas should become apparent. Over time, the development of a comprehensive (though not exhaustive) database will address the key risks facing the society and the risks for which particular management efforts are under way or envisioned.

Research on Risks

The risk profile will provide a guide for targeting research and ordering risk assessments. Certain large, or potentially large, risks that have hitherto not undergone study, for example, ought to command priority on the risk assessment agenda. Beyond this essential filling of gaps, a strategy of broad-based or sectoral assessment of classes of risks – industrial, occupational, natural or environmental risks – would train attention on the common characteristics of a particular class of hazards and the patterns of risks within a class, as well as the ways in which both are changing (or not changing) over time.

Similarly, a well-developed risk profile would disaggregate the national risk profile by regions and by urban/rural setting, and thereby provide a more accurate account of the risk in different settings. Even risk profiles of different types of villages – in effect, integrated 'hazardousness-of-a-place' studies (Hewitt and Burton, 1971) – hold promise as a novel genre of village study. Such mid- and micro-level examinations of risk might well serve as the basis for more in-depth analyses of those risks that will be the focus of future risk-management efforts. The prospect for constructing hazard maps may be a useful outcome of the risk profile.

Risk acceptability and standard-setting

The national risk profile might also contribute to the ongoing debates over 'risk acceptability' and the transfer of potentially hazardous technology. A crucial ingredient in the judgements about importing new technology is the capability to place the associated risks into the broader contexts of the risks that already exist in the receiving country and of the benefits attendant on the acquisition. The profile will add perspective on the increments of risks that will occur, as well as on the contribution of the imported technology to a reduction in risks from other sources.

Monitoring and evaluation

The national risk profile will identify certain risk areas for which databases exist for monitoring how particular risks are changing over time. It will also drive the selection of indicators to be used in monitoring. As time-series data become available, evaluations of new events, technologies or management initiatives will determine their impacts upon trends and distributions of risks. Additions to the database through monitoring and evaluation will advance the progressive development of the risk profile toward a level of becoming a genuine reflection of risk in the developing country at hand. Furthermore, risk profiles for individual countries will enhance the global monitoring efforts currently under way (UNEP, 1987).

Management priorities

Management structures should be prime beneficiaries of the national risk profile. An initial application, the mapping, onto major domains of risk, of national legislation on environmental and risk-related issues, will identify areas that enjoy reasonably adequate coverage, as well as those that have eluded government action.

Similarly, institutions and their scopes of jurisdiction should undergo comparison with the profile. Both the style and the character of institutions should be commensurate with the problems and priorities of individual countries (Whyte and Burton, 1980, p142). As work acquires sophistication, a parallel profile of risk-reduction opportunities ought to ride tandem with the profile of major risks. Not that all risks are amenable to reduction through government intervention. The profile of risk-reduction opportunities may well look quite different from the profile of risk. It may prove instructive to compare these opportunities to the national allocation of resources to various risk problems. Are the most pressing problems being addressed, or, as the US Environmental Protection Agency admits in its own self-study (USEPA, 1987), does the discrepancy between profile and priority prevail? Are any prime opportunities for reducing overall risks being neglected? How might a reallocation and coordination of managerial resources and efforts enhance the potency of risk-reduction measures?

Social equity

A recurring concern in international development involves ensuring equity among various segments of the population with access to new resources. Numerous projects and technologies, it is apparent, have benefited only a few and have increased inequities in society. The risk profile, if it contains disaggregation by relevant geographical and social divisions, will provide relevant information on current allocations and burdens of risks. Pesticide use in developing countries is a case in point (Repetto, 1985). The population as a whole reaps few benefits from the application of pesticides to the cash crops grown primarily for large estates. Rather, the owners of large farms prosper at the expense of exposed workers and contaminated water, soil and food supplies (Medawar, 1979, p21). This social anatomy of risk can provide one useful framework to gauge how particular policy decisions are likely to reverberate on the risk conditions of different populations within society.

Geographers have noted the uneven distribution of risks in time and space (Zeigler et al, 1983, p43). National, regional and local maps of such distributions are apt to yield a graphic demonstration of inequities, as well. Moreover, maps of observed, perceived or potential risk or hazard zones will allow decision makers to redress blatant inequities and to anticipate likely ones. National profiles will facilitate the construction of maps, such as that of 'potential technological hazard zones' in India (Karan et al, 1986, p196), which will display the diverse location and distribution of risks within a given country at a given time and in the future.

PROFILES IN PERSPECTIVE

National risk profiles, even were they easy to come by, would not provide a panacea for managing risks in developing countries. At the same time, however, even incomplete profiles hold some promise for easing the management burden. The more complete the inventory and classification of risks, the easier it will be to establish priorities, set standards and devise strategies for monitoring and evaluation. Decision makers in all countries need information that is as current, accurate and comprehensive as possible. For those trying to make intelligent choices on a developing nation's technological future, adequate information is imperative (Ashford and Ayers, 1985, p895). For those trying to achieve an appropriate level of environmental protection, a national inventory of natural resources and critical environmental problems is in order (Train, 1985, p676).

Does the construction of a risk profile warrant priority on any national agenda? After all, few developed countries can lay claim to anything approaching such a profile. A nation beset by poverty, hunger and disease surely has more pressing problems.

Ironically, the very context of adversity presses for stronger rather than weaker controls: 'The need to protect natural resources to ensure sustainable development suggests that these countries should adapt higher, rather than lower, standards of protection' (Train, 1985, p677). Sustainable development requires a recognition of the inextricable links and complex interactions among people and their environments (Clark, 1986, p5). Thus, a context of poverty and disease may exhibit a lower tolerance for industrial pollution (WRI, 1984, p26).

The national risk profile comprises a local initiative; but its import is global. What the *World Resources* annual reports for 1986 and 1987 (WRI and IIED, 1986, 1987) do in terms of 'an assessment of the resource base that supports the global economy', the national risk profile will do for assessing the spectrum of risks that threaten the well-being of people and what they value. The 'intricate risk mosaic' (Foster, 1980, p43) of one country will probably invite cross-national comparisons, may stimulate imitation and will certainly guide the decision-making process and the setting of priorities.

11 Risk and Criticality: Trajectories of Regional Environmental Degradation

Roger E. Kasperson, Jeanne X. Kasperson and B. L. Turner, II

INTRODUCTION

In its influential report, the World Commission on Environment and Development (WCED) called for a global risk-assessment programme to buttress and extend the work of the United Nations Environment Programme (UNEP) (WCED, 1987). This chapter reports on an international project centred at Clark University in the US that has explored the causes and consequences of growing environmental risk over a 50- to 70-year period in nine regions distributed throughout the world. The nine regions are Amazonia; the Eastern Sundaland region of Southeast Asia; the Ukambani region of southeastern Kenya; the Nepal Middle Mountains; the Ordos Plateau of China; the Aral Sea; the southern High Plains of the US; the Mexico City region; and the North Sea. The authors begin by considering the notion of criticality and by developing definitions and a classification of environmentally threatened regions. Research teams were assembled for all nine regions and studies were conducted. In this chapter, the authors review the development of concepts and methods used in these studies and the major cross-cutting findings that emerged. They argue that a growing disjuncture exists in the studied regions between the rapid rates of environmental degradation and the slow pace of societal response, threatening environmental impoverishment and loss of options for future generations, as well as the escalating costs of substitution in resource use and risk mitigation efforts.

The causes and consequences of human-induced environmental degradation are concentrated in regions and places distributed throughout the planet where they threaten life-support systems and the well-being of

Note: Reprinted from *Ambio*, vol 28, Kasperson, R. E., Kasperson, J. X. and Turner, B. L., 'Risk and criticality: Trajectories of regional environmental degradation', pp562–568, © (1999), with permission from the Royal Swedish Academy of Sciences

current and future generations. During the 1980s, Russian geographers created 'red-zone' maps of 'critical' environmental situations (Puchachenko, 1989; Mather and Sdasyuk, 1991, pp159–175), and the National Geographic Society (1989) compiled a world map of 'environmentally endangered areas'. At about the same time, biologist Norman Myers (1988) pioneered his concept of 'hot spots of biodiversity' at risk. Threatened areas throughout the globe captured extensive discussion at the Earth Summit in Rio de Janeiro in 1992, and the Intergovernmental Panel on Climate Change (IPCC) has probed the risks associated with prospective climate change and the potential concentration of adverse impacts in highly vulnerable regions around the world (Watson et al, 1998).

The WCED called for a Global Risk Assessment Programme to complement the monitoring and assessment function of UNEP by providing 'timely, objective and authoritative assessments and public reports on critical threats and risks to the world community' (WCED, 1987, p325). The WCED envisioned four principal functions:

1   identify critical threats to the survival and security of people, globally or regionally;
2   assess the causes and likely consequences of those threats;
3   provide authoritative advice on what must be done to avoid, reduce or adapt to these threats; and
4   constitute an additional source of advice to governments and intergovernmental organizations on policies and programme implementation to address these threats.

Since the WCED's report, other initiatives on global environmental change, such as the IPCC and the work of various international centres of integrated assessment, have enlisted risk concepts and methods in their assessments. And the SysTem for Analysis, Research and Training (START) Programme has undertaken a regionally structured series of activities to build environmental risk assessment and management capabilities throughout the globe (Fuchs, 1999). These efforts have pointed to the need for regional studies of long-term environmental risk to supplement, test and deepen studies of growing environmental degradation at the planetary scale.

The Project on Critical Environmental Zones, an international and interdisciplinary effort centred at Clark University in the US, took up this very problem.[1] Armed with a common research protocol, research teams examined nine regions in which large-scale, human-induced environmental changes purportedly threatened the sustainability of the existing system. They sought to determine whether the attributes and indicators of critical environmental zones allow the formulation of a definition of the term 'criticality', as well as assessments of its causes and consequences, so that different regional assessments can reach comparable conclusions. Thus, the United Nations University Press has already published an overview

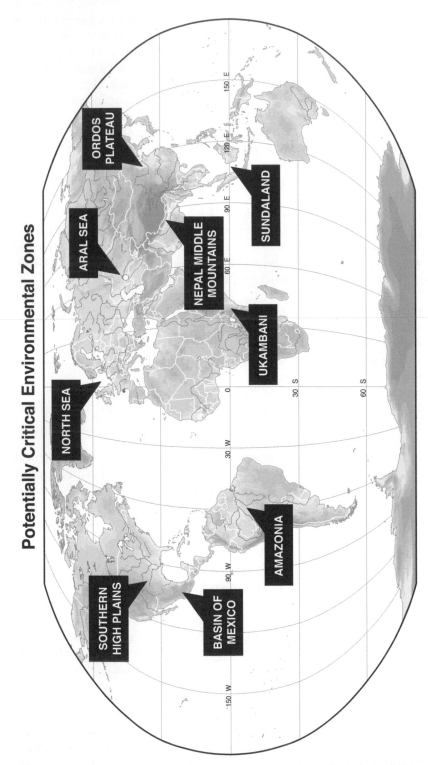

Figure 11.1 *Nine environmentally threatened regions*

volume, *Regions at Risk: Comparisons of Threatened Environments* (Kasperson et al, 1995), and regional studies of Eastern Sundaland (Brookfield et al, 1995) and Amazonia (Smith et al, 1995), Mexico City (Ezcurra et al, 1999), the Ordos Plateau of China (Jiang, 1999) and the Southern High Plains of the US (Brooks and Emel, 2000). All entail assessments of growing regional environmental risk and vulnerability. This chapter reports on the approach and major findings emerging from this decade-long effort.

THE NINE REGIONS

In selecting nine regions distributed throughout the world (see Figure 11.1), the project sought to capture a range of physical environments, human habitats and resource-use systems, population density, and types of political economy. Drawing these together in a selection matrix, we used our nine case studies to cover a substantial measure of the range of the matrix, but not all possibilities. Multidisciplinary research teams were created for each, with team leaders – in all but one case – drawn from the region. Common questions, research designs and study protocols were developed in project team meetings.

Amazonia

Amazonia (Smith et al, 1995) encompasses the Brazilian section of the Amazon River's drainage basin, the world's largest continuous expanse of tropical forest. The sensitive nutrient cycling in the upland soils and the regional hydrological cycle are intimately tied to these forests, which have lost an area the size of France over the past 20 years and continue to suffer the ill effects of deforestation, estimates of which have recently doubled, owing to logging and fires (Nepstad et al, 1999). The economic viability of the small farms that have appeared, the conversion of those croplands to cattle ranching and the toxic pollution of the region's rivers by mining heighten concerns about the sustainability of human activities in this delicate ecosystem.

Sundaland

The partly submerged sub-continental spur that extends southeastward from Asia is known as Sundaland (Potter et al, 1995). Eastern Sundaland refers to the portion that contains the island of Borneo and the eastern Malay Peninsula. Home to one of the largest expanses of tropical forest in the world, this region has long suffered the joint assault of the international timber industry and various development projects. Many of the same environmental impacts that have altered conditions in Amazonia also apply here; but the massive scale of the logging operations and the impacts of the associated forest fires question the ability of the diverse forest biota to recover and the long-term sustainability of the timber industry and its current practices.

Ukambani

The Ukambani region (Rocheleau et al, 1995) of southeastern Kenya occupies the semi-arid, east-facing slopes that separate the highlands proper from the coastal plains. Historically, people cultivated the better pockets of this land, allocating the drier sections for seasonal grazing of livestock. Recent population growth has expanded the pockets of permanent cultivation down the slopes and into the more arid lands. Many of the farmers involved in shifting their plots downslope are marginal smallholders who find it difficult to till the land without causing large-scale soil erosion. The accompanying intensification of grazing patterns portends a degraded and possibly desertified future, although some researchers point to recovery in progress (Tiffen et al, 1994).

Nepal Middle Mountains

The Nepal Middle Mountains (Jodha, 1995) constitute a broad strip of well-watered but rugged land that runs through the centre of the country to make up the heart of Nepal. Rapid population growth and changing rules of resource allocation have stressed the increasingly smaller and fragmented land holdings. The marginality of many of these households and the residents' continuing search for supplemental off-farm income suggest that this sensitive high-energy slope environment is at significant risk. More landslides and greater soil loss could well contribute to deforestation and increased sedimentation in the drainage basin of the Ganges.

Ordos Plateau

The Ordos Plateau (Jiang et al, 1995; Jiang, 1999) is the large, arid sandstone upland located in the great bend of the Hwang Ho (Yellow) River in Inner Mongolia. Long home to nomadic pastoralists, the plateau has witnessed, especially over the past 50 years, a heavy influx of Han Chinese farmers who have converted the short grasslands into cultivated land. Government-sponsored development of a woollen industry in the region has promoted the intensification of livestock production. These changes in land use, combined with the natural desiccation of the region, have contributed to large-scale soil loss and sandification.

The Aral Sea

Once the fifth largest freshwater lake in the world, the Aral Sea (Glazovsky, 1995) was an oasis in the deserts of central Asia. Today, the state-sponsored large-scale irrigation schemes for cotton production that began diverting the two main rivers that fed the sea in 1960 have produced an ongoing environmental catastrophe. The sea has lost about one half of its surface area and 70 per cent of its volume. As the water volume has dropped precipitously, the lake has become saline. A once-thriving fishing industry has died, and rainfall in the region has declined drastically. The chemical pesticides and fertilizers used extensively on the irrigated cotton fields have so polluted soils and potable water supplies that agricultural productivity has decreased and human health has suffered dramatic declines.

The Llano Estacado

The Southern High Plains (Brooks and Emel, 1995, 2000) extend over the western half of the Texas panhandle and adjacent portions of New Mexico in the US. Traditionally the home of cattle ranching, this region of arid, short-grass prairie was transformed for large-scale agriculture after the introduction of groundwater irrigation. Farmers primarily planted cotton, and their increased demands for water from the underlying Ogallala aquifer have produced severe depletion of groundwater. These changes, along with growing competition in national and international cotton markets, have devastated many family farms, degraded formerly cropped lands and heightened the risk of a long-term regional collapse.

The Basin of Mexico

The Basin of Mexico (Aguilar et al, 1995; Ezcurra et al, 1999), a mile-high interior basin overwhelmed by greater Mexico City, will soon be the largest metropolitan area in the world, with nearly 22 million people confined in 5000 square kilometres. Government policy has long focused on concentrating wealth and industry within the region, thereby exacerbating population growth and spawning industrial activities that have polluted the air and water. Today, the surface water in Mexico City is highly polluted, and deteriorating air quality is hazardous to human health.

The North Sea

A large and densely settled population rings the shorelines of the North Sea (Argent and O'Riordan, 1995). As the technological capacity to extract natural resources such as oil continues to grow, the region's natural systems are under stress. The use of the sea as a sink for an ever-growing amount of effluents places additional stress on the North Sea. Fish stocks and other renewable resources hover on the edge of overexploitation as the many different ways in which people use the sea conflict with one another.

DEFINING AND CLASSIFYING CRITICALITY

Conceptualizing environmental criticality is a challenge owing to the wide range of perspectives, many of them conflicting, from which it is possible to view the environmental threat or regional environmental change. To simplify, we identify two polar positions – the geocentric and the anthropocentric perspectives – that undergird assessments of criticality or endangerment (Turner et al, 1990b). In this distinction, the geocentric perspective defines criticality in terms of physical changes or ecological dimensions, and criticality is reached when human-induced perturbations have so altered the biophysical system that a different system results. The anthropocentric perspective, by contrast, views technological and social change as sufficient to overcome human-induced changes or degradation of the biophysical system.

The valuable features of each approach and the contradictions between them highlight the need for an integrative, holistic approach to conceptualizing environmental threats and assessing regional environmental change. Experience suggests that addressing rates and types of change in nature–society relations is essential to understanding environmental transformations, as is attention to the fragility of the natural system, the societal vulnerability to perturbations and the ability to respond to environmental threats. A purely geocentric perspective ends up designating as 'endangered' or 'critical' regions that successfully maintain large populations and high standards of living and are likely to do so far into the future. Correspondingly, a narrowly anthropocentric perspective is apt to generate unrealistic assessments based on optimistic assumptions about human ability to adapt to any environmental problem or to overcome any physical limits. Neither is adequate for the balanced approach to critical environmental situations needed to guide thoughtful analysis or judicious strategies aimed at sustainability.

An integrative perspective on criticality recognizes the essential role of the environment in sustaining human life, but realizes at the same time that not all elements of the environment are essential or equally important. It must also appreciate the enormous adaptation and mitigation potential of human management and response. It should explicitly acknowledge continuing human adjustments and adaptations to environmental change, look beyond specific natural resource-use systems to the broader realm of environmental interactions, take a long-term perspective, and place regions squarely in their global context. Figure 11.2 suggests such an integrative approach to environmental threat, recognizing that both human driving forces and natural variability cause environmental stress and perturbations – just as human systems shape social and economic vulnerabilities and natural systems ecological fragility that interact with these stresses to produce adverse impacts on human beings, other species and the physical environments in which they live.

We offer a set of definitions and concepts to guide an integrative approach to understanding environmental risk and criticality. We begin by differentiating 'criticality' from other situations that denote lesser degrees of environmental risk. We recognize four conditions:

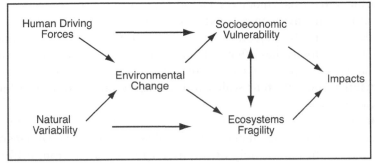

Figure 11.2 *Integrated risk assessment of regional environmental change*

1   Environmental criticality refers to situations in which the extent and/or rate of environmental degradation preclude the continuation of current human use systems or levels of human well-being, given feasible adaptations and societal capabilities to respond.

2   Environmental endangerment refers to situations in which the trajectory of environmental degradation threatens in the near term (this and the next generation) to preclude the continuation of current human use systems or levels of human well-being, given feasible adaptations and societal capabilities to respond.

3   Environmental impoverishment refers to situations in which the trajectory of environmental degradation threatens in the medium to longer term (beyond this and the next generations) to preclude the continuation of current human use systems or levels of well-being and to narrow significantly the range of possibilities for different future uses.

4   Environmental sustainability refers to situations in which nature–society relations are so structured that the environment can support the continuation of human use systems, the level of human well-being and the preservation of options for future generations over long time periods.

Critical, endangered, impoverished and sustainable regions are regions (as defined earlier) characterized by these situations. This classification, in principle, allows a mapping of regions of the world that are at environmental risk into states or stages along a continuum between criticality and sustainability.

TRAJECTORIES OF RISK AND REGIONAL DYNAMICS OF CHANGE

Analyses are needed of the dynamics of environmental endangerment if regional differences are to be taken seriously and comparative analysis is to move forward. By 'regional dynamics of change', we refer to the relationships that exist among the factors that together shape the changing nature of human–environment relationships and their risk effects within a particular region. By 'trajectories of risk', we refer to the trends among these relationships as they affect regional risk and sustainability over time. The analysis of regional dynamics requires successive examinations of relationships from different scales and vantage points and over differing historical periods.

Our own regional analyses suggest that the conditions of impoverishment, endangerment and criticality, and the regional dynamics that cause them, take widely different forms and arise from different circumstances in each regional context. No simple evolutionary pattern or set of regional dynamics of change holds true across all nine regions. In particular, the relationship between growing environmental degradation and changes in the wealth and well-being of inhabitants varies markedly from region to region. Figure 11.3 suggests some cases, although many others are possible. Case 1 represents the situation often assumed in which increasing environmental degradation causes wealth and well-being to decline – in other words, growing criticality. But in case 2, a shift from an agricultural to an industrial economy allows continuing increases in wealth and well-being in the face of continuing environmental degradation (although, presumably, this cannot continue indefinitely). In case 3, continued exploitation of resources supports increasing accumulation of wealth, but continued deterioration of environmental quality eventually results in 'bite-back' upswings in environmentally induced disease and downward trends in human well-being.

In fact, case 2, however counter-intuitive, reflects a very common pattern in our nine regional studies and highlights the complexity that tangles regional studies of risk. First, much time typically elapses between the onset of degradation and the depletion of natural resources or the overwhelming of environmental sinks (such as oceans or the atmosphere, where contamination can be absorbed). A significant time lag often delays the induction of effects (as in the 20–30-year latency period for many cancers). Environmental degradation can stretch over lengthy time periods during which the region's population becomes wealthier, healthier and generally better off. Moreover, when a particular environmental component is degraded or exhausted, individuals and societies shift to other productive systems and environmental assets that offer equal or better rewards.

In these situations of increasing human wealth and well-being in the face of, indeed with the assets derived from, the degradation of the regional environment, the long-term value or 'natural capital' of the environment may be sequestered for the benefit of current users, who may be drawing down that capital instead of living off natural 'interest' (Turner et al, 1990b). Thus, the case 2 trajectories show current generations drawing down nature in ways that diminish the range of economic options

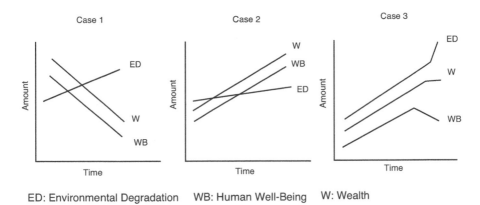

ED: Environmental Degradation   WB: Human Well-Being   W: Wealth

Figure 11.3 *Contrasting simple regional trajectories of environmental change*

for the future. Accordingly, trajectory analysis is useful in assessing rates of environmental change and how they may affect long-term environmental risk and future options, including those involving loss of life-support capabilities.

Constructing regional trajectories of risk helps to illuminate the nature and rates of change in human–environment relations. But the analysis of such trajectories also requires assessments of the regional dynamics of change. To begin, it is important to conduct a basic analysis that relates environmental degradation and human well-being with the driving forces usually implicated in environmental change. The selection of key driving forces of change for analysis needs to vary from region to region and should reflect their relative importance to environmental change and human well-being within a particular region. The analysis itself should seek to explain not only the trajectory of each of these variables, but the causal relationships among them.

The degree of spatial linkage between the region and other regions and the global economy is a major attribute of regional context that structures the impact of external driving forces and changing regional vulnerabilities. Much evidence suggests, for example, the growing dependence of many agricultural economies in the developing world upon fluctuations in world market prices and shifting demand in distant markets (Raskin et al, 1966; Blaikie and Brookfield, 1987; Turner et al, 1990b). In such cases, environmental degradation and emerging overall regional risk can be explained only with detailed attention to the restructuring of such economies over time and the way in which such changes interact with differential patterns of human vulnerability and well-being in the region.

Significant disaggregation of risk and outcome trends is necessary in order to understand regional dynamics. Figure 11.4 suggests models of the types of analyses that can help to clarify the interactions among internal regional changes and patterns of growing environmental degradation. Case A is one of increasing social and economic polarization, attendant upon

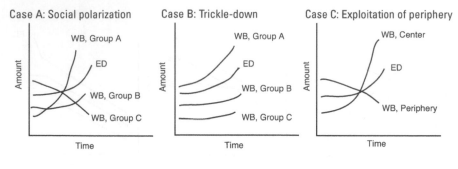

ED: Environmental Degradation   WB: Well-Being

Figure 11.4 *Three cases of regional social processing causing or interacting with environmental degradation*

growing environmental degradation. As analyses of global and national inequalities suggest, such polarization can become a major driver of environmental degradation and risk. In case B, the benefits reaped from the drawing down of regional resources are more equitably distributed. Such a 'trickle-down' effect may help to avoid situations in which impoverishment of marginal groups drives severe environmental degradation; but it may simultaneously create a political situation highly resistant to interventions to mitigate the upward trajectory of environmental damage. Case C provides insight into centre–periphery relationships that may structure accumulation at the centre of capital derived from environmental or natural resource exploitation in the periphery.

Finally, the nature of societal responses in the face of emerging trajectories of environmental change requires careful attention as an element of the regional dynamics. One key aspect of this analysis will centre upon the differential responses at various scales or managerial levels. Such assessments can characterize the types, number and effectiveness of responses undertaken at these various levels, as well as the options and structural constraints under which they operate. This analysis may begin with the environmental manager closest to the situation (the 'farmer first' approach) or, alternatively, with state policies and actions (Chambers et al, 1989; Scoones and Chambers, 1994). Risk analysts would more likely begin with the structure of the hazard or the consequences of environmental change and work backwards, or upstream, to examine the driving forces of change by using various models of social and political change and policy formation (Kates et al, 1985).

A key consideration will be the comparison between trajectories of growing environmental damage and changing societal capability to intervene to mitigate the damage and to head off the regional trend towards growing environmental criticality. The tendency for societal responses to lag seriously behind emerging environmental endangerment has received extensive treatment from Meadows and colleagues (Meadows et al, 1972, 1992). The importance of neglect and lag times in societal response merits detailed attention in analyses of regional dynamics of change. The regional trajectory to environmental criticality will not emerge without warning, of course. As environmental degradation proceeds, various events and indicators of change will prefigure the movement to greater damage and greater future risk. Such events and indicators will constitute a stream of 'risk signals' that can alert society and various managers to impending damage (see Chapter 7, Volume I). How such warning systems work, how they connect to and inform management systems, and how and why appropriate responses do or do not occur are questions fundamental to understanding trajectories of regional change and diagnosing the outcomes that emerge along the trajectories (Kasperson et al, 1995).

MAJOR FINDINGS

No single set of regularities emerges from comparison of the regional trajectories of environmental change and associated interpretations to explain the diversity of the regional cases and the complexities implicit in the slopes and shapes of the risk trajectories. Whereas most regions shared similar long-term decreasing sustainability, each merits examination within its particular historical, landscape and societal context. The observations that follow suggest several cross-cutting issues worthy of further attention; but they do not pinpoint one or several factors that dominate the regional dynamics of change or uncover the 'smoking guns' of growing environmental risk within these areas.

GLOBAL ENVIRONMENTAL CHANGE THROUGH A REGIONAL LENS

Analyses of global environmental change abound with allusions to long-term world population growth, aggregate loss of global biodiversity, world energy use and non-renewable resources, and international summit conferences and institutions to combat these large-scale changes. Environmental problems take on a different hue at the regional scale. It is not, of course, that the issues and their causes, so apparent at the global scale, are not present in the world's regions; they are. But the change depicted in the smooth averaged curves that portray global trends shows up in much sharper relief in the world's regions. And the mosaic and processes of change often appear very different from the averaged or aggregated global trends. Regional landscapes, economies and cultures reappear and can provide new insights and richness that challenge broader assumptions and interpretations.

The trajectories of change in our nine threatened areas provide a warning, one that supplements those discoveries or events such as stratospheric ozone depletion or mounting carbon dioxide (CO_2) levels in the atmosphere at the global scale. In nearly all of these regions, trajectories of change are proceeding to greater endangerment, sometimes rapidly so, while societal efforts to stabilize these trajectories and to avert further environmental deterioration are lagging and are generally only ameliorating the damage, rather than decisively intercepting the basic human driving forces of change.

This is not to say that disaster is imminent in most of these regions (although we judge the Aral Sea already to be an environmental catastrophe and the Basin of Mexico as rapidly nearing criticality); it is to say that the trajectories of change in most of these regions are rapidly outpacing societal responses, that these trends are environmentally impoverishing the future populations who will occupy these regions, and that the trajectories of risk point to an escalation of long-term costs of regional substitution, adaptation and remedial measures. At some indeterminable point in the future, these trends will eclipse regional societal capabilities to head off widespread damage and declines in human well-being.

Symptoms of emerging criticality

Since environmental change is ubiquitous throughout the world, it is essential to identify and to diagnose symptoms of emerging environmental criticality in the 'noise' of overall environmental change. Most current discussions of global environmental change focus on rapid land-cover change (especially deforestation), on global climatic change and ozone depletion, and on loss of biodiversity. At the regional scale, however, it is not at all clear that these are the dominant problems. Rapid land-cover change is characteristic of many of our nine cases and, more generally, occurs in many regions throughout the world undergoing intensive human occupation. Although such changes often lead to short-term losses, regeneration and shifts in human uses allow continuing adaptations and productive use of the regional environment even in the face of dire predictions.

Among our nine regions, those with the steepest trajectories to criticality are arid or semi-arid areas where water resources depletion threatens continued human habitation and well-being. The Aral Sea, owing to extensive mismanagement of regional water resources, is the most striking case. But two more prosperous situations – the Llano Estacado region of the southern High Plains in the US and the Basin of Mexico – reveal long-term trends of water depletion and dependence that are unsustainable over the long term, and that portend a rising potential for disruptions in life-support systems and for eventual regional collapse. This pattern seems to reflect the distinctive qualities of water as a resource: its indispensability for many human activities, combined with the tremendous, almost prohibitive, cost of importing it over any substantial distance when local supplies prove inadequate, as well as its central role in many human conflicts (Gleick, 1998).

A rich tapestry of human causation: IPAT Plus

At the regional level, the pattern of human causation is richly variegated. Rarely can a single dominant human driving force be discerned that explains the historical emergence of environmental degradation or that captures the complexity of regional change. Nor do the various grand theories, whether they arise from welfare economics, political economy, neo-Marxist thinking, global dependency models or development theory, provide satisfying broad interpretations, although all offer some insight. So what we term the regional dynamics of change – the interplay among the trends of environmental change, vulnerabilities and fragility, human driving forces, and societal responses, require scrupulous examination within their cultural, economic and ecological contexts. And the most satisfying interpretations invariably recognize the shifting complexes of driving forces and responses over time, tap diverse social science theory and lie firmly grounded in careful empirical work.

Human-induced environmental change, it has been argued, is largely the product of changes in population, affluence and technology (Ehrlich and Holdren, 1971; Holdren and Ehrlich, 1974). Accordingly, one way to analyse the human forces at work is to apply the IPAT formula, expressed as:

$$I = P.A.T$$

where I = impact, P = population, A = affluence and T = technology. In this formula, quantitative values are provided for the terms on the right (P, A, T) in order to derive a measure of impact.

In this discussion, impact refers to environmental degradation, population provides a measure of base resource needs, affluence is measured according to per capita consumption, and technology indicates the efficiency of resource production and consumption. This formula provides a broad measure of the demand for resources and the impact this demand has on the environment as expressed through the processes of production and consumption. Population, affluence and technology are obviously important causal factors in regional change. The processes at work in driving environmental change in our nine regions cannot be explained, however, by looking only at population, affluence and technology. Other forces equal or supersede the impact of these three forces in shaping the environmental trajectories of a given region. Clearly, it is necessary to factor into the IPAT formula an additional or 'plus' factor that permits a variety of categorizations. In the case of these nine regions, however, it usually manifests itself in government policy or some combination of social conditions involving institutions, poverty and vulnerability, and cultural beliefs.

The state plays a critical role in initiating and managing the trajectory of resource use in a given region: witness the far-reaching impacts in the regions controlled by the command economies of the former Soviet Union and China and the quasi-command economy of Mexico. The development of the Aral Sea irrigation systems was neither an accident nor the result of the hidden hand of the market. Central planning ministries in Moscow deliberately created these systems, controlling what they produced, how they produced it, the sociopolitical structure of the production process and the reward structure for production. The Aral Sea irrigation scheme was designed, in large part, to provide a marketable resource (cotton) that could be sold for hard currency. Moscow saw the total global market value of cotton production as the critical issue and the drawdown of the Aral Sea as a necessary price.

Economic development schemes in the Ordos Plateau and the Basin of Mexico provide similar examples. Central planning ministries in Beijing promoted a woollen industry in the Ordos Plateau and collaborated with Japanese firms to design and manage the enterprise. The growth of this industry stimulated the intensification of livestock production, as well as coal mining. Government policies also played a key role in concentrating wealth, industry and services within or near the Basin of Mexico. These inducements attracted an influx of rural poor and rich alike to Mexico City, and exacerbated the massive population growth that is draining already precarious natural resources.

In Amazonia, the Nepal Middle Mountains, Ukambani and the Basin of Mexico, poverty severely limits people's access to resources, technologies and entitlements that could reduce the impacts of environmental degradation. Lacking sufficient capital to purchase the kind of inputs required to make cultivation sustainable, the Amazonian farmer either resorts to land expansion or sells the land to a larger rancher. Squeezed on tiny plots in Nepal or marginal lands in Ukambani, from which sufficient income is difficult or impossible to derive, many men migrate to work elsewhere for part of the year, draining needed household labour from the farms. The upkeep of terraces and other management tasks that sustain land suffer as a result. In the Basin of Mexico, poverty adds to the pollution of water as people who lack access to adequate sanitation dump untreated wastes into the water supply.

Social values towards the environment and associated behaviours also extensively shape the regional dynamics of change. People's belief in the need to concentrate activities in prime locations, such as the Basin of Mexico, speaks to a primate city set of values (Pezzoli, 1998). Social identity and religion also encourage population growth, particularly in areas such as the Aral Sea Basin, the Nepal Middle Mountains and the Basin of Mexico. In the Aral and North Sea areas, as well as in the Basin of Mexico, faith in technological fixes accompanies a general underestimation of the adverse impacts that attend concentrated economic development. Many believe, in accord with the anthropocentric perspective, that technological development and market-induced innovation are sufficient to overcome scarcity and degradation. In those regions where elites monopolize political power and patronage is routinely accepted as part of the governance system, institutional corruption flourishes, as it has in the Aral Sea Basin and in Sundaland.

Many driving forces, it is clear, are rooted deeply in political and economic systems. They interact with population, affluence and technology to push regions toward criticality. The broader social conditions in which these forces operate shape the specific ways in which they interact and drive environmental risk. Regional explanations of environmental change need to place the impacts of population, affluence and technology in a larger, more inclusive explanatory framework.

Delay and overshoot

In their seminal work *The Limits to Growth* (Meadows et al, 1972) and the subsequent *Beyond the Limits* (Meadows et al, 1992), Meadows and colleagues argue that human society has a tendency towards overshoot, in which environmental changes occur rapidly, signals of such changes are late, distorted or denied, and societal responses are slow. As a result, environmental degradation overshoots responses, creating the potential for collapse of some kind.

Our regional cases provide sufficient evidence to bolster this argument, while adding new perspectives to the interpretation of causation. In all nine regions, societal responses have been delayed and ineffective, often badly so. This is particularly the case with the seven marginal regions where signals of environmental endangerment have typically been ignored, suppressed or devalued, and environmental risk has been allowed to continue to grow. The widespread nature of delay and of half-hearted or unimplemented responses appears to be due less to a lack of perception or awareness than to competing values or concentrated elite power. As such, it provides little basis for optimism that, in the first decade of the new millennium, existing national policies, programmes and institutions will prove adequate to the task of meeting mounting environmental deterioration in many regions of the world. To put it another way, most of our regions are on trajectories headed toward endangerment or criticality, with diminishing time for creating effective responses if deteriorating situations are to be stabilized.

It is further evident that societal responses in most of the regions have focused on 'downstream' measures – mitigative effects designed to ameliorate damage and to cushion the impacts of environmental deterioration – rather than on interventions focused on changing basic driving forces. The reason is that 'upstream' interventions unavoidably interfere with the driving forces of change and, thus, with other desired state and societal priorities. So even in the Aral Sea and the Basin of Mexico, where environmental degradation has reached or approaches criticality, and where adverse human health impacts are already apparent, interventions (whether from regional governments or the international community) aimed at altering the basic driving forces have yet to be implemented with determination.

Delays in societal response to environmental risks of various kinds often stem from inadequacies in the signalling and knowledge base or the high levels of associated uncertainty (see the discussion of signals in Chapter 8, Volume I). Thus, the lengthy emergence of response to global warming and the loss of global biodiversity are connected in no small part to the large uncertainties and long time-spans associated with potential impacts in the former and the scant database for the latter. Risk research attests to assessment failures as a recurring source of delayed or even maladaptive responses and of nasty surprises (Svedin and Aniansson, 1987; Schneider et al, 1998). But the evidence from our nine regional studies suggests that state policies aimed at economic growth, lack of political will, willingness to tolerate damage in marginal areas and among vulnerable peoples, and widespread political corruption, rather than public perceptions and awareness of environmental threat, are the more deeply seated sources of growing environmental deterioration.

CONCLUSIONS

The regional dynamics of environmental change across the globe reveal a widespread disjuncture between the fast variables of environmental change and the leisurely pace of societal response. Interestingly, the global scale reveals a more mixed picture in which societal responses to such change as stratospheric ozone depletion, global warming and industrial accidents have often been quite rapid, if less than totally effective (see, for example, the discussion of responses to the Bhopal accident in Chapter 5, this volume). Nevertheless, signals of growing environmental risk have been diagnosed with considerable speed and coping actions have been initiated. But the trajectories of change in the nine regions add weight to the argument for overshoot. Only in Amazonia, the North Sea, the Ordos Plateau, parts of Sundaland and, perhaps, Ukambani do responses appear to have some potential for stabilization, or at least a significant 'flattening' in the trajectory towards greater endangerment or criticality in the near future. Meanwhile, the dominant situation remains one of divergence in the rates of environmental change versus societal response that promise increasing environmental impoverishment and loss of options for future populations, as well as escalating costs for the substitution and mitigative efforts that must eventually occur. The primary causes for disjuncture lie less in inadequacies in scientific diagnosis, understanding and assessment than in the planet's sociopolitical structures, institutions and value systems.

NOTE

1   This project involved multidisciplinary teams of researchers from six continents. In all but one instance, case-study leaders were from the region being examined. Our discussion of each region draws directly upon the work of the contributors to *Regions at Risk* (Kasperson et al, 1995). The comparative assessments and conclusions reached in this chapter are, however, entirely our own.

12 Assessing the Vulnerability of Coastal Communities to Extreme Storms: The Case of Revere, Massachusetts, US

*George E. Clark, Susanne C. Moser, Samuel J. Ratick,
Kirstin Dow, William B. Meyer, Srinivas Emani,
Weigen Jin, Jeanne X. Kasperson, Roger E. Kasperson and
Harry E. Schwarz*

Introduction

The impacts of hazardous events are usually unevenly distributed among and within nations, regions, communities and groups of individuals. Vulnerable groups are those who are likely to suffer a disproportionate share of the effects of hazardous events. For the purposes of this chapter, we draw on a growing literature of vulnerability studies (reviewed by Dow, 1992, 1993) to define 'vulnerability' to hazards as people's differential incapacity to deal with hazards, based on the position of groups and individuals within both the physical and social worlds. Building on this literature, we see vulnerability, as we discuss in Chapter 14 of this volume, as a function of two attributes:

1   exposure (the risk of experiencing a hazardous event); and
2   coping ability, subdivided into resistance (the ability to absorb impacts and continue functioning) and resilience (the ability to recover from losses after an impact).

In this chapter, we use coping ability as an antonym of social vulnerability. Exposure (the risk of experiencing a hazardous event) as a concept is well explicated in the 'technical' risk literature (see, for example, Renn, 1992,

Note: Reprinted from *Mitigation and Adaptation Strategies for Global Change*, vol 3, Clark, G., Moser, S., Ratrick, S., Dow, K., Meyer, W., Emani, S. Jin, W., Kasperson, J. X. Kasperson, R. E. and Schwartz, H., 'Assessing the vulnerability of coastal communities to extreme storms: The case of Revere, Masachussetts, US', pp59–82, © (1998), with kind permission of Kluwer Academic Publishers

Table 12.1 *Sources of vulnerability themes*

Age	Bolin (1982)
	Bolin and Klenow (1983)
	Drabek and Key (1984)
	Quarantelli (1991)
	Rossi et al (1983)
Disabilities	Parr (1987)
Family structure and social networks	Bolin and Bolton (1986)
	Drabek and Key (1984)
Housing and the built environment	Bolin and Bolton (1986)
	Bolin and Stanford (1991)
	Godschalk et al (1989)
	USNRC (1984a)
	White and Haas (1975)
Income and material resources	Bolin and Bolton (1986)
	Bolin and Stanford (1991)
	Drabek and Key (1984)
	Perry and Lindell (1991)
	Quarantelli (1991)
	Rossi et al (1983)
Lifelines	USNRC (1984a); Platt (1991)
Occupation	Bolin (1982)
Race and ethnicity	Bolin and Bolton (1986)
	Drabek and Key (1984)
	Perry et al (1983)
	Perry and Lindell (1991)
	Rossi et al (1983)
	Trainer and Bolin (1976)

pp58–61) and is, in general, the most widely cross-disciplinarily understood dimension of vulnerability. For this reason, we do not comment further on exposure except to say that it, too, is partly socially constructed in that existing land use and daily commuting patterns, to name but two exposure variables, are social and temporal phenomena. We use a simplified measure of exposure – a floodplain map – in order to focus on the more explicitly social variables that together determine coping ability. Coping ability, the ability to either absorb impacts (e.g. by exiting the scene or dealing with the hazard in place) or to recover from them (e.g. though insurance, cash reserves or other means), is influenced by a large list of variables identified by sociologists, geographers, political scientists and other investigators (see Table 12.1 and the section on 'Methods').

This chapter does not pretend to be a review of the vulnerability literature or a catalogue of all of the past and future definitions and approaches to vulnerability. However, there are several authors who should be mentioned for those who wish to explore the terminology in depth. Timmerman (1981) provided an early look at the concept and origins of the term 'vulnerability' and sparked a need to explore and clarify the term, as well as our understanding of the phenomenon. Liverman (1990) focused the debate by providing a catalogue of the 'conditions and variables ... important in determining vulnerability to global environmental change'. Wisner (1993a) looked at vulnerability in the developing world, examining the concept of marginality and the role of the state in vulnerability. Blaikie et al (1994) also focused on the developing world, analysing the forces that combine to cause a crisis. Chen (1994) examined vulnerability on a relatively large scale in light of global environmental change. Dow and Downing (1995) commented on the state of vulnerability research in the wake of the Social Science Research Council-sponsored First Open Meeting of the Human Dimensions of Global Environmental Change Community at Duke University in June 1995. Cutter (1996) broadly reviews the definitions of vulnerability and the types of studies carried out since 1980. Hewitt (1997) reviews the vulnerability literature with an emphasis on vulnerability as a form of powerlessness that is socially reproduced.

Exposure and coping ability as co-determinants of people's vulnerability to hazards are of particular interest to hazard managers inclined to address behavioural, managerial, institutional and other human activity-related issues that change the likelihood of severe impacts from hazards. Short of affecting the environmental hazard itself, hazard managers seek points of intervention in the causal chain between a hazardous event and the downstream human consequences. In the hazards literature of the past two decades, this search has been supported through causal modelling inspired by the classic work of Hohenemser et al (1985), which has since then also found entry into the global change literature (e.g. Hohenemser et al, 1985; Moser, 1997; Clark et al, 2001; and see Chapter 2 in this volume) which illustrate that this view of hazards as processes strongly implies three possible strategies for hazard control:

1   prevention of hazard events;
2   prevention of hazard consequences once events have taken place; and
3   mitigation of consequences once these have occurred.

The parallels to the search for possible responses to global climate change are obvious in spite of the variegated terminology:

1   prevention of further increases in the atmospheric concentration of greenhouse gases;

2   mitigation of impacts (bounding of first-order environmental changes); and
3   adaptation to the impacts (higher-order responses to impacts) (Bruce et al, 1996; Houghton et al, 1996; Watson et al, 1996).

Understanding the causal linkages among the social and physical processes that interact to produce a hazard, on the one hand, and illuminating the causal linkages across scales from the global to the local, on the other, are parallel challenges that need to be addressed in order to prepare adequately for the management of climate change-related environmental hazards (e.g. Turner et al, 1990b; Cash, 1997; Global Environmental Assessment Project, 1997).

Here we focus on the former challenge largely because the likely magnitude of local consequences from global climatic changes are far from predictable and the causal mechanisms that would link global processes to local impacts are still inadequately understood. In the absence of such greater ability to predict local consequences of global climate change, it is still possible and (as a no-regrets strategy) advisable to address human vulnerability to environmental perturbations. We thus adapt the hazards causal model discussed in Chapter 2 by introducing vulnerability via its two principal dimensions: exposure and coping ability.

Some attempts have already been made to link vulnerability and the causal structure of hazards. Watts and Bohle (1993) use a tripartite model of property relations, political power and economic power. While this model is useful in exploring the political economic dimensions of the problem, it addresses neither the spatial aspects nor the physical environment. Cutter (1996) posits a synthetic 'hazards of place' model that recognizes both the physical and the social in relation to the causal processes of hazards.[1] While Cutter integrates the physical and social into a model of hazard processes, our incorporation of vulnerability into the hazards causal model of Hohenemser et al (1985) accomplishes this integrative step, while at the same time emphasizing the spatial distribution of vulnerability and, through the notion of hazard as process, allowing the identification of points of action in order to combat the effects of the hazard.

The crux of vulnerability to global environmental change is as follows: people stand to experience impacts from hazards of global change of varying degrees that fall along a spectrum from positive to negative, based on their position in the social and physical worlds. The problem, then, is twofold:

1   How do we begin to analyse and understand the differential potential for harm in a local context?
2   How do we incorporate vulnerability into our understanding of hazards as a type of interplay between social and physical phenomena in order

to have a way of discussing vulnerability's implications and of adjudicating between different policy and management options?

In order to answer the first question, we need to explore the social and spatial distribution of vulnerability to global change hazards. In order to answer the second question, we examine the implications of this research for causal modelling of hazards and for storm preparation and response. We do so in the context of a small US coastal city, Revere, Massachusetts (see 'Setting'), as an example of the type of coastal communities and the types of management challenges faced by coastal zone and hazard managers in developed countries. Clearly, we make no claim that Revere is generally representative, especially not in an international comparison. We do believe, however, that Revere is far from alone with regard to the problems of on-the-ground storm hazard and sea-level rise management faced along developed shorelines. Comparative studies of coastal communities in other countries are invited. Our analysis offers one approach to address vulnerability, not at the large scale at which economists and other workers often assess problems, but at the scale at which hazards managers commonly need to implement hazard reduction strategies.

SETTING

Revere, Massachusetts (population of 42,786 in 1990), located just north of Boston, is exposed to flooding and wave damage on three sides, from the Atlantic Ocean to the east and from tidal rivers to the north and south. We chose this city for our study because of:

- the availability of extensive detailed economic and topographic baseline data (Hunt, 1990);
- the large portion of flood-prone residential area in the city; and
- the range of economic circumstances, from working-class to affluent (despite Revere's local reputation as a blighted community; Vigue, 1997), in the floodplain and community as a whole.

The threat of accelerated global sea-level rise, which is projected to be between 13cm and 94cm by 2100 (Warrick et al, 1996), portends increased flood and wave damage. Massachusetts also has a recent historical 2mm of annual relative sea-level rise due to subsidence, which if extrapolated over the Warrick et al (1996) study span could amount to an additional 20cm by 2100. Wind, snow and ice damage may also increase, depending upon the highly uncertain effects of global climate change on storm frequency and intensity. Recent experience underlines the significance of the storm hazard to Massachusetts coastal communities, whether climate change and accelerated sea-level rise occur or not. The area weathered four especially severe coastal storms during the 19-month period between August 1991 and March 1993: Hurricane Bob in 1991 and three severe extra-tropical storms, regionally known as 'northeasters' – the Halloween Northeaster or No-name Storm of 1991, the Blizzard of December 1992 and the Blizzard of March 1993, which affected the entire eastern seaboard of the US. Most recently, the area endured a major snowstorm in January 1996 that left 0.6m of new snow on eastern Massachusetts, set records for total snow depth in the area and put the region on the way to a new record for seasonal snowfall (Nealon and Brelis, 1996). The damage from each of these storms pales in comparison to the flood damage from the Blizzard of 1978, when federal assistance for coastal Massachusetts totalled US\$38 million and Red Cross spending in Revere alone totalled US\$400,000 (Corps of Engineers, 1979; Hunt, 1990). Each event also caused serious disruption beyond its monetary toll.

THE PROBLEM

Such impacts and the potential for worse ones with climate change indicate the value of a better understanding of how hazardous events and human populations interact. Aggregate estimates of damage of the sort just cited, though useful, do not address the point emphasized in vulnerability analysis: that the impacts of hazardous events are unevenly distributed among and within the exposed populations.

METHODS

The literature on vulnerability identifies many elements that contribute to differential ability to cope with hazards. Aspects that are frequently mentioned are age, disabilities, family structure and social networks, housing and the built environment, income and material resources, lifelines (including transportation, communication, utilities and other services), occupation, and race and ethnicity. Studies addressing each of these themes are listed in Table 12.1. Less explicitly dealt with in the vulnerability literature are transience, immigration and education levels; but the significance of all three can be inferred from discussions of the importance of hazard perception and experience (e.g. Mitchell, 1984). Table 12.2 illustrates how various attributes may influence the ability to deal with and recover from storms. We chose census data to represent these attributes because they are widely available and familiar to local managers. Data from the 1990 census on 34 variables reflecting themes from the literature were assembled at the block group level (see Table 12.3). The block group is a census unit containing approximately 1000 people; this unit is chosen for analysis because it is the smallest for which relatively complete socioeconomic data are available. We acknowledge that vulnerability varies on small scales and even at the household level.[2]

Table 12.2 *Key themes and possible exposure, resistance and resilience scenarios*

Age	Young children entail an extra burden of child care, which may be disrupted during storm events (exposure). Parents may lose work time if their daycare is closed down. The same may apply to adults in elder care. Both young and old populations also may be unable to resist storms or respond on their own, although the vulnerability of elders is decreased by their wealth of experience
Disabilities	Disabilities can hinder taking action in any of the phases of vulnerability
Family structure and social networks	Large families may be harder to care for or keep track of in a storm and its aftermath (exposure, resistance, resilience). Yet, if enough are working, large families could be a benefit in sharing recovery costs. Social networks play a role in disaster warning, perception and behaviour (exposure, resistance). Strong social networks may help to bear the cost of rebuilding as well (resilience)
Housing and the built environment	The spatial distribution of the built environment is highly inertial, establishing patterns of peoples' location, use and travel (exposure). Likewise, the quality of construction can determine the ability of individuals or groups to successfully ride out a storm (resistance)
Income and material resources	Income allows spending on prevention items, such as retrofitting the house or relocating furnaces to higher floors in a flood. Money or vehicles may also enable a fast exit from a hazardous environment (exposure, resistance). Of course, money also allows rebuilding to proceed (resilience). On the other hand, those with a lot of wealth have a lot to lose (exposure)
Lifelines (which includes transportation, communication, utilities, emergency response and hospitals)	Telephones and the media provide advance warning of storms. Transportation can provide a way out of the hazard for evacuation or a way in if caught in rush hour (exposure). Utilities of all sorts are necessary for resistance and resilience. Hospitals and emergency response certainly provide resistance to the storm and help to those who have already fallen victim (resilience). Emergency response crews can also lessen exposure by helping people to evacuate
Occupation	Some occupations, such as fishing, tend to be located in harm's way (exposure). Once equipment is ruined, it may be months and a whole fishing season away before insurance or relief payments begin, so opportunity is lost. Self-employed people often have poor documentation of business receipts and therefore may have difficulty establishing the record necessary to receive recovery aid (resilience)
Race and ethnicity	Minorities may encounter discrimination when seeking post-storm aid (resilience), which may change the ability to prepare for the next storm (resistance). They may be confined by real estate discrimination to certain hazard-prone neighbourhoods (exposure)

Table 12.3 *Census variables analysed*

WRKPRNT	Children 17 and under whose resident parents or guardians all work, expressed as a percentage of the population
AMERINPC	Percentage of people who are American Indian, Eskimo or Aleutian
ASIAPIPC	Percentage of people who are Asian or Pacific Islander
BIGFAM	Households with seven people per occupied housing unit
BLACKPC	Percentage of population who are black
BLD.39PB	Percentage of housing units that were built prior to 1939
BLIZIMIG	Percentage of the population which was foreign-born and came to the US between 1980 and 1990 (approximation of post-Blizzard of 1978)
CARPLPC	Percentage of people who carpool to work
CARSPCAP	Cars per person
CHILD5PC	Percentage of people aged five and under
COMMUTE	Average travel time to work (minutes), not including those who work at home or those who do not work
DISABLPC	Work-disabled people per capita
FEMALEPC	Percentage of population who are female
FISHERPC	Percentage of population employed in fishing, agriculture or forestry
HISPNCPC	Percentage of the population who are Hispanic
HOMELESS	Percentage of population counted in shelters or on the street
HOMEVALU	Median value (US dollars) of owner-occupied homes
IMMOBLPC	Percentage of the population who are physically immobile
LANGISOL	Percentage of the population who live in a household without at least one English-speaking adult or older child
LOCAREPC	Percentage of the population who have a low capacity for self-care
MTG35.PH	Percentage of owner-occupied households with mortgages 35 per cent or more of household income
NEWCMRPC	Percentage of the population who moved in from 1989 to 1990
NEWIMMIG	Percentage of the population who is foreign-born and entered the US between 1987 and 1990
NODIPL 18	Adults with educational levels less than a high school diploma
NURSHMPC	Percentage of people who live in nursing homes
NWCMSTRM	Percentage of the population who moved in between 1980 and 1990 (approximation of post-Blizzard of 1978)
OLD65.UP	Percentage of people aged 65 and more
PCINCOME	Per capita income in 1989
PHONEPH	Percentage of occupied housing units with no telephone
POVRTYPC	Percentage of people with incomes below the federally defined poverty line
PUBTRANS	Percentage of the population who travels to work on public transportation
RACEOTPC	Percentage of race not white, black, native American Asian, or Hispanic
RENTINC	Median gross rent as a percentage of median household income
SLFMPLPC	Percentage of people who are self-employed

Table 12.4 *Groupings of vulnerability factors*

Literature themes	Census variables	Factor name
Factor grouping 1		
Income/resources	Low income	'Poverty'
	High federal poverty	*Variance explained: 25%*
Race/ethnicity	Hispanic	
	'Other' race	
	Black	
Education	Few high school diplomas	
Lifelines	Fewer cars	
Factor grouping 2		
Transience	Newcomer since 1980	'Transience'
		Cumulative variance explained: 24%
	Recent newcomers	
Housing and the built environment	Few high mortgages	
Factor grouping 3		
Disabilities	Immobile	'Disabilities'
	Low self-care	*Cumulative variance explained: 42%*
	Disabled	
Factor grouping 4		
Race/ethnicity	Asian and Pacific	'Immigrants'
		Cumulative variance explained: 49%
Lifelines	Public transport	
Immigrants	Immigrants since 1980	
	Recent immigrants	
Factor grouping 5		
Age	Few elders	'Young families'
		Cumulative variance explained: 55%
Family structure	Many children less than five years' old	
	Working parents	

Nevertheless, the block group is a practical unit in advising local officials on the allocation of resources. We used factor analysis as an objective way to simplify our multivariate data set. Factor analysis allows researchers to identify and cluster variables that measure essentially the same underlying theme. Factorial ecology during the 1960s and 1970s, for example, showed that many descriptive attributes of urban populations could be grouped into a few factors, but that these factors were not further reducible. These factors displayed differing and characteristic spatial patterns so that city form was not simply a matter of overall ring or sector patterns of all phenomena, but a mosaic resulting from the superimposition of socioeconomic status, family status and ethnicity. A factor analysis of coping abilities shows how far, and in what combinations, the many variables suggested by the literature measure separate or similar characteristics in the chosen block groups. We identified five most important factors from the analysis (see Table 12.4) and gave each an appropriate name. The factor that accounted for the largest amount of variance included issues of income and material resources, race and ethnicity, education, and lifelines (transportation, utilities and emergency response – see Platt, 1991). For want of a more inclusive term, and because per capita income had the strongest association of all the variables, we named this factor '*poverty*'.

The remaining factors, in order of variance explained, were '*transience*', associated most with the people new to the area since a record-setting blizzard in 1978; '*disabilities*', which included measures of immobility, ability to care for oneself and work disability; '*immigrants*', which reflected the concentration of recent Asian immigrants in certain parts of the study area; and '*young families*', which represented families with young children without a parent at home, high numbers of young children and a low proportion of elderly residents.

Four more factors, which showed lesser significance but had eigenvalues of greater than one, included the following groups of census variables: few homeless and high percentage of women; fishing occupations, nursing homes, self-employment; Native Americans; few telephones; and rent as a high percentage of income. For lack of a compelling reason to exclude them, they were incorporated within the composite vulnerability indices described below.

The factor analysis thus underlines the complex and multidimensional character of hazard vulnerability. Low coping ability (high social vulnerability) cannot be reduced to a single variable such as income or unfamiliarity with the area and its hazards. At the same time, the results of the factor analysis suggest that the multidimensional complexity of coping ability may be represented by far fewer factors than the original list of 34 variables.

RESULTS

We mapped the factor scores for each individual factor onto a base map of block groups for the town of Revere. Figure 12.1 is a map of factor loadings for the 'poverty' factor. Remember that 'poverty' as used here includes a complex of several census variables. Similar maps were created for each factor. Mapping each of the factors independently provides valuable information; however, it is also useful to combine the multidimensional factors into a single scalar measure of social vulnerability in order to provide an overall assessment. We use two methods for obtaining the scalar index of coping ability:

less "poor"

"poorer"

missing data

Notes: The coastline is to the east, with tidal rivers to the north and south. Boston lies directly south along the coast. In this and following maps, north is at the top of the page.

Figure 12.1 *Factor scores on the 'poverty' factor, Revere, Massacusetts*

1　*averaging*, which provides an absolute index; and
2　*data envelopment analysis* (DEA), which produces a relative measure.

The method which practitioners would choose to use in a study depends upon a number of circumstances, including confidence in weighting of factors and the end use of the index.

The most common way to combine factors would be to create an index based upon a weighted average, where the weights reflect the importance of each of the factors to the activities or decisions that need to be made. One difficulty in implementing this method centres on the way in which the weights are obtained, which often requires subjective assessments of importance. Another characteristic of this technique is that averaging may obscure high values on one of the factors when it is combined with other factors whose scores are low. For some end uses of an index – for example, emergency response – this may not be desirable since extreme values may indicate where resources are needed most. Creating a scalar index using DEA (see Charnes et al, 1978; Haynes et al, 1993) has the same mathematical structure as the weighted average except that the weights are obtained for each block group objectively through the use of an

Notes: Composite social vulnerability using the average method is on the left; using the DEA method, it is on the right. The block group on the coast at 'A' appears more vulnerable using the DEA method because its inhabitants are highly transient. The block group on the coast at 'B' appears more vulnerable using the DEA method because it contains a large number of disabled inhabitants.

Figure 12.2 *Average versus data envelopment analysis (DEA) overall social vulnerability, Revere, Massachusetts*

optimization model. In this method, no subjective a priori evaluations need to be made about the weights. In addition, when DEA is used, block groups that have high values on only one factor may still be identified as vulnerable. The DEA index technique provides a relative measure of vulnerability, ranking block groups on the basis of their comparative degrees of social vulnerability. Note that the DEA measure can change significantly if new block groups are added to the analysis, making this technique sensitive to the set of block groups chosen.

Figure 12.2 shows the average social vulnerability for Revere on the left (in the absence of other information all factors were equally weighted) and the DEA social vulnerability index on the right. Darker areas are calculated to have a higher degree of social vulnerability. As expected, there are several block groups that appear more vulnerable in the DEA map due to the dampening of high values in the weighted average. For example, one block group (A) is flagged because of a high factor score on transience and appears more vulnerable in the DEA map. Block group (B) has a high factor score for disabilities; therefore, it also is more vulnerable in the DEA map. For the remainder of the discussion, we focus on the DEA map, using a three-category (or low, medium and high vulnerability) version of the DEA map when discussing overall social vulnerability.

In order to understand how social vulnerability interacts with physical exposure to the hazard, we adapted a Federal Emergency Management Agency (FEMA) Insurance Rate Map of flood zones for Revere (see Figure

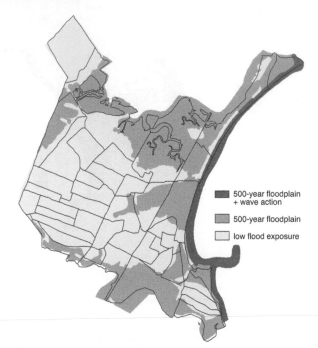

Figure 12.3 *Physical flood exposure, Revere, Massachusetts*

12.3). This adapted map has three different risk zones: one of no risk (light grey), one 500-year flood risk zone (medium grey) and one zone subject to both 500-year flood risk and wave action (dark grey). (In our study area, the 100- and 500-year floodplains are coincident for all practical purposes. Either term could be used.)

It is important to note that although a smaller risk is assumed for people who do not live in the floodplain, they may also suffer the effects of severe storms. They may be affected by wind, rain, snow and ice damage, and their routes of travel, daycare centres, other lifelines and places of business may also be affected by flooding.

ANALYSIS

Finding all of those areas that are both physically high risk *and* socioeconomically in less of a position to cope with the hazard allows us to display the interaction of physical risk with socioeconomic dimensions of resistance and resilience. Analysis of how each block group of Revere scores on the physical and socioeconomic scales lets us create a new, overall vulnerability map of Revere (see Figure 12.4). The legend shows increasing physical vulnerability in darker shades of grey on the Y-axis and increasing social vulnerability in denser cross-hatching on the X-axis.

The immediate conclusion that can be drawn from these maps, which show how vulnerability varies spatially, is that the threat of physical

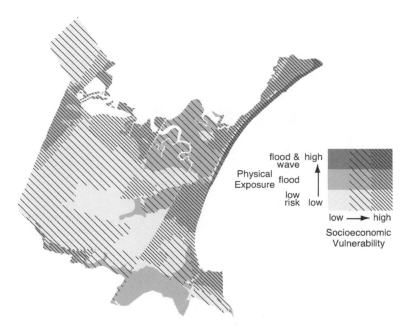

Notes: Note that the upper left legend category, the combination of highest physical exposure and lowest DEA socioeconomic vulnerability, does not occur in Revere.

Figure 12.4 *Overall social (DEA) and physical vulnerability, Revere, Massachusetts*

vulnerability is differentially compounded by social vulnerability. Maps at this level of generality are quite useful in pointing out areas that need more in-depth attention. For example, coastal zone and emergency managers have to focus their finite time, personnel and fiscal resources when deciding upon hazard mitigation projects. This choice may be guided by physical vulnerability only (e.g. 100-year flood zones on FEMA maps) or they may be guided, as suggested here, by a combination of physical and social vulnerability (e.g. choosing all combinations of medium to high physical and medium to high social vulnerability; or choosing only the areas of greatest physical and social vulnerability; and so on). In order to then find out what makes an area socially vulnerable, we have to go back to the factor level and, perhaps, ultimately even the variable level. The underlying social situation that makes a particular area vulnerable can then inform the strategy and choice of mitigation effort. For example, if an area scores high on the 'disability' factor, emergency evacuation plans could be adjusted to ensure earlier warning and evacuation, and the availability of transportation for the physically less able or mobile. In other words, the type and degree of intervention can be adjusted to the special needs in specified flood-prone coastal areas. It should be apparent that the scale of analysis is important in determining the use of a map, and that a gradation of scales can be used as a sifting

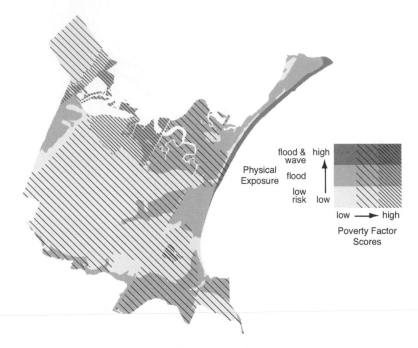

Notes: Note that the upper right legend category, the combination of highest poverty factor scores and highest physical exposure, does not occur in Revere.

Figure 12.5 *Poverty factor scores and physical vulnerability, Revere, Massachusetts*

tool, informing where to invest time and resources as more focused studies and actions are planned.

In our study, we go one step back down the scale to the factor level. Figure 12.5 is a simplified (three-category) version of a poverty factor map cross-tabulated with the FEMA flood zone map. Figure 12.5 shows why, in a particular area, people are so vulnerable and how they intersect with the three levels of physical risk.

IMPLICATIONS FOR CAUSAL MODELLING OF HAZARDS

More succinctly describing the components of vulnerability through factor analysis allows us to add the concept of vulnerability to a model of how hazards happen. The causal model discussed in Chapter 2 outlines the circumstances that must coincide at the intersection of the human and physical systems in order for a hazardous event to take place. Furthermore, their model is constructed in a way that highlights possible points of appropriate and timely intervention in order to avert the consequences – opportunities to expand the 'practical range of choice' (White, 1961; Wescoat, 1987).

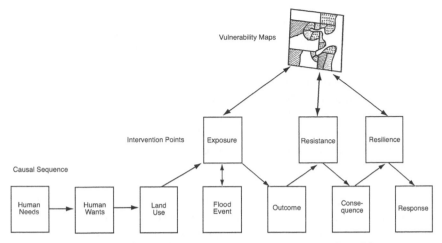

Figure 12.6 *Vulnerability and the hazards causal model*

Recognizing and incorporating within the causal model the spatial distribution of the physical and social components of vulnerability (in this case the factors of poverty, transience, immigrants, disability and young families) will help interested parties focus attention on problems of social justice, a major issue in emerging hazards and global environmental change research (Wescoat, 1993).

According to the causal model, hazards emerge through a sequence of choices and occurrences: human needs, human wants, choice of technology, initiating events, outcomes, exposure and consequences. A reinterpretation of the hazards causal model to incorporate vulnerability based on this research appears in Figure 12.6.

In this adaptation of the hazards causal model from technical to environmental hazards, we substitute 'land use' for 'choice of technology'. In doing so, we do not intend to imply that land use or social situation is a 'choice' for everyone, merely that landscapes are socially adapted phenomena and need to be thought of as such rather than as *tabulae rasae* upon which environmental hazards play themselves out.

In our model here, exposure, resistance and resilience each function as a filter or amplifier for the impacts of the hazardous event at different stages. Where these components of vulnerability are seen to worsen the hazard impact, they identify places where progress could be made in hazard management.

Exposure is determined by both existing land use and by the characteristics of the storm event, such as timing, duration and intensity. At the same time, the storm as a hazardous event implies an interaction with human systems, so the arrow goes both ways between exposure and flood event, producing an intersection, known as an outcome. For example, floodwaters rise to a given height, which potentially affects a given number of houses, businesses, cars, people, etc. The consequences (the damages to life, health, well-being and property) of an outcome are then determined by people's resistance to the hazard. Finally, the ability to respond (to recover from the event and prepare for the next) is determined by the resilience of the population or individual.

Another aspect of this model is that a map of each factor can be used to inform each stage of vulnerability (exposure, resistance and resilience). To use the poverty example, poor people may be constrained by the real estate market to a given neighbourhood which has poorly maintained drainage or physical protection from the hazard (exposure). In terms of resistance, poor people may lack the money or equipment to buy or make storm-proofing adjustments to their homes. For resilience, they may not have been able to purchase insurance to let them rebuild. To continue the filter/amplifier metaphor of the causal model, 'turning down' a vulnerability factor such as poverty in a given area makes the outcomes of hazardous events become more favourable.

Structuralist practitioners (such as Hewitt, 1983, 1997; Susman et al, 1983; Watts, 1983; Blaikie, 1985; Wisner, 1993a) might argue (somewhat correctly) that the causal model addresses only the proximate causes of vulnerability, as opposed to systemic root causes such as global capitalism, colonialism, racism and so on. They might further argue that we merely point out poverty rather than question why poverty exists here and elsewhere in the first place.

We agree that these are necessary efforts. However, we seek to emphasize intervention points in the causal chain that can be addressed within the scope of coastal policy makers, managers and community organizers. We also argue that we, too, attempt to promote a somewhat sophisticated view of poverty as including elements of economics, racism, education and transportation (see Table 12.4). Furthermore, we encourage the identification and rectification of more upstream causes of vulnerability as local analysis is performed.

Concerted interagency and community efforts to forge better links between emergency management, public works, labour and employment agencies, and private organizations are useful pathways of action at the community level, even if they do not fix, for example, the exploitation of labour through global capitalism. By speaking to the scale, areas and processes over which local leaders have some control, we hope to leave room for action and even optimism on a grass-roots level.

IMPLICATIONS FOR COASTAL HAZARD MANAGEMENT

Such practical damage reduction programmes might include contingency plans for weather-robust public transportation to and from shelters, insurance centres, places of employment and regional rail networks since transportation access issues score highly on each of the poverty, disability and immigrant factors. Interpreters and victim advocates might be used to help those covered by the immigrants and transience factors to communicate with or to be made aware of available community resources. Many potential initiatives can be identified through vulnerability analysis, and the more 'proactive' or 'upstream' the step taken, the greater the downstream benefit. Reduced exposure through improvement or elevation of housing stock means fewer people needing to resist or overcome effects of storms. Increased resistance means less need for recovery, and so on.

It is no accident that maps of vulnerability feature prominently in this framework. Based on our factor analysis, they represent a succinct understanding of the distribution of a large number of physical and social variables, which is crucial to a more complete picture of the hazards process. On a practical basis, vulnerability maps similar to the ones produced in this study have potential benefits for both crisis management and planning for future contingencies, including potential impacts of global climate change. For example, disaster management resources can be more easily and swiftly directed to the locales of greatest need, and resources can be applied according to the specific factors that render a certain population socially vulnerable. Vulnerability maps thus allow for the focusing of limited resources on areas of highest priority or of potentially greatest improvement.

Likewise, these vulnerability maps should be very useful for generating scenarios. One can model a '100-year coastal storm' (or any other hazardous event) and see who is affected in various ways. Thus, vulnerability maps could be used not only as a response tool to intervene in an ongoing event, but also as an impact assessment tool that is more sophisticated than most current global change impact assessment models, and hence as a planning tool that can be employed even in spite of lacking localized information about changing sea level and storm climate.

By focusing attention and action on the current ability of communities to deal with coastal hazards on a local scale, this chapter suggests that there are many opportunities to improve upon hazard management that are beneficial in the present and that leave vulnerable areas in a better position for the future if hazards become exacerbated by climate change. Localities using this type of 'no regrets' approach would be enabled to take action against hazards in spite of a high level of uncertainty about the future.

ACKNOWLEDGEMENTS

This research has been supported by the National Oceanographic and Atmospheric Administration (NOAA) Climate and Global Change Program, Human Dimensions Section, and by the Department of Energy through the Northeast Regional Center of the National Institutes for Global Environmental Change (NE NIGEC). An earlier version of this chapter was presented at the First Open Meeting of the Human Dimensions of Global Environmental Change Community at Duke

University, 3 June 1995. Opinions expressed herein are those of the authors and not necessarily of the supporting agencies, authors' employers or other organizations. This chapter is dedicated to the memory of Harry E. Schwarz, colleague, teacher, mentor and friend. He is greatly missed.

NOTES

1   Although Cutter cites Hewitt and Burton (1971) as the inspiration for her term 'hazards of place', it should be noted that Hewitt and Burton use the term 'hazardousness of a place' differently: not to introduce 'vulnerability' or to emphasize causal modelling, but simply to mean the inclusion of multiple hazard types in the analysis of hazards at a given location.

2   Revere has 44 block groups; we extracted and analysed the data for a larger area consisting of Revere and the adjoining municipalities of Malden, Lynn and Saugus, with 225 block groups in all. The larger scope allows us to have robustness of analysis as well as a sense of how Revere (with the most flood-prone residential area) compares to other towns in the watershed. At the same time, our focus on Revere allows us to keep the focus on social variation within one community rather than institutional differences between neighbouring jurisdictions.

13 Border Crossings

Jeanne X. Kasperson and Roger E. Kasperson

INTRODUCTION

In a world careening towards a single global economy and a relentlessly interactive world communication system and popular monoculture, it is apparent that, ever increasingly, the sources of health and environmental risks that confront individual nation states and other political units lie beyond their political boundaries. Technological change – such as the creation of genetically modified organisms (GMOs) and innovations in cybernetic systems – proceeds with widening scales of impact and repercussions beyond national borders, and the effects are becoming ubiquitous worldwide (French, 2000). As symbolized by the political turmoil at the World Trade Organization (WTO) meetings in Seattle in the US in 1999, international trade, finance and resource regimes are assuming greater importance in the allocation of global risk. International risk management accords, such as the Kyoto Protocol to the United Nations Framework Convention on Climate Change (UNFCCC) and the harmonization of regulation in the European Community, portend significant international intrusion into the structure of national industrial systems and policy. Meanwhile, political change reveals divergent trends as greater integration pertains in some arenas (e.g. the European Community), while greater political fragmentation prevails in many others (e.g. the new states of central Asia), amid a concurrent weakening of national state structure and a strengthening of non-governmental civil society.

Despite these trends, risk scholars have accorded transboundary risks only limited attention. A host of questions surrounds the incidence, impacts, equity problems, societal responses and management interventions for the control of transboundary risk. This chapter focuses on the complex dynamics of how the direct biophysical and economic risks associated with transboundary activities interact with social, psychological

Note: Reprinted from *Transboundary Risk Management*, Linnerooth-Bayer, J., Löfstedt, R. and Sjöstedt, G. (eds), 'Border crossings', Kasperson, J. X. and Kasperson, R. E., pp207–243, © (2001), with permission from Earthscan and political processes to send signals about these border crossings and how they should be managed. We also explore how social institutions and management authorities process or might process these signals. Following an analytic framework developed in 1988 by Clark University researchers and Paul Slovic and his colleagues at Decision Research (see Chapter 6, Volume I), we envision this enquiry as an analysis of the 'social amplification and attenuation' of transboundary risks. We then draw upon this analysis to address the implications of the risk dynamics involved for the communication and management strategies.

As to whether transboundary risks pose distinctive challenges, events over the past two decades have provided dramatic testimony to their far-reaching potential effects and associated management conundrums. The Bhopal accident was not only an Indian tragedy but a wrenching trauma for the global chemical industry as well (see Chapter 5 in this volume). In the US, the Emergency Planning and Community Right-to-Know Act of 1986, passed as Title III of the Superfund Amendments and Reauthorization Act (SARA), and the extraordinary post-Bhopal efforts of the US Chemical Manufacturers Association (CMA) were direct results of the accident and its aftermath. Such risk-monitoring and risk-communication requirements were already formally in place in Europe; but the horror of Bhopal kept the implementation of the Seveso Directive high on the policy agenda. Moreover, the subsequent accidents at Sandoz and Chernobyl dispelled any perception that Bhopal could not have happened in Europe. The Chernobyl accident, for its part, removed an innocence concerning the possible reverberations and scale of a nuclear plant accident, and perhaps for technological accidents more generally. The transnational character of the Chernobyl fallout, and in particular the international incidence of hot spots, surprised all of Europe, bared in sharp relief the inadequacy of protective strategies, and may also have stamped indelible imprints upon public attitudes, in Europe and elsewhere, concerning nuclear power technology (Hohenemser and Renn, 1988; Renn, 1988). International efforts to clean up the Baltic have highlighted the difficulties in implementing international regulatory regimes in countries with disparate economic priorities and capabilities even when broad consensus exists at the policy level (SEI, 1996; Greene, 1998), a situation that undoubtedly presages experience at the close of the 20th century with the faltering of the Kyoto Protocol and not-so-concerted international efforts to avert global climate change. The controversy during the 1990s with bovine spongiform encephalopathy (BSE), so-called mad cow disease, has demonstrated that the social amplification of risk surrounding the handling of the risks involved in an exported food commodity (British beef) can result in dramatic transboundary responses that threaten not only a pillar of a national economy, but also stigmatize a national industry and erode social trust in experts and national regulatory authorities (Powell et al, 1997; ESRC, 1999; Granot, 1999). Finally, the ongoing debate over GMOs, and especially the differences between European and North American views of the risk involved, signals how differences in national risk assessment and regulatory practices can interact with broader cultural values and historical political tensions to place transboundary risks at the centre stage of debate and conflict.

Here, using the social amplification of risk as a framework of analysis and drawing upon empirical evidence from recent examples of transboundary risk conflicts, we explore the nature of these growing challenges. We argue that, as with other risk experience, this is not a single problem or challenge, but that transboundary risks come in different complexes and forms. Distinguishing among them can provide greater clarity about the social amplification-of-risk processes that result and the feasibility and likely effectiveness of strategies of managerial intervention, including risk communication, public participation and precautionary approaches.

WHAT IS DIFFERENT ABOUT TRANSBOUNDARY RISKS?

At first glance, transboundary risks are not much different from other risks. The dangers of industrial accidents, the movement of a radioactive plume, or the damage to health associated with ecological exposure to acid deposition do not change because differing political jurisdictions are present. Similarly, the unintended effects of the release of GMOs upon non-target species or upon agricultural systems will cross political jurisdictions. Threats to humans and their environments, in short, neither observe sovereignty nor respect political authority. In this respect, they evince some attributes of common property resources (Cutter, 1993, pp67–68). In *World Risk Society*, Beck (1999) goes so far as to claim that 'misery can be marginalized, but that is no longer true in the age of nuclear, chemical and genetic technology' and refers to the 'border-transcending magnetism' of the risks of these technologies.

But on closer inspection, this simple view of risk dissipates. In fact, even for the direct potential impacts on health and ecosystems, location often matters greatly. Hazardous facilities may be intentionally located at the edges of political jurisdictions, either to export the risk to others (the classic example of this being the downstream contamination of rivers by upstream users), to capitalize on more permissive safety standards in an adjoining political system, or to exploit margins and peripheral regions because they are viewed as expendable and of little importance to political elites and central authorities. The US–Mexico borderland accommodates a remarkable constellation of hazardous *maquiladoras* drawn to the Mexican side of the border by laxer safety standards and cheaper labour (Sanchez, 1990), whereas the depletion of the Aral Sea for irrigation to benefit the Soviet state through much of the past century represents a clear case of intentional exploitation of a peripheral region (Glazovsky, 1995). Our recent study of long-term environmental change in nine threatened regions around the world found that peripheral location and marginality near borders, whether geographical or sociopolitical, were major contributors to environmental degradation and to tardy and ineffective mitigative responses from central authorities and elites (Kasperson et al, 1995; see also Chapter 11 in this volume).

Indeed, the nature of transboundary risks is such that the spatial separation between the area generating the risk and the areas exposed to potentially harmful consequences may exacerbate potential vulnerabilities, thereby increasing risk. As Beck observes, 'Beyond the walls of indifference, danger runs wild' (Beck, 1992, p46). If those at risk reside in one nation state and are unaware, or less aware than their neighbours, of the threats posed by hazardous facilities or human activities in another nation state, they may be ill-prepared for an accident. Emergency communication and response systems may run up against language and cultural barriers, and border-management systems may also be lacking or underdeveloped. All of these deficiencies certainly complicated the Chernobyl accident. To the extent that knowledge of the risk is lacking, communications are vulnerable to boundary obstacles, and unless precautions have been taken for emergency preparedness, an accident or release is likely to produce greater adverse consequences than would otherwise be the case. Put another way, since risk is a joint product of perturbations and events, on the one hand, and the vulnerability of those at risk, on the other, transboundary situations may easily enlarge and exacerbate existing health and environmental risks.

Among the most telling attributes of transboundary risks is their great potential for the social amplification or attenuation of risk. If the source country of the risk and the recipient country have overlays of past conflicts, cultural differences or ongoing tensions, even minor potential 'exports' of risk may generate widespread media coverage, societal attention, public concern and protests. Ordinary citizens, in short, may have extraordinary risk aversion to even the most minor risks emanating from another party whom they regard as hostile, uncaring or just plain untrustworthy. Distrust centred on what is perceived to be a lack of competence by those managing the risk or a lack of concern for the health and safety of those across the border will surely heighten the volatility of risk questions and debate. The merest hint that the risk producers are cornering the benefits while transferring the risks to neighbours who receive no benefits is likely to rekindle old tensions and perhaps even generate new demands for compensation, greater risk control or both. All of the above are dramatically at work in public concerns over the 'Frankenfoods' and 'terminator technology' involved in the export of genetically modified (GM) foods by US corporations to European consumers.

Meanwhile, risk attenuation may also be at work. Where significant benefits accrue to the risk-source region while the risk consequences fall on extra-regional peoples or places, managers of the exported risk may be less driven to expend scarce resources for risk control and minimization. Transboundary risks may even become bargaining chips in political negotiations surrounding broader and more complex arrays of interests, and in national differences in the framing of risk problems, assessment procedures and alternative constructs of precaution. The public may also be less concerned if it knows that the consequences of an uncertain accident are likely to fall elsewhere. Thus, it is quite possible, perhaps even likely, that the processes of risk amplification and attenuation will be simultaneously at work with transboundary risks, with different intensities or effects apparent at different scales and layering of simultaneous amplification and attenuation, so that misperceptions, miscommunication, distrust and conflict arise more readily and prove to be quite resistant to resolution or accommodation.

CLASSIFYING TRANSBOUNDARY RISKS

What defining qualities of transboundary risks shape societal perceptions and responses? How can we usefully sort the apples and oranges in the 'basket' of transboundary risks?

Historically, the Organisation for Economic Co-operation and Development (OECD) has distinguished two principal classes of international pollution:

1   *Upstream–downstream pollution*: upstream (or upwind) countries benefit from the natural export downstream (or downwind) of polluted water (or air), and downstream (or downwind) countries suffer from receiving it. Winners and losers are determined by fortune of location in relation to natural environmental flows and pathways. Upstream donors have little incentive (other than political goodwill) to control their pollution. Downstream recipients have no control over the pollution they receive, and they generally have a weak political bargaining position and must rely on goodwill, international pressure or some trade-off bargaining with the polluter country.

2   *Reciprocal pollution*: the costs and benefits of polluting processes are scattered throughout a number of countries, including the source country or countries. This most commonly occurs where the countries involved have unrestricted access to common resources such as air or oceans (which may be shared in practice, although in principle they are continuous, homogeneous and indivisible). This is the most common form of international pollution; but effective solutions are generally elusive because unresolved conflicts and disagreements over the distribution of social and economic costs and benefits commonly persist (OECD, 1972).

Of course, a much broader array of transboundary risk situations abides beyond this simple dichotomy and its characteristic focus on one specific type of risk (pollution).

A closer look suggests that a robust classification needs to address a host of special considerations, including equity and fairness, differences in environmental standards, problematic institutional issues, and an apparent potential for the social amplification (or attenuation) of risk. Whether the risk that crosses the border is a legitimate visitor or an unwelcome trespasser, and whether the very crossing implies the explicit consent of the country that experiences adverse risk consequences are important matters. Other distinctions – including the types of release of pollutants, potential major accidents, and differing environmental effects and transport media – merit consideration as well. It has been suggested, for example, that a taxonomy of transboundary risks might treat types of risks, transport media, country characteristics and national versus common property resources (Linnerooth-Bayer, 1996, p170).

Previous social studies of risk (e.g. Kasperson and Stallen, 1991; Rolén 1996; Stern and Fineberg, 1996) have clearly affirmed the importance of social and political context in risk assessment. Indeed, any meaningful discussion of analysing and managing transboundary risk must begin with the explicit recognition that this is neither a single problem nor a single-context type of situation. Recognizing that all sorts of transboundary risks abound, we focus in this chapter primarily upon those risks that cross national political boundaries. In keeping with the tenor of the volume as a whole, then, we regard transboundary risks as risks that arise when human activities in one or more nation states threaten current or future environmental quality, human health or well-being in at least one other nation state.

We next recognize four different transboundary risk situations, in which the differences entail important aspects of social context (see Figure 13.1). *Type 1, border-impact risks* involve activities, industrial plants or developments in a border region that affect populations or ecosystems in the border region on both sides of the boundary. Although such risks, whether from accidents or routine releases, threaten inhabitants or ecology in this region, they usually do not involve long-distance transport or displacement of the risk. This type of transboundary risk typically is bi-national, although it may involve more than two countries if multiple boundaries happen to coalesce, as in the Upper Silesia–northern Bohemia or the Aral Sea regions. Upstream–downstream problems or developments along a shared natural resource are not uncommon (Park, 1991). The development of the Gabcikovo–Nagymaros hydroelectric power stations on the border between Slovakia and Hungary illustrates this type of transboundary risk well, and the nature of the dispute between the two countries suggests the array of issues and dynamics of conflict that this type of transboundary risk often generates (Fitzmaurice, 1996).

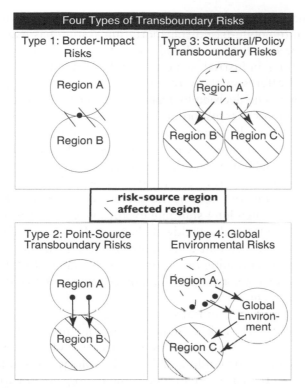

Figure 13.1 *A fourfold classification of transboundary risks*

Type 2, point-source transboundary risks involve one or several clear point sources of potential pollution or accident that threaten an adjoining country or region (or several countries or regions). The concentration of the risk source at one or several clearly identifiable locations, whether situated close to the border or away from it, provides a sense of problem or focus for perception of risk. The Chernobyl accident, of course, was the archetype of such a risk situation; but Lithuania's Ignalina nuclear power plant and its threat to other Baltic states and the Bärseback nuclear plant in southern Sweden, once operating a scant 20km from Copenhagen, exemplify this type of transboundary risk (Löfstedt, 1996a, 1996b). Indeed, the Swedish government's 1997 decision to close the Bärseback plant was motivated in no small part by the continuing public concerns in Denmark over the potential effects of a nuclear accident.

Type 3, structural/policy transboundary risks differ from the foregoing categories in that they involve less identifiable and more subtle and diffuse effects associated with state policy, transportation or energy systems, or the structure of the economy. Such risks typically cross boundaries uninvited; but they can enjoy varying degrees of hospitality, complicity and consent from the affected countries. Here the stakes, and the risks, are much more dramatic; but access to the risk issues is often more indirect and opaque. Elsewhere, we have described these risks as one type of 'hidden hazards' (see Chapter 7, Volume I). The US decision to promote an automobile-based transportation system has had enormous implications for air pollution and acid precipitation in the US and Canada. The coal-burning industrial structure in England has had extensive environmental effects in the form of widespread acid precipitation in northern Europe. The decision by France to become the nuclear state *par excellence* has imposed transborder risks on the rest of Europe. The forthcoming decisions by China as to the burning of its high-sulphur coal (and under what technological conditions) will have far-reaching implications throughout Asia and, of course, for the global environment itself. And the BSE controversy illustrates how risk management of potentially contaminated beef in the source country (the UK) and the transfer of risk through long-standing export patterns to European (and other) countries unleashed a worldwide ban and devastation of the British beef industry and subsequent controversy over the lifting of the ban and the assessment of continuing risk (ESRC, 1999; Granot, 1999). Here the conflict drew from a history of previous and ongoing tensions between the UK and the continent.

Although ubiquitous and almost routine, risks in this class share formidable hurdles for risk communication and management. Issues concerning the mix of energy systems, the structure of the economy, and the public policies surrounding industrial ecology, consumption and environmental protection are traditionally the prerogative of nation states. Equally, the risks are typically pervasive but opaque, and truly hidden, hazards. How to communicate and manage such risks involves a quantum level of difficulty that exceeds the ambition of national risk programmes and usually entails international negotiation and the construction of international risk regimes.

Type 4, global environmental risks present a still higher bar for risk assessment and management. What distinguishes this class of risks is that human activities in any given region or country, or set of regions and countries, affect many or all other countries or regions, often remote from the source country, through alterations of the global 'systemic' environment (Turner et al, 1990b). Nearly always, multiple and diffuse sources combine to alter aspects of the global environment through complex pathways of biophysical change in which the exact nature of interactions and causes is highly uncertain. Similarly, although potential effects may be sometimes dramatic and possible to pinpoint by particular geographic distributions (e.g. effects of sea-level rise on the Maldive Islands or Bangladesh), more typically (as in climate change) the potential impacts are murky, the spatial resolution of precise effects is poor, and winners and losers are difficult to discern. In short, this category of transboundary risk, which rudely intrudes upon national risk agendas, embraces many of those issues most difficult to accommodate.

Increasingly, international regimes and institutions are occupying a central place in producing, structuring and regulating this class of risks.

The European Community, through its harmonization of economic, trade and environmental policies, is assuming wide authority as a risk allocator and manager. The WTO's efforts to reduce trade barriers are not only opening new markets worldwide, but are also generating concerns in many countries over potentially far-reaching impacts on the industrial and agricultural structures, food supply and patterns of consumption, as well as on the environment, many of which may not be to the advantage of the less developed countries. The implementation mechanisms of the Kyoto Protocol – especially the Clean Development Fund, emissions trading and joint implementation – carry wide potential for international control over national developmental pathways, prompting some developing countries and non-governmental organizations (NGOs) to charge 'environmental colonialism' (Agarwal and Narain, 1999). These increasingly important shapers of worldwide risk patterns are difficult to classify on the involuntary/voluntary transboundary risk continuum. In the political turmoil that beset the 1999 WTO meetings, a primary locus of public controversy was the degree of coercion that some see the organization exercising over smaller states, as well as its alleged exclusion of non-state parties from processes that are rapidly becoming more open arenas for multiple interests and advocates beyond the traditional economic representatives from nation states.

THE SOCIAL AMPLIFICATION OF RISK

The concept of risk amplification and attenuation proceeds from the thesis that risk events, defined very broadly, interact with psychological, social, institutional and cultural processes in ways that can either heighten or dampen perceptions of risk and shape the risk behaviour of institutions, groups and individual people (see Figure 13.2). Behavioural responses, in turn, generate secondary and tertiary social or economic consequences, which extend far beyond direct harm to human health or the environment to include significant indirect impacts, such as liability, insurance costs, loss of confidence in institutions, stigmatization, or alienation from community affairs (Kasperson, 1992; also Chapter 6, Volume I).

Such secondary effects can (in the case of risk amplification) trigger demands for additional institutional responses and protective actions, or, conversely (in the case of risk attenuation), place impediments in the path of needed protective actions. In this usage, 'amplification' is a generic term referring to both intensifying and attenuating signals about risk. Thus, alleged 'overreactions' of people and organizations receive the same attention as alleged 'downplaying'.

Risk, in this view, is, in part, biophysical threats of harm to people and, in part, a product of social experience and the social processing of risk signals. Hence, all hazardous events are 'real': they involve not only transformations of the physical environment or human health as a result of continuous or sudden (accidental) releases of energy, matter or

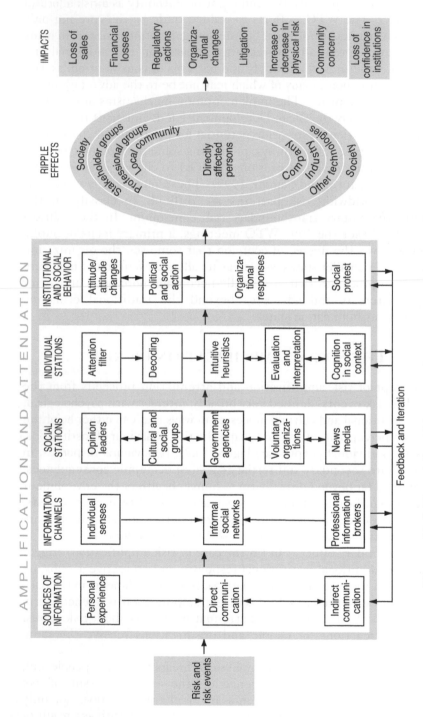

Figure 13.2 *The social amplification and attenuation of risk: A conceptual framework*

information, or reports on such transformations, but also perturbations in social and value structures. These events remain limited in the social context unless they are observed by human beings and communicated to others (Luhmann, 1986, p63). The consequences of this communication, the subsequent reframing of risk signals at other social stations, and ensuing social interactions may lead to other physical and socioeconomic transformations, such as changes in technologies, changes in methods of land cultivation, changes in the composition of water, soil and air, or effects on institutions. The experience of risk is, therefore, both a matter of physical harm and the result of culture and social processes by which individuals or groups acquire or create interpretations of risks. These interpretations provide rules of how to select, order and explain signals that emanate from the risk experience. Additionally, each cultural or social group selects certain risks and adds them to its strand of worry beads to rub and burnish even as it shuts out other risks as not meriting immediate concern.

The amplification process can take many different paths. It may start with a physical event (such as an accident), claims by environmental groups, or a report on an environmental or technological risk. Some organizations and individuals also, of course, actively monitor the experiential world, searching for risk events related to their agenda of concerns. In both cases, individuals or groups select specific characteristics of these events or aspects of the associated depictions and interpret them according to their perceptions and mental schemes. They also communicate these interpretations to other individuals and groups and receive interpretations in return. Social groups and individuals process the information, locate it on their agenda of concerns and may engage in risk-related behaviour. Some may change their previously held beliefs or gain additional knowledge and insights and be motivated to take action. Others may use the opportunity to compose new interpretations that they send to the original sources or other interested parties. Still others may find the new information as confirming long-held views of the world and its order.

The individuals, groups or institutions who collect information about risks communicate with others and through behavioural responses act, in our terminology, as *amplification stations*. It is obvious that social groups or institutions can amplify or attenuate signals only by working in social aggregates and participating in social processes; but individuals in groups and institutions act or react not merely in their roles as private persons, but also according to the role specification associated with their positions. Amplification may therefore differ among individuals in their roles as private citizens and in their roles as employees or members of social groups and organizations.

Social institutions and organizations also occupy a primary role in society's handling of risk, whether internal or transboundary in nature, for it is in these contexts that many risks are conceptualized, identified, measured and managed (Short, 1992, p4). In post-industrial democracies, large organizations – multinational corporations, business associations and government agencies – largely set the contexts and terms of society's debate about risks. These organizations vary greatly in their goals for and commitments to risk management. Freudenburg (1992, pp13–14) has called attention to breakdowns in internal organizational communications as a contributor to the bureaucratic attenuation of risk, as occurred in the minimization and suppression of a potential risk alert during the early signals from BSE cases. Other studies (e.g. Kasperson and Kasperson, 1993) reveal that large corporations develop very different kinds of organizational cultures that shape their ability to identify and assess the risks of their products and determine if and how these risks will be communicated to other social institutions and the public.

Risk issues are also an important element on the agendas of NGOs concerned with environmental and health issues. The nature of these groups is important in the framings and social constructions of the risk problem and the types of rationality brought to interpretation and to preferred management strategies. To the extent that risk becomes a central issue in a political campaign or a source of contention among social groups, it will be brought to greater public attention and often imbued with value-based interpretations. Polarization of views and escalation of rhetoric by partisans typically occur, and new recruits are drawn into the conflict. These social alignments about risk disputes often outlive a single case and become anchors for subsequent risk episodes, setting the stage for new risk controversies.

Role-related considerations and membership in social groups shape the selections of information that individuals regard as significant. People frequently ignore or attenuate interpretations or signals that are inconsistent with their beliefs or that contradict their values. If the opposite is true, they may well magnify, or amplify, the interpretations, signals or both. The process of receiving and processing risk-related information by individuals is well researched in the risk perception literature (Slovic, 1987; Freudenburg, 1988). But this is not sufficient: individuals also act as members of cultural groups and larger social units that co-determine the dynamics and social processing of risk. In this framework, we term these larger social units *social stations of amplification*. Individuals in their roles as members or employees of formal organizations, social groups or other institutions not only follow their personal values and interpretative patterns, they also perceive risk information and construct the risk 'problem' according to cultural biases and the rules of their organization or group (Johnson and Covello, 1987).

Both the information flow depicting the risk or risk event and the associated behavioural responses by individuals and other social stations of amplification generate secondary effects that extend beyond the people directly affected by the original hazard event or report. Secondary impacts include such effects as enduring public perceptions and attitudes, impacts on the local or regional economy, political and social pressure, social disorder, changes in risk monitoring and regulation, increased liability and insurance costs, and repercussions on other technologies. Secondary impacts are, in turn, perceived by social groups and individuals so that additional stages of amplification may occur to produce higher-order impacts. As a result, the impacts may spread, or 'ripple', to other parties, distant locations or future generations. Each order of impact will not only disseminate social and political impacts, but may also trigger (in risk amplification) or hinder (in risk attenuation) positive changes for risk reduction. The concept of social amplification of risk is, hence, dynamic, taking into account the continuing learning and social interactions resulting from social experience with risk.

The analogy of dropping a stone into a pond (see Figure 13.1) illustrates the spread of these higher-order impacts associated with the social amplification of risk. The ripples spread outward, first encompassing the directly affected victims or the first group to be notified, then touching the next higher institutional level (a company or agency) and, in more extreme cases, reaching other parts of the industry or other social arenas with similar problems. This rippling of impacts is an important element of risk amplification since it suggests that the processes can extend (in risk amplification) or constrict (in risk attenuation) the temporal and geographical scale of impacts.

INEQUITY AND DISTRUST: Contributors to the social amplification of transboundary risks

Since the distinctiveness of transboundary risks, we have argued, arises in part from their unusual potential for social amplification and attenuation, some exploration is needed of the social properties that affect how social institutions and cultural groups process such risks. Two major factors that appear to have particular capability for contributing to social amplification and attenuation are the risk inequities that are involved in the 'export' of risk and the distrust that may underlie or accompany risk incidence across boundaries. We begin with some conceptual discussion of the issues that each poses.

Inequity

Challenging complexities and ambiguities surround inequity problems that are embedded in transboundary risk conflicts. The causes of environmental degradation lie rooted in a variety of basic driving forces, such as agriculture, industrial ecology, population growth and urbanization. The transforming economies of Eastern and Central Europe add additional complex problems of social and economic justice. Beneficiaries and motivations are often difficult to discern, and interactions with the physical environment may be poorly understood and highly uncertain. Those likely to be most affected across political boundaries or in other states are often only dimly perceived, and little may be known of their values, capabilities or living circumstances. Past international conflicts may add an overlay of sociopolitical isolation and tensions.

Then, too, there are the complexities and ambiguities attached to the notion of equity itself (Kasperson, 1983a). Equity means different things to different people. Although it is often conceived as the 'fairness' of a particular arrangement, the standards and underlying principles of fairness vary. Some see it as a concordance between benefits and burdens; others, as an allocation of burdens to those best able to absorb or deal with them. Some view equity as primarily concerned with the substantive outcomes of an activity or project; others perceive equity as concerned with the procedures used to make the allocations. What is clear is that equity involves both matters of fact and matters of value, and it is an artifact of culture. And if equity is to be confronted explicitly in fashioning international initiatives for the management of transboundary risk, it will need to capture the diversity of values underlying different perceptions of responsibilities and goals. Whether or not a single principle or set of principles can win endorsement across political boundaries remains in question, and debates about inequity hold great potential for entering into, and significantly contributing to, the social amplification of risk in transboundary risk debates. Even though the precautionary principle enjoys wide acceptance in the UK and on the continent, the BSE controversy spawned a myriad of framings of this principle and its application. Similarly, it also highlighted major differences in approaching uncertainty, especially between the UK and Germany.

Transboundary risk controversies may well pose several different types of equity problems; so some conceptual clarification and background are germane. To begin, equity may be defined as the fairness of both the process by which a particular decision or policy is enacted and the associated outcomes. This suggests that any full analysis of equity needs to consider both distributional and procedural equity.

The primary equity issue in transboundary risk situations is likely to be the spatial (i.e. geographical) pattern of benefits and eventual harms associated with a particular set of activities. Thus, economic activities can, in principle, be compared as to their spatial pattern of harmful and beneficial impacts among states or other jurisdictions. In simple cases, these empirical patterns serve as a basis for making inferences about the obligations and responsibilities (if any) that beneficiaries have for those harmed and the adequacy of legal structures and institutional mechanisms for meeting these responsibilities. In the siting of a hazardous waste facility, for example, it is increasingly common to enlist a variety of means to compensate local host communities for bearing estimated risk impacts.

But assessing the geographical distribution of potential transboundary impacts will often be sufficiently complex and uncertain to challenge even carefully developed empirical studies. Estimating such inequities will involve, for example, defining the proportion of lake acidification in Sweden resulting from coal burning in Poland, placing a value on the resulting ecological damage, estimating the probability that an accident at the Ignalina nuclear plant will cause radiological damage in Sweden, defining the distribution of future beneficial and adverse economic impacts that will result from the Gabcikovo–Nagymaros project, or judging the damage attributable to unintended changes in species composition associated with the use of genetically modified organisms in agriculture. In a study of the distributional equity outcomes associated with the West Valley nuclear waste reprocessing plant in the US, Kates and Braine (1983) found themselves able to provide only a qualitative set of judgements about the magnitude of effects on local, state and national political jurisdictions. There are very few examples of careful empirical assessments of distributional outcomes associated with the kinds of transboundary risks elaborated in Linnerooth-Bayer et al (2001). Mounting such assessments across political boundaries may prove intractable, as differences in concerns over the risks, variations in assessment methodologies and conflicting approaches to estimating uncertainty exist and will greatly complicate this task and the likely acceptance of findings.

Scoping out the equity assessment is also problematic and likely to spark debates, even about what issues and what history are relevant for equity discussions. Equity, it can be argued, cannot be assessed solely within a particular regime or policy arena. It is inappropriate to divorce coal burning from other types of environmental change and past inequities involved in the previous transboundary interactions of peoples, nations and economics. Indeed, it will not be surprising if the type of transboundary equity problem that generates greatest concern is the one that has received the least attention – what we refer to as cumulative geographical inequity. New inequities that correlate with other past inequities suffered by disadvantaged societies, societies in transition or marginal groups are particularly pernicious because their effects are likely to be synergistic and not simply additive. Previous inequities are also almost certain to have increased the vulnerability of some groups to new projects. Thus, it should not be surprising when countries of Central and Eastern Europe and those in the developing world object vehemently to admonitions from the Nordic countries, Western Europe or the US to reduce future fossil-fuel emissions in order to preserve or improve the well-being of the rest of Europe or the globe as a whole. Cumulative geographical inequities, in short, may well be expected to form the core of many debates over transboundary risks and enter into the suitability or preference of particular policy options. Moreover, those inequities may be highly relevant to strategies for managerial intervention – particularly determining who should pay for reducing transboundary risks and where control should be vested in the regulatory regime.

The *procedural equity* of the processes by which transboundary environmental problems have arisen and by which they may be resolved differs from distributional equity issues. Here the concern is with the adequacy and appropriateness of the decision processes and political regimes that have created the risks. People living today, it can be argued, were not part of the considerations involved in decision making or the self-interest that externalized damage and burdens on other places, groups and generations. Established international institutional mechanisms for addressing such procedural inequities are unavailable, undeveloped or highly contentious. As a result, achieving procedural equity will be one of the most complex problems surrounding the management of transboundary risks.

Distrust

Addressing transboundary risks that are feared, that overlay previous histories of conflicts and other inequalities, and that contain complex equity considerations calls for high levels of social trust across transboundary participants so that management interventions can go forward. *Social trust*, in our usage, refers to the expectation that other persons, institutions or states in a social relationship can be relied upon to act in ways that are competent, predictable and caring (see Chapter 2, Volume I). *Social distrust*, equally, is the expectation that other persons, institutions or states in a social relationship are likely to act in ways that are incompetent, unpredictable, uncaring and, thus, probably inimical.

Four key dimensions of social trust, in our view, enter into range and depth of trust-related behaviour, and all are highly relevant to the social amplification processes surrounding transboundary risk situations. Each dimension can play an important role in the development and maintenance of social trust; none is sufficient in itself, however, to ensure the existence of such trust. These four dimensions are as follows:

1   *Commitment*: to trust implies a certain degree of vulnerability of one individual or group to another, to an institution, or to another social and political system. Thus, trust relies upon perceptions of an uncompromised commitment to a mission or goal (such as protection of the public health) and fulfilment of fiduciary obligations. Perceptions of commitment rest on perceptions of objectivity and fairness in decision processes and on the provision of accurate information.
2   *Competence*: trust is gained only when an institution or another government in a social relationship is judged to be reasonably competent in its actions over time. Although expectations may not be violated if these institutions or governments are occasionally wrong,

consistent failures and discoveries of unexpected incompetence and inadequacies can lead to a loss of trust. With regard to the cases treated in this volume, risk managers and institutions responsible for transboundary risks must show that they are technically competent in their mandated area of responsibility.

3   *Caring*: perceptions that an institution or government will act in a way that shows concern for and beneficence towards trusting individuals are critical. Perceptions of a caring attitude are an especially important ingredient where individuals must depend upon authorities in other societies who control the generation of risk. Obviously such perceptions are difficult to achieve even within a nation state; trust in those who have no formal accountability to individuals cannot be expected to come easily.

4   *Predictability*: trust rests on the fulfilment of expectations and faith. Consistent violations of expectations nearly always result in distrust.

Public perceptions, specific situational contexts and general societal factors all play important roles in the development of social trust. Thus, the nature of an individual's past interaction with organizations, with social institutions as a whole and with other societies involved in the transboundary risk will be major influences on the existence of social trust. The degree of trust or distrust that prevails, in turn, will play a major role in the dynamics of social amplification and attenuation of the transboundary risk issues.

SOCIAL AMPLIFICATION OF TRANSBOUNDARY RISKS: Cases in point

Löfstedt and Jankauskas (2001) ask: why are the Swedes and the other European nations so concerned about Ignalina? Indeed, the question is quite relevant, for despite this Lithuanian nuclear power plant's having design similarities to the Chernobyl plant, the Swedish Nuclear Inspectorate had, by 1995, already allocated 90 million Swedish krona (SEK) to upgrading safety at Ignalina, which funded improvements in fire protection, early-warning radiation systems, monitoring of metal strength, enhancement of operating procedures and maintenance, and establishment of a Lithuanian nuclear inspectorate modelled on the Swedish system. And another SEK 70 million had been allocated for the 1995–1996 budget year. Meanwhile, the plant itself sits a full 700km from the Swedish coast, which, the Chernobyl accident notwithstanding, affords a substantial cushion of distance from accident risks.

The high level of public concern is, apparently, an artifact of the social amplification of the risk. It is quite evident that even in advance of a serious (or perhaps even minor) accident that would drive levels of public concern and reaction even higher, significant risk amplification has already occurred. The authors report, for example, substantial public worry in Marmo, Sweden, over environmental problems – especially acid rain and industrial wastes – originating in Eastern Europe. One in five respondents in an open-ended solicitation of problems cited 'unsafe nuclear power plants'. Concerns were even higher about coal burning and acid rain, and fully 92 per cent of the respondents supported Sweden's giving environmental aid to Eastern Europe, with a primary motivation being to protect the Swedish environment from damage.

These concerns among the public appear to outstrip apprehensions among Swedish policy makers; but significant worry exists there as well, particularly among those at the national level. As one environmental spokesperson put it: 'We are heavily affected by transboundary environmental problems. I am especially worried about unsafe nuclear power stations and acid rain. Look what happened with Chernobyl' (cited in Löfstedt and Jankauskas, 2001, p46). The social dynamics that have created the high levels of concern among the public and policy makers alike cannot be judged from the data presented, particularly in the absence of any extensive recent media coverage. And, of course, virtually all Swedes know for a fact that Swedish lakes have registered dramatic adverse effects as a result of coal burning in Poland. The linkage of the Ignalina plant to the disastrous impacts of Chernobyl, threats of terrorism at the plant and the pollution of the Baltic more generally all may play a role. Elements of social distrust are also undoubtedly present, as Swedish officials and the public have reservations concerning the competence of Lithuanian managers to operate the Ignalina plant safely. But exactly how media coverage, distrust, concern over nuclear plant risks and the memorability of the Chernobyl accident interact to shape Swedish perceptions of the transboundary risk posed by the Ignalina plant remains opaque at present.

In contrast to the Swedish response, elements of both social amplification and attenuation of risk are apparent in Lithuania. The Lithuanian media have, particularly since Lithuanian independence, given extensive attention to the problems of the Ignalina plant and the risks that it poses. All respondents in the survey of public perceptions in Kaunas had heard of the Ignalina plant, and imagery centred on the problems and perceived lack of safety of the plant. The reality of a Russian-built plant in Lithuania clearly enhanced the negative imagery. Distrust of the responsible authorities was also quite apparent: 63 of 100 respondents in Kaunas reported a lack of trust, although different dimensions or bases for the distrust appeared to be present. Moreover, since Kaunas respondents were concerned about environmental problems generally, it may well be that both the Ignalina plant problems and coal burning are linked to this more generalized concern.

Despite these elements of social amplification in the media, negative imagery and the distrust of the responsible authorities, attenuation of the risk has also apparently occurred. In an open-ended question of concerns about environmental problems, nuclear power scored well down the list. Generally, Kaunas respondents were more concerned about local problems than transboundary risks, a pattern also apparent among the Lithuanian policy makers interviewed. Most Kaunas respondents opposed closing the Ignalina plant, with many citing the need for electricity as a primary reason. And since the Russians were blamed for imposing the plant on Lithuanians to begin with, many saw the Lithuanians as hapless victims (of Russian transboundary incursions), rather than villains responsible for exporting risk to Sweden.

The Gabcikovo–Nagymaros hydroelectric power plant controversy (Vari and Linnerooth-Bayer, 2001) illustrates several other features of the social amplification of transboundary risks. Environmental activism played a critical role in this controversy, with 60,000 people participating in a 1990 protest rally, and with environmentalists from Austria, Czechoslovakia and Hungary coalescing into a coordinated political force. Meanwhile, the controversy itself was part of a larger political evolution in Hungary, since it offered a primary means of expressing opposition to the government. Indeed, the amplification potential of the Gabcikovo–Nagymaros conflict waxed and waned in relation to broader political changes in Hungary (Fitzmaurice, 1996). It is clear that social amplification of risk can be heavily influenced by the extent of its incorporation within broader social and political movements.

And by ideology, as well. Vari and Linnerooth-Bayer (2001) show that the public debate over the risks of the dam project became linked to two ideological paradigms. In Hungary, the dam system became so much a symbol of the old regime that opposition became entwined with broader political reform. In Slovakia, in contrast, the project became a symbol of national pride and economic progress. The paradigm divergence provided a political overlay contributing to greater polarization in perception and social valuation of the risk, whereas the lack of international institutions made the emerging conflicts difficult to overcome or resolve.

Transboundary air pollution, an archetypal type 3 case in our classification, displays a number of distinctive assessment and management challenges associated with its patterns of social amplification. Given the importance that notions of equity and fairness take on in a situation with a high national stake in the economic activities generating the pollutants and the far-reaching nature of the effects, as with acid deposition, it is not surprising to witness significant influences on the types of assessments undertaken and the underlying rationale for taking management actions. In the case of acid rain in Europe, the source and type of assessments became very important. The Regional Acidification INformation and Simulation (RAINS) model was adopted as the basis of assessment in no small part because it emerged from an international research institute, IIASA (the International Institute for Applied Systems Analysis), and so could be viewed as more politically neutral than a model generated within one of the contending nations. Similarly, given the diversity of value perspectives pervading the transboundary context of acid rain, a management approach that had the appearance of being based in natural sciences, rather than being a choice among contending values or conceptions of equity, easily won favour. This approach allowed contending countries to justify adoption of the critical loads targets on quite different bases and rationales – a 'bright line' on which consensus could be won – including substantial impact reduction (science) sought in Scandinavia, the appearance of avoiding action in the UK and Germany, and ideological reasons in the Soviet Union. In short, the complexity and intensification of this type of transboundary situation may well shape the types of assessment and management approaches likely to be successful.

The 1986 Chernobyl accident, in our classification a dramatic example of type 2 transboundary risks, showed significant amplification and attenuation of the risk occurring simultaneously among different elements of the public and different groups within particular nations. Differing patterns of social trust and credibility undoubtedly abetted the marble-cake mix of amplification and attenuation. According to one study (Peters et al, 1987), 60 per cent of all Germans surveyed viewed the federal government and other official institutions as having been totally or partially trustworthy. Public opinion polls in France and Italy revealed that 70 per cent or more of the public distrusted and lacked confidence in official government information sources. In Sweden and The Netherlands, in contrast, trust levels appear to have been higher than in many other parts of the European Community. As Otway et al (1987) point out, the competing and often conflicting information provided by adversarial interest groups and stakeholders compounded the distrust problem and contributed to public misperceptions and confusion. Exploiting Chernobyl either to protect or attack national nuclear energy programmes produced a great range of conflicting information and interpretations of the accident and fallout threats (Hohenemser and Renn, 1988). In the welter of discordant communication, unreliable sources often commanded as much or more credibility as reliable sources. During the unfolding of the post-accident responses, industry quickly lost much of its credibility, and governmental sources were not far behind.

The accident suggests that conflicting stances by officials in different nations or political jurisdictions in providing risk information, explaining the risk and recommending protective actions in transboundary risk settings can easily spawn public misperception, confusion and distrust. Yet consistency in management responses to risk across national political boundaries will be difficult, perhaps impossible, to achieve. It is also clear that anticipatory or pre-emergency communication and education are essential in securing a basic level of preparedness among the potentially affected public. Models for such preparedness programmes, including necessary communication systems, have been well developed for major industrial accidents, including nuclear mishaps (Sorensen and Mileti, 1995). Although it is a generic risk communication issue, the need to put transboundary risk and protective measures into context was driven home in the aftermath of the Chernobyl fallout. Quantitative information on risk is unlikely to have the desired effects without the means to help the public understand the data and to connect the data to protective actions. Finally, it is apparent that in many transboundary settings the public will distrust official information sources across boundaries regardless of how well they do their job. Creative measures to maintain or increase trust and confidence before and during an emergency may need to be a part of transboundary risk communication programmes.

The more recent experiences with the so-called mad cow disease scare in the UK and the ongoing controversy over GM foods are highly suggestive of both the growing importance of transboundary risks and how volatile they may be (Tait, 1999). The BSE case provides additional confirmation of how transboundary risk settings can complicate risk management and drive social amplification of the risk as a result of differences in risk assessment practices. Wynne and Dressel (2001) show, for example, that the UK's approach to uncertainty demands the specification of the risk mechanism or damage pathway and concrete identification of uncertainties, whereas practices on the continent, and especially in Germany, accord more credence to abstract 'theoretical' risks. Furthermore, continental traditions have involved substantial integration of the social and the natural sciences, whereas assessment practices in the UK have been strongly rooted in the physical sciences and empirical analyses. This setting facilitated the rapid resort to characterizations in the UK of European risk managers, and especially those in Germany, as 'unscientific' and prone to 'overreaction' and 'hysteria'. In turn, continental perceptions emphasized the lack of precautionary procedures in the UK, given a food supply risk fraught with poor knowledge and high uncertainty.

Whereas the BSE case entails considerable 'risk amplification mirroring', as noted above, it also suggests the propensity of conflict over transboundary risks to escalate latent patterns of historical tensions and distrust into overt conflict. In doing so, it points to the high potential of transboundary risks to drive stigmatization processes. So while British consumers' trust in their own officials' management of BSE risks underwent a dramatic decline between 1991 and 1995, the 1996 European Union (EU) ban on the movement of all live cattle of British origin unleashed nationalistic loyalty and a fury of anti-EU sentiment in the UK and, subsequently, a powerful backlash from European countries. The furor of media coverage and loss of credibility of British risk managers drove further escalation of risk concerns and caused extensive damage to the 'identity' of British beef, whereas the rising political mistrust and misunderstanding undermined whatever basis might have existed for creating common approaches to risk management. Some of these same connections between transboundary risks and social amplification processes are apparent in the controversy over the export of GM foods from the US to Europe. Specifically, transboundary effects are apparent in the risk assessment process, patterns of distrust and stigmatization of GM foods. As Levidow (2001) notes, European regulatory approaches perforce have had to proceed in the face of divergent accounts of the risk and differing notions of sustainable agriculture (Persley and Siedow, 1999). Debates have involved varying assessments of the scope of 'indirect adverse effects' of GM crops, with Denmark, France and the UK all moving towards greater use of value judgements. Even here, however, individual nations have interpreted the precautionary principle in different ways to justify their own policy stances on the use of GMOs or living modified organisms (LMOs). And surely the fact that the US, and an American corporation (Monsanto), has been in the lead in the use of GM foods builds upon historical antipathy to the 'invasion' of US fast foods and popular culture on the European continent. But it is also the case that these intercontinental conflicts become more intractable as a result of the difference in US regulatory approaches to GM foods and agricultural strategy, in which the US emphasizes earlier approval and a less precautionary strategy than that prevailing in Europe. Then, too, the transboundary setting facilitates the entrance of a diversity of highly contentious value positions (Stirling and Mayer, 1999, found, for example, no fewer than 117 criteria emerging in judgements about GM crops and other alternatives). Meanwhile, by the 1980s a maturing European political movement had already succeeded in stigmatizing GM foods, so that current labelling strategies represent not only a means for making risk more voluntary, but a powerful means for extending the visibility of the stigma (i.e. the marking).

From the exploration of empirical evidence concerned with the social amplification and attenuation of transboundary risks, we identify several propositions that merit testing in further research:

- Compared with other risk problems, transboundary risks have distinctive properties likely to intensify the social amplification and attenuation of risk, enlarging the gaps between expert assessment and public perceptions and making them more volatile and difficult to manage.
- Because transboundary risks occur across national risk management programmes with varying approaches to, and practices of, assessment, values will pervade assessment procedures and interact with other dimensions of political conflict.
- A 'mirror' structure in social processing of risk is likely to be a common pattern in such situations, with social attenuation in the risk-source region and social amplification in the risk-consequence region.

- This 'mirror' structure will frequently be accompanied, and further complicated, by a scale-related 'layering' of intensity of effects, with attenuation most pronounced at the local scale and amplification most pronounced at the national scale.
- Equity problems and social distrust will play a more powerful role in the social amplification of transboundary risks than they do with most other types of risk problems, and risk amplification may reveal an intensification of asymmetrical influence of trust-building and trust-destroying events.
- Contextual effects, particularly those involving previous political animosities and ongoing sociopolitical movements, will more extensively shape the social dynamics of risk amplification and attenuation than with other types of risk.

The following section considers the implications of the attributes of transboundary risk for risk communication and management.

MANAGING AND COMMUNICATING TRANSBOUNDARY RISKS

Since, as we note above, communicating to the public within a country or across borders about the same transboundary risk raises a number of different problems, it is appropriate to distinguish among these problems. We enlist our fourfold categorization of transboundary risks to examine communication and management challenges, as summarized in Table 13.1.

Border-impact risks

Although border-impact risks take many forms, they typically have their greatest impacts in the border regions. They also commonly occur in frontier areas on the periphery of states or political jurisdictions, areas often viewed as marginal or expendable by the state or political elites at the centre. Disputes over natural resources present at the boundary are common, as the many international conflicts over upstream–downstream uses of rivers, diversion of waters, rights to fishing grounds located just outside territorial limits or border locations for hazardous industries suggest. Borders will continue to be targets for the siting of hazardous industrial facilities because of either their peripheral situation or their status as the meeting ground that divides differing legal or economic systems (and, thus, many involve differing regulatory structures or contrasting labour costs). Disputes over border developments have a strong potential for amplification of risks through connections to national political movements and agendas, as occurred at the Gabcikovo–Nagymaros hydroelectric power plant project on the Hungary–Slovakia border (Vari and Linnerooth-Bayer, 2001). Linkages with national politics may provide an overlay to escalating polarization in risk perception and

Table 13.1 *Communicating risks across borders: A profile of differentiated challenges*

Type of transboundary risk	Nature of hazard	Affected public	Key communication issues
Border-impact risks	Hazards usually localized in border regions; conflicts over resource use, allocation or depletion; impacts often near term in nature	Concentrated among border peoples, with more diffuse impacts at regional or national levels; potential perception of threats to national security	Hostile relations between boundary states may impede communication; past history important; language barriers; interests of ethnic subgroups may be present
Point-source transboundary risks	Clearly identifiable risk source(s); distance decay of effects may occur; accidents and toxic releases dominant problem	Layered impacts, some in proximate location, others in more distant locations across borders; benefits dissociated from risk, raising equity problems	Identifiable and visible risk sources; emergency preparedness and warning systems important; trust issues often involved; risks may be strongly socially amplified
Structural/policy transboundary risks	Diffuse and opaque risk sources; long delay in emergence of hazards and lag times in effects common; diffuse management responsibility	Highly distributed within risk-source state and affected transborder areas; future generations may be major risk bearers	Few rights for redress currently exist; difficult to link risks with economic structure or public policy; linkages to future impacts often opaque
Global environmental risks	Highly variable though potentially catastrophic; long time lines; widely distributed risk sources; large uncertainties; effects on global commons	Often remote in space and time; winners and losers often opaque; conflicting values and priorities surround risks and benefits; risks overlay other socioeconomic problems; vulnerable groups bear most of risk.	Few established transborder communication links; hazards accumulate slowly; episodic extreme events; equity and distrust problems exacerbate effects

social response to the risk, thereby adding to the obstacles to arriving at bi-national or multinational agreement over the project.

Concerning risk communication about this category of risks, certain problems (e.g. language barriers, cultural differences and institutional fragmentation) common to many transboundary risks are likely to be present. Institutional structures, regulations and standards will often vary on opposite sides of a border. The strategy, level and content of communication will often need to diverge. Political groups within each country may exploit boundary risk issues for national agenda-setting.

Much will depend upon the nature of relations between the two adjoining states and the nature of relations among groups or interests within the boundary area itself. The risk debate may easily mirror or take on ingredients of these ongoing relations, especially if long-standing tensions and conflicts are in play. Issues of distrust or inequity in such situations may generate intense amplification. On the other hand, if bi-national relations have long been amicable, as is true of the Great Lakes management regime in North America (Beierle and Konisky, 1999), then an innovative participatory approach to transboundary risk regime-building may become possible.

Point-source transboundary risks

The Chernobyl experience typifies, if in extreme form, this class of transboundary risk. A few dominant sources of risk, readily identifiable, define the risk situation. From a management perspective, efforts can focus on ameliorating the risk. It may even be comparatively easy to identify winners and losers. From a communications perspective, consultations and negotiations with the peoples at risk from the concentrated risk source are advisable.

As with all transboundary risks, significant impediments may be present. Difficulties in communicating across jurisdictions will still prevail. Lack of trust in the safety management at the facility may intensify public concerns over hazards and generate significant social amplification of the risks. Past tensions or hostilities may erode even well-intentioned communication efforts by the risk-source country. Nonetheless, of the four types of transboundary risks we have identified, this is likely the easiest to manage, although risk communication discussions in the aftermath of Chernobyl suggest this is not always the case.

Structural/policy transboundary risks

Arguably, this may be the most difficult and formidable arena for transboundary risk management and communication. First, the sources and causes of the hazards are typically multiple, diffuse, indirect and opaque. The structure of an economy focused on heavy manufacturing, using high inputs of energy and exercising few environmental controls, as occurred with Soviet policy in Russia and Eastern Europe, has pervasive and long-term effects. Agricultural systems supported by high chemical inputs may have to endure long-term environmental effects and, probably, reduced crop yields as well. The extensive import by China of highly inefficient and polluting automobiles from Singapore and elsewhere has environmental ramifications in Korea, Japan and globally. National population policies, despite their evident roles as 'drivers' of environmental change, remain the prerogative of individual nation states. GM foods regulated and exported through existing international agreements represent a new transboundary risk issue of high volatility and one almost certain to remain contentious far into the future.

Such transboundary matters did not preoccupy international risk-management regimes during the second half of the 20th century. Indeed, the Science Advisory Board of the US Environmental Protection Agency (USEPA, 1987) concluded that further progress on environmental protection in the US depended upon a shift of focus from 'end-of-pipe' clean-up to intervention in the structure of the economy and basic social and economic policy. But these issues will be very much on national and international environmental agendas in the 21st century. Since national economic policies and structures have been extensively protected by national sovereignty, they have thus far largely escaped extra-national intervention. But as the environmental interdependence of nations becomes progressively more visible, questions of national security, as has already happened, will be progressively linked with environmental cooperation. This will inevitably cause states to confront both basic policies and structures of economy and trade policies, and the growth of non-governmental actors and global civil society may well challenge existing national and international regimes and institutions in these risk arenas. These transboundary risks, in short, may have a 'forcing' potential for institutional change.

Meanwhile, transboundary risk issues remain formidable in their own right. Linking environmental and human health effects with different mixes of energy systems, transportation choices, social-welfare arrangements, aging populations or urbanization policies entails a level of assessment beyond the capability of most current approaches to risk analysis. Imagine communicating about risk 'residuals' that involve such a complex array of economic, social and health effects. Putting risk 'in context', in the current fashion, will be an interesting challenge to the assessors. Communicating the assessment results and linking them to policy choice and to diverse and far-flung publics will likely confound the most optimistic of risk communicators. Yet, arguably, this may be the most important of the transboundary risk problems.

Global environmental risks

The distinctive property of these risks, which share much in common with the previous class of transboundary risks, is that they produce their effects through alteration of the global risk system as a whole. For the purpose of this chapter, we treat what elsewhere we term 'cumulative global environmental risks' (Turner et al, 1990b) as falling within our type 3 risks discussed above.

Again, the major risk sources are associated with basic population and economic structures and policies. But the environmental effects are registered through alterations of basic biogeochemical flows, particularly when human alterations produce larger fluctuations than those attributable to natural variability. The obstacles confronting risk communication are imposing: the risk sources are widely distributed and historically cumulative; the links between human activities and changes in climate, atmosphere, oceans and biosphere generally are opaque and poorly understood; large uncertainties complicate the pattern of effects; the links between risk sources and affected peoples are often remote in space and time; and management options lie embedded in larger problems of growing global inequalities and conflicting views of responsibility and values. The ramifications, or 'signals', of emerging risks are often episodic, shrouded by expert debate and disagreement and prone to unwelcome future surprises.

Communicating about such risks is already under way in various countries. Interestingly, a highly interactive global mass media and internet system may work to overcome the plethora of hurdles that might impede recognition by the world's people of their common stake in addressing these issues. As a Gallup poll of members of the public in 24 countries unexpectedly demonstrated, concern about environmental degradation throughout the world is much more pronounced than might have been expected (Dunlap et al, 1993). And there are indications that threats to the planetary environment as a whole may constitute a special category of concerns to members of the public throughout the world. Perhaps, as our study of nine environmentally threatened regions throughout the world abundantly suggests (Kasperson et al, 1995), the particular problem of communicating such risks lies less with evoking public concern than with mobilizing and articulating such concern through national political systems focused on differing economic and political goals and possessing very different political cultures.

Managing transboundary risks, therefore, clearly entails different complexes of problems. Although significant progress in national programmes and international legal regimes is evident in particular regions and risk domains, current efforts are but the early stage in what is a rapidly growing collection of international regimes (Victor et al, 1998). As with many other international environmental risk problems, the rate of the mounting challenge will continue to pose daunting challenges to these emerging international responses. At the same time, it is apparent that an international imperative to communicate at least certain types of risks across national borders is routinely interwoven within the international fabric of risk responsibilities. At the end of January 2000, delegates from 135 countries produced a new biosafety protocol to regulate the transboundary movement of GMOs, now LMOs (Blassing, 2000). The *Cartagena Protocol on Biosafety* (*IER*, 2000; see also www.biodiv.org/biosafe/protocol) speaks to a global awareness and willingness to share the risks (and benefits) of an eminently transboundary risk. The document may well validate a claim that 'a plural world citizenship is soaring with the wind of global capitalism at its back' (Beck, 1999, p17). To be sure, 'informed consent' and democratic principles figure prominently. The tremendous growth in NGOs worldwide and communications technology, especially the internet, is providing infrastructure to this changing ethic. Yet the rapid growth in transboundary risks is clearly substantially outpacing managerial responses and policy innovations.

14 Vulnerability to Global Environmental Change

Jeanne X. Kasperson, Roger E. Kasperson,
B. L. Turner, II, Wen Hsieh and Andrew Schiller

INTRODUCTION

In its wake-up call to a world confronted by accelerating changes in the planetary environment, the World Commission on Environment and Development (WCED, 1987) in *Our Common Future* described a tapestry of global problems in which ecological and sociopolitical problems were deeply meshed. Global environmental threats and poverty were so interrelated, the WCED argued, that a global risk assessment programme was needed to discern the roots of the stresses emanating from human activities, a new environmental ethic would be required, and an integrative attack on the dynamics of environmental degradation and poverty was essential.

Various assessments since 1987 have confirmed this basic diagnosis. Taking climate as an example, from the work of the Intergovernmental Panel on Climate Change (IPCC), we know that humans have altered the climates in which they live, and the evidence is growing that human activities are responsible for most of the warming (IPCC, 2001). A litany of effects that will unfold over the coming decade is likely to impose widespread stresses and perturbations on human and ecological systems, including an ongoing rise in global average sea level, increases in precipitation over most mid and high latitudes of the northern hemisphere, increased intensity and frequency of droughts, floods and severe storms, and unforeseen abrupt changes and extreme climatic events. We also know that the related effects will be strongly concentrated in particularly vulnerable regions as a complex array of other stresses – including growing populations, poverty and poor nutrition, accumulating

Note: Reprinted from *Global Environmental Channge: Understanding the Social Dimensions*, Dickmann, A. and Dietz, T. (eds), 'Vulnerability to global environmental change', Kasperson, J. X., Kasperson, R. E., Turner, B. L., Hsieh, W. and Schiller, A., forthcoming © (2005), with permission from MIT Press

atmospheric and water contamination, gender and class inequalities, the ravages of the AIDS epidemic, and inept or politically corrupt governments – also acts on these regions and shrinks capacities to cope. The ecosystems likely to be most vulnerable will be those already at the limits of their range, those in which barriers to species migration preclude or impede redistribution, those at the interface of ecotones (e.g. estuaries, coral reefs, seasonally dry forests and some fire-adapted systems), artificially simplified agro-ecosystems that contain less genetic diversity to adapt, and those ecosystems likely to experience further human-induced demands as a result of reduced production in other ecosystems upon which humans rely. Ecosystems highly dependent upon social infrastructure or political stability are also likely vulnerable – where purposeful human actions that could be disrupted by climate change (e.g. artificial burning regimes to mimic natural processes and white rhinoceros protection activities) have supplanted natural processes. The coalescence and interaction of social and ecological vulnerabilities may result in spirals of degradation that reinforce and even accelerate mutual vulnerabilities and damage. In short, the people, ecosystems and regions already beset by a struggle to cope with a concatenation of stresses largely beyond their control are likely to bear most of the burden of climate change.

This should not be surprising, of course, based on what is already known about the changing global pattern of natural disasters and their effects on humans (Abramovitz, 2001; IFRC, 2001; UNISDR, 2002). Data from Munich Re reveal that 'great' natural catastrophes, those that overtax the ability of the region to help itself and that necessitate interregional or international assistance, increased from 20 during the 1950s to 84 during the 1990s (Munich Re, 2001, p15). During the period of 1991–2000, natural disasters affected some 211 million persons per year, seven times the number affected by human conflict. Most revealing is the role of vulnerability in this toll. Of all those killed and affected by natural disasters, 98 per cent were in developing countries, fully 86 per cent in Asia, but only 1 per cent in Europe. Meanwhile, the *World Disasters Report 2001* (IFRC, 2001, p165) notes that of the 2.3 million people reported as killed by conflict during 1991–2000, over three-quarters were from nations of low human development. Although economic losses from natural disasters are concentrated in wealthy societies, the lower losses in developing countries carry far larger and longer-term negative effects on political, social and economic productivity and stability (Munich Re, 2001, p20).

This intertwining of risk and vulnerability is a familiar theme in the field of risk analysis, the practitioners of which have long defined risk as the joint product. So, in 1942, Gilbert F. White wrote a PhD dissertation in which he proposed to reduce damages from floods by decreasing exposure of people to floods and increasing their capacities to anticipate and cope with floods (White, 1945). A long tradition of research on natural hazards (Mitchell, 1989; Mileti, 1999) has examined sources of vulnerability, extended by greater attention over the last decade to social and economic structures and global processes predisposing human societies to high risk (Blaikie et al, 1994; Bohle et al, 1994; Ribot, 1995, 1996; Cutter, 1996; Adger and Kelly, 1999). Studies of technological risk have examined the proneness of technological and industrial systems to failure (Perrow, 1984, 1999; La Porte, 1996; La Porte and Keller, 1996; Vaughan, 1996), the vulnerability of critical societal infrastructure to natural disasters and human terrorism (Haimes, 1990, 1998), and the factors propelling societies to greater energy insecurity (Stobaugh and Yergin, 1979; Yergin and Hillenbrand, 1982; Yergin, 1991; Khatib, 2000; UCS, 2002). Meanwhile, concerns over food security and threats to small agriculturalists and fishers have sparked interest in the vulnerability of livelihood systems, especially in the South (Chambers and Conway, 1992; Scoones, 1996; Carney, 1998; Carney et al, 1999; Vogel, 2001). And the emerging field of ecological risk assessment has a developing methodology, derived from human risk analysis, focused on characterizing and reducing human-driven risks to ecosystems, including toxic wastes and pollution (Suter, 1990, 1993).

This range of studies suggests one elemental insight largely missed in the evolving vulnerability literature: although writings typically champion a 'right' model or approach, in fact vulnerability, as we detail in Chapter 16, Volume I, refers to widely different situations, differing complexes of stresses, varying complexes of predisposing vulnerability factors, and dissimilar sociopolitical and community contexts. Searching for the 'right' theory or model for vulnerability is reminiscent of the prolonged sterile debate over the 'right' political model for community politics until it finally dawned that there was in fact no single correct model – that different places had quite different power structures and that sound comparative studies required multiple models (Clark, 1974).

At the same time, vulnerability has clearly 'arrived' and become something of a buzzword, a highly visible topic in a wide expanse of global change and sustainability studies and assessments and in the agenda of international food and health organizations (FIVIMS, 1999, 2000, 2001). It is also a prime issue in the post-11 September 2001 effort on terrorism (Talbott and Chanda, 2001; Kennedy, 2002). For hazards research, Mitchell (2001, pp87–88) ascribes some of the appeal of vulnerability as an analytical concept to 'its rhetorical connotations as well as its acuity as an intellectual device'. Observing the mounting attention to vulnerability in studies of natural hazards and disasters, a recent review implicates increased vulnerability as overwhelming the effective use of knowledge. The IPCC produced a special volume from its second assessment on regional vulnerabilities to climate change (Zinyowera et al, 1998) and put vulnerability centre stage in priority (if not actual work) in its third assessment (IPCC, 2001; McCarthy et al, 2001). The United Nations Environment Programme (UNEP) and the United Nations Development Programme (UNDP) are actively developing vulnerability assessment procedures, indicators and guidelines. Although the World Bank and regional banks have rediscovered poverty and undertaken extensive anti-poverty programmes, they have yet to integrate them within any broader framework of vulnerability analysis. Meanwhile, the International Federation of Red Cross and Red Crescent Societies (IFRC), in its strategic *Work Plan for the Nineties*, takes up the challenge of 'improving the situation of the most vulnerable' (IFRC, 1999b, p8). The Famine Early Warning System (FEWS) regularly produces two types of vulnerability assessments: 'current vulnerability assessments' of various populations' abilities to meet their current food needs and 'food security and vulnerability profiles' of long-term food security issues (www.fews.org). The Millennium Ecosystem Assessment (2003) has targeted vulnerability of human-ecological systems as a priority area of study. The International Human Dimensions Programme on Global Environmental Change (IHDP), taking stock of the centrality of vulnerability issues in its various initiatives, has also accorded vulnerability high priority as a cross-cutting issue (Lonergan, 1999; *IHDP Update*, 2001). The new international sustainability science initiative now gathering momentum has adapted vulnerability as an archetypal problem and accorded it major priority (USNRC, 1999; Kates et al, 2000; Research and Assessment Systems for Sustainability Program, 2001).

These are high expectations and priorities. What is the capability of current scholars and analysts, and the state of accumulated knowledge, to deliver? In the discussion that follows, we argue that:

1   The existing research and assessment cupboard is filled with lots of things, but it is unacceptably cluttered and bereft of an integrative framework of theory and analysis.
2   The truly integrative analysis envisioned by the World Commission on Environment and Development (WCED) has been undermined by the failure to frame the analytic subject as coupled social–ecological systems.
3   Ideological squabbling in the social sciences has created an unproductive divergence between 'structure' and 'agency' in the framing of basic problems.
4   The requisite funding sources and programme for a sustained attack on an overriding set of issues of global importance have been lacking.
5   As a result, the cumulative progress over the past decade on understanding and analysing vulnerability to global environmental change has fallen far short of what could and should have been achieved.

That said, it is also the case that the carry-over from the current state of scholarly knowledge into the arena of assessment and practice has been disappointingly short of what has been possible.

The goals, particularly for the next stage of work, must be realistic. Systematic studies of vulnerability cannot be expected to result in a quantitative understanding equivalent to that of the forcing functions in the scientific arena of global environmental change; but they can provide substantial qualitative understanding of the vulnerability of particular ecosystems, peoples and places.

DEFINITIONS

The term 'vulnerability' derives from the Latin root *vulnerare*, meaning 'to wound'. Accordingly, vulnerability in simple terms means 'the capacity to be wounded' (Kates, 1985). Chambers (1989) elaborated this notion by describing vulnerability as 'the exposure to contingencies and stress, and the difficulty in coping with them'. Building upon Cutter (1996), Box 14.1 arrays the numerous meanings that have been attached to the concept of vulnerability. Here we define vulnerability simply *as the degree to which a system or unit (such as a human group or a place) is likely to experience harm due to exposure to perturbations or stresses*. It is apparent from relating the notion of vulnerability to a basic structural model of hazard (see Chapter 12 in this volume) that three major dimensions of vulnerability are involved in the evolution of hazard:

1 *exposure* to stresses, perturbations and shocks;
2 *sensitivity* of people, places and ecosystems to the stress or perturbation, including their capacity to anticipate and cope with the stress; and
3 *resilience* of the exposed people, places and ecosystems – that is, their ability to recover from the stress and to *buffer* themselves against and *adapt* to future stresses and perturbations.

These dimensions also suggest the types of indicators needed to develop maps of vulnerability or 'hot spots.' Box 14.2 provides definitions of key concepts relevant to the discussions that follow in this chapter.

FOUNDATIONAL THEORY: Sen, Holling and Chambers

Although the vulnerability literature encompasses a diverse collection of writing and contributors, it also exhibits a remarkable degree of commonality among a few basic concepts and approaches. This commonality arises from the influential work of the economist Amartya Sen, the ecologist C. S. Holling and the developmental theorist Robert Chambers. It is important, therefore, to review briefly these ideas before distinguishing among the major existing conceptual approaches to vulnerability.

Box 14.1 Selected definitions of vulnerability

Gabor and Griffith (1980)
Vulnerability is the threat (from hazardous materials) to which people are exposed (including chemical agents and the ecological situation of the communities and their level of emergency preparedness). Vulnerability is the risk context.

Timmerman (1981)
Vulnerability is the degree to which a system acts adversely to the occurrence of a hazardous event. The degree and quality of the adverse reaction are conditioned by a system's resilience (a measure of the system's capacity to absorb and recover from the event).

Susman, O'Keefe and Wisner (1983)
Vulnerability is the degree to which different classes of society are differentially at risk.

Kates (1985)
Vulnerability is the capacity to suffer harm and react adversely.

Pijawka and Radwan (1985)
Vulnerability is the threat or interaction between risk and preparedness. It is the degree to which hazardous materials threaten a particular population (risk) and the capacity of the community to reduce the risk or adverse consequences of hazardous materials releases.

UNDRO (1988)
Vulnerability is the degree of loss to a given element or set of elements at risk resulting from the occurrence of a natural phenomenon of a given magnitude.

Bogard (1989)
Vulnerability is operationally defined as the inability to take effective measures to insure against losses. When applied to individuals, vulnerability is a consequence of the impossibility or improbability of effective mitigation and is a function of our ability to detect the hazards.

Chambers (1989)
Vulnerability refers to exposure to contingencies and stress, and difficulty in coping with them. Vulnerability thus has two sides: an external side of risks, shocks and stress to which an individual or household is subject; and an internal side which is defencelessness, meaning a lack of means to cope without damaging loss.

Mitchell (1989)
Vulnerability is the potential for loss.

Liverman (1990)
Distinguishes between vulnerability as a biophysical condition and vulnerability as defined by political, social and economic conditions of society. Vulnerability is defined both in geographic space (where vulnerable people and places are located) and in social space (who in that place is vulnerable).

Downing (1991b)

Vulnerability has three connotations: it refers to a consequence (e.g. famine) rather than a cause (e.g. drought); it implies an adverse consequence; and it is a relative term that differentiates among socioeconomic groups or regions, rather than an absolute measure of deprivation.

Dow (1992)

Vulnerability is the differential capacity of groups and individuals to deal with hazards based on their positions within physical and social worlds.

Smith (1992)

Risk from a specific hazard varies through time and according to changes in either (or both) physical exposure or human vulnerability (the breadth of social and economic tolerance available at the same site).

Alexander (1993)

Human vulnerability is a function of the costs and benefits of inhabiting areas at risk from natural disaster.

Cutter (1993)

Vulnerability is the likelihood that an individual or group will be exposed to and adversely affected by a hazard. It is the interaction of the hazards of place (risk and mitigation) with the social profile of communities.

Watts and Bohle (1993)

Vulnerability is defined in terms of exposure, capacity and potentiality. Accordingly, the prescriptive and normative response to vulnerability is to reduce exposure, enhance coping capacity, strengthen recovery potential and bolster damage control (i.e. minimize destructive consequences) via private and public means.

Blaikie et al (1994)

By vulnerability we mean the characteristics of a person or group in terms of their capacity to anticipate, cope with, resist and recover from the impact of a natural hazard. It involves a combination of factors that determine the degree to which someone's life and livelihood are put at risk by a discrete and identifiable event in nature or in society.

Bohle et al (1994)

Vulnerability is best described as an aggregate measure of human welfare that integrates environmental, social, economic and political exposure with a range of potential harmful perturbations. Vulnerability is a multi-layered and multidimensional social space defined by the determinate, political, economic and institutional capabilities of people in specific places at specific times.

Cannon (1994)

Vulnerability is a measure of the degree and type of exposure to risk generated by different societies in relation to hazards. Vulnerability is the characteristic of individuals and groups of people who inhabit a given natural, social and economic space, within which they are differentiated according to their varying position in society into more or less vulnerable individuals and groups.

Dow and Downing (1995)
Vulnerability is the differential susceptibility of circumstances contributing to vulnerability. Biophysical, demographic, economic, social and technological factors such as population ages, economic dependency, racism and age of infrastructure are some factors that have been examined in association with natural hazards.

Cutter (1996)
Vulnerability is conceived as both a biophysical risk as well as a social response, but with a specific areal or geographic domain. This can be geographic space, where vulnerable people and places are located, or social space – who in those places is most vulnerable?

Vogel (1998)
Vulnerability is, perhaps, best defined in terms of resilience and susceptibility, including such dimensions as physical, social, cultural and psychological vulnerability and capacities that are usually viewed against the backdrop of gender, time, space and scale.

IFRC (1999a)
Vulnerability can be defined as the characteristics of a person or group in terms of their capacity to anticipate, cope with, resist and recover from the impact of a natural or man-made hazards.

UNEP (1999)
Vulnerability is a function of sensitivity to present climatic variability, the risk of adverse future climate change and capacity to adapt ... the extent to which climate change may damage or harm a system; vulnerability is a function of not only the system's sensitivity, but also its ability to adapt to new climatic conditions.

Adger (2000)
Individual and collective vulnerability and public policy determine the social vulnerability to hazards and environmental risks, defined here as the presence or lack of ability to withstand shocks and stresses to livelihood.

Cutter et al (2000)
Broadly defined, vulnerability is the potential for loss of property or life from environmental hazards.

IPCC (2001)
Vulnerability is defined as the extent to which a natural or social system is susceptible to sustaining damage from climate change. Vulnerability is a function of the sensitivity of a system to changes in climate and the ability to adapt the system to changes in climate. Under this framework, a highly vulnerable system would be one that is highly sensitive to modest changes in climate.

Source: adapted and updated from Cutter (1996, pp531–532)

Box 14.2 Definitions of key words

Vulnerability: the degree to which a person, system or unit is likely to experience harm due to exposure to perturbations or stresses.

Exposure: the contact between a system and a perturbation or stress.

Sensitivity: the extent to which a system or its components is likely to experience harm, and the magnitude of that harm, due to exposure to perturbations or stresses.

Resilience: the ability of a system to absorb perturbations or stresses without changes in its fundamental structure or function that would drive the system into a different state (or extinction).

Stress: cumulating pressure on a system resulting from processes within the normal range of variability, but which, over time, may result in disturbances causing the system to adjust, adapt or be harmed.

Perturbation: a disturbance to a system resulting from a sudden shock with a magnitude outside the normal vulnerability.

Adjustment: a system response to perturbations or stress that does not fundamentally alter the system itself. Adjustments are commonly (but not necessarily) short term and involve relatively minor system modifications.

Adaptation: a system response to perturbations or stress that is sufficiently fundamental to alter the system itself, sometimes shifting the system to a new state.

Hazard: the threat of a stress or perturbation to a system and what it values.

Risk: the conditional probability and magnitude of harm attendant on exposure to a perturbation or stress.

Amartya Sen and entitlement theory

'Why hunger?', Sen asks. Given the enormous expanse of productive power in agriculture, he notes, it is certainly possible to guarantee adequate food for all, and yet chronic hunger and severe famine persist. In 1977, Sen debuted his theory of *entitlements*, which he elaborated shortly thereafter in *Poverty and Famines* (Sen, 1981). Put succinctly, the work is a deep analysis of the crucial roles of human endowment and exchange entitlements. Sen ascribed the causal roots of entitlement failures as far removed from food production, residing instead in the social and economic system that governs the rights of people to exercise command over food and other necessities of life.

Sen contends that famine can occur without any loss in food availability (although often a reduction in food availability coalesces with entitlements' failure to create famine conditions). In his view, access to food (food entitlement) arises from the ability to command food through legal and customary means. This theory of entitlement includes production of one's own resources, as well as exchanges and transfers through labour and markets in an exchange economy. According to Sen (1977), in an exchange economy, whether a family will starve depends, on the one hand, upon its *endowments* – what it has to sell (products, labour) – and upon whether it can sell what it has (opportunities for exchange) and at what prices (what the market will bear). The risk of starvation also depends, on the other hand, upon how much a family has to pay for food compared with its endowments. Thus, aspects of the exchange economy create family-specific abilities or lack of them (based on endowments), as well as externalities to which a family must respond (the price it must pay for food). Together, these elements create a 'space' of famine vulnerability within which a household may find itself.

Exchange *entitlements* depend not merely upon the relevant exchange rates, but also upon market imperfections and other institutional barriers and upon the actual ability to sell or buy the commodities in question (Sen, 1981). Importantly, exchange entitlements depend upon various institutional arrangements, in any given social or economic system, that affect people's command over commodities. Such institutional arrangements are based on rules governing 'the rights that people ... have to exercise command over food and other necessities' (Sen, 1981, p375). The importance of exchange entitlements lies in their involvement at critical points in the chain of famine causation. In this sense, Sen's entitlement theory provides a structure for analysing famines rather than presenting a particular theory of 'ultimate' explanation.

Entitlements, Sen maintains, represent the set of alternative commodity bundles that a person can command in a society using the totality of rights and opportunities on hand (Sen, 1984). The rights and opportunities, including what one owns and what one earns, he terms 'endowments'. In Sen's theoretical framework, endowments serve to create entitlements that broaden one's power over commodities. Entitlement mapping is a key way to outline and understand the relation between endowments and entitlements. It provides an analytical schema of how entitlements arise from endowments and, moreover, how such entitlements reside at critical points in the chain of famine causation or avoidance.

Entitlement theory provides a powerful entry into the ways in which social relations, economic systems and individuals create disasters out of moderately risky situations. The structures within economic systems, institutions and society more generally become a primary tool to explain why certain people – and not others – experience disasters in particular areas at particular times. Entitlement theory also suggests that much of vulnerability is socially controllable. In other words, coping and responses to perturbations are often more instrumental in determining the eventual toll of the disaster than is the magnitude of the perturbation.

Various researchers have extended or expanded Sen's theory. Leach et al (1999), for example, see Sen's original theory as too restrictive since it focuses primarily upon the command over resources derived through market channels, reinforced by formal legal property rights. Instead, they argue, many ways of gaining access to and control over resources exist beyond the market, and many ways of legitimating such access and control reside outside of formal legal institutional mechanisms (Gore, 1993). Furthermore, these routes of access and control can be more important influences on entitlements than the formal legal institutions that serve to dampen or magnify the vulnerability of different people at various times. Accordingly, Leach et al (1999) extend Sen's entitlement analysis to emphasize the role of social institutions in mediating environment–society relationships at various scales. They argue that a set of interacting and overlapping institutions, both formal and informal, which are embedded in the political and social life of an area, mediate access to and control over resources. In this context, social institutions act as social contracts and relationships, including government and legal institutions, informal social relations and kinship networks.

C. S. Holling and resilience

Theoretical and empirical ecology has inherited much from classical physics and, according to Holling (1978), this legacy has created the tendency to focus on constancy in defining and understanding ecological systems. Ecosystem stability has often been assumed and change has had to be explained (van der Leeuw, 2000). Holling (1986) maintains that we can better understand ecosystems if we shift the emphasis towards assuming change and then try to explain stability. He argues that ecosystems continually confront the unexpected, so that the constancy of the system's behaviour or condition state is less important than the persistence of the relationships that define the system. Consequently, Holling's theory of ecological resilience focuses on the existence of those relationships. As he puts it, there are two sciences – the science of parts where analysis centres upon specific processes that affect specific variables, and the science of the integration of parts (Holling, 1996a). Individuals die, populations disappear and species become extinct; but ecosystems endure only if relationships among ecosystem elements – both biotic and abiotic – persist. With the concept of dynamic stability and resilience theory, he sees sets of disturbances at different temporal and spatial scales not as problems but as an integral part of the development and dynamics of ecosystems. Disturbances open up opportunities for ecosystem renewal and reorganization, development and evolution (Holling, 1996b).

Ecosystem resilience – the capacity to buffer or absorb disturbance – is fundamentally important because it makes system reorganization after disturbance possible (Nyström and Folke, 2001). Because the ability of ecosystems to reorganize after disturbance allows them to continue providing life-support services to humans and other species, resilience is the key to sustaining the flow of life-support services, even as ecosystems evolve and change (Folke et al, 2003).

Resilience is predicated on redundancy and the overlap of function in systems. Resilience refers to the size of the valley, or basin of attraction, around a state, which corresponds to the maximum perturbation that can be taken without causing a shift to an alternative stable state (Scheffer et al, 2001, p591). Simplification of ecosystems can cause a loss of resilience. Human-dominated ecosystems, which now make up a large part of the Earth, tend to be simpler and are often more efficient at producing goods. Consequently, however, they are also likely to be less resilient than non-human-dominated systems. Loss of resilience increases the likelihood of ecosystems shifting into entirely different states (Nyström et al, 2000; Jackson et al, 2001). Abrupt shifts among very different stable domains are plausible in regional ecosystems.

Human-dominated systems may reduce resilience because such systems are invariably managed to suppress, circumvent or remove many natural disturbances, while foreign disturbances are introduced. Paine et al (1998) have termed these fundamental changes in disturbance regimes 'compounded perturbances'. Management of human–environment systems for efficiency with elemental changes in disturbance regimes could lead to greater surprise and shocks, and potentially even interruption of life-support services. Indeed, Holling stresses the inherent unknowability and unpredictably to sustaining the foundations for functioning systems of people and nature (Holling, 1996a, 1996b, 1997).

In Holling's world of oscillating and ever-changing systems, the notion of resilience to perturbations is a key determinant of the persistence of systems. Resilience determines the persistence of relationships within a system and is a measure of the ability of a system to absorb perturbations before the system changes its structure (Holling and Meffe, 1996, p330). In this definition, resilience is a property of the system, and a system's persistence (versus its probability of extinction) is the outcome of its resilience, and the 'golden rule' of natural resource management is to strive to retain critical types and ranges of natural variation in ecosystems (rather than reducing or controlling such variation) (Holling and Meffe, 1996, p334).

Holling's theory of resilience has been highly influential in vulnerability studies in the human and ecological sciences; but, unfortunately, few efforts in the vulnerability stream of work have fully exploited the detail, nuances and more recent developments of the theory. Although the term 'resilience' is now routinely used to describe the capacity of people and societies to cope with and recover from impacts, ideas of stability versus resiliency, brittleness, co-evolution of society and ecology, reorganization, and renewal – all central concepts in resilience theory – still await full incorporation by vulnerability analysts outside the Resilience Alliance. Yet, clearly this work is a strong motive force towards integrated ecological–social analyses.

Robert Chambers: Coping strategies, empowerment and development

Robert Chambers, a development theorist, has written perceptively on issues of vulnerability in a development context and particularly enriched the analysis of coping and adaptability. Distinguishing between poverty and vulnerability, Chambers (1989, p1) characterizes vulnerability as 'defencelessness, insecurity, and exposure to risk, shocks and stress'. Accordingly, in his oft-cited view, vulnerability has two sides: 'an external side of risks, shocks and stress to which an individual or household is subject; and an internal side which is defencelessness, meaning a lack of means to cope without damaging loss' (Chambers, 1989, p1). He also takes a broad view of such losses, including their cumulative nature, as entailing becoming physically weaker, economically impoverished, socially dependent, humiliated or psychologically harmed.

Chambers (1997) also emphasized that participatory assessment approaches enable the vulnerable to express their multiple realities. As these realities are local, complex, diverse, dynamic and unpredictable, the participatory approach uses multiple, subjective indicators of poverty status that emerge out of the experience of the poor, collected through participatory techniques. Chambers argues that participatory assessment methods seek to empower the vulnerable – women, minorities, the poor, the weak and the disadvantaged – to enact reversals in who has power. As such, participatory research methods offer a better means for assessing poverty and capturing what people themselves identify as its principal dimensions and indicators. Moreover, they help to identify what the poor have rather than what they do not have, and in so doing focus on their assets, including tangible assets such as labour and human capital, productive assets such as housing, and largely invisible intangible assets such as household relations and social capital. To reduce vulnerability, people can make certain substitutions among different assets. For example, households that keep children in school rather than send them out to work are poorer in income terms; however, in the longer term, their strategy can reduce vulnerability through consolidating human capital as an asset.

The main issue here is that large stocks of one kind of assets may be of little use; the more assets people command in the 'right mix', the greater their capacity to buffer themselves against external shocks (see also Moser, 1998, and Bebbington, 1999). Chambers (1997) also distinguishes three different types of agriculture worldwide, each of which has a different vulnerability to global environmental change:

1   'first agriculture', with high-technology inputs, in industrialized countries;
2   'second agriculture': the Green Revolution areas, in developing countries; and
3   'third agriculture', with low mechanization but crucially important to the poor, in the South.

The last-mentioned, which includes most of the poorest and most food-insecure people of the world, is, Chambers argues, the critical group for mounting attacks on global patterns of vulnerability.

Chambers also accords particular attention to the diversity of vulnerability and coping strategies of the poor. Contrary to popular misconception, he sees such strategies as complex, diverse and activist in the face of threat. Most poor people, he argues, do not put their eggs in one basket but act to reduce risks, to increase their adaptability and to seek greater autonomy, a line of argument, it might be noted, that is prevalent in the literature on pastoral nomadism (Johnson and Lewis, 1995). Like hazard theorists, he sees the essence of such coping as creating and maintaining wider options, particularly through the willingness and ability of household members 'to do different things in different places at different times' (Chambers, 1989, p2). Generally, poor people use complex investment strategies in coping pools, seek to diversify their portfolios of assets in order to better handle varying contingencies and bad times and, over time, to strive to minimize irreversible losses. All the same, as Chambers notes, external interventions during crises often come too late, after poor people have become poorer by disposing of productive assets or taking on debts or obligations that threaten the security of their livelihoods.

Noting the lack of a developed theory and accepted indicators and methods of measuring vulnerability, Chambers (1989, p6) set out key research needs to guide vulnerability work:

•   developing 'simple and sure' methods for enabling vulnerable people to analyse their conditions and identify priorities;
•   developing and testing indicators of vulnerability;
•   assessing the modes, costs and benefits of reducing vulnerability and preventing impoverishment, compared with recovery actions;
•   assessing and comparing coping strategies under stress, including sequences of response, thresholds of different types of responses, and the value and use of different assets;
•   examining the effects of civil disorder on vulnerability and coping strategies, including effects on economy and household strategies;
•   evaluating relief and development policies, including ways of strengthening people's current coping strategies; and
•   delineating the effects on adult disability and death on household viability, strategies and behaviour.

This priority list has lost none of its relevance over time and still demarcates essential lines of enquiry for advancing our understanding of vulnerability and, particularly, of coping under stress.

CONCEPTUAL APPROACHES

Analyses of vulnerability reflect a wide diversity of approaches, ranging from those narrowly based in ecology, to others centred on vulnerability as a component of hazards, to yet others that see vulnerability primarily as an expression of political economy. These largely divergent perspectives on vulnerability have yielded a highly fragmented literature, unduly vitriolic disciplinary and ideological debate, and sparse empirical testing and application of the competing frameworks. Only recently has movement to bridging these differences become apparent, particularly in the work of the Resilience Alliance (Gunderson et al, 1995; Berkes and Folke, 1998; Folke et al, 2003), but elsewhere as well. Here we overview the major conceptual approaches to vulnerability that have emerged over time and their salient elements.

Hazard and risk analysis

The creation of a field of systematic hazard analysis is a phenomenon of only the past 50 years; prior to that, individual dangers and disasters were examined and probed, but comparative analyses and formal methods of analysis were few. Gilbert F. White's *Human Adjustment to Floods* (White, 1945) was a particular benchmark in stimulating subsequent work, owing to its careful analysis of the growing toll of floods in the US and its concern to examine the range of human behaviour and coping exercised in managing risks. Not until several decades later, however, did White (1961) develop a broader conceptual approach aimed at enlarging the array of coping measures used in efforts to reduce the toll of natural hazards upon society.

Stimulated by White's early work, extensive research on natural hazards during the 1960s and 1970s created a set of concepts and formats for analysing natural hazards and human responses to them (White, 1974; White and Haas, 1975). The basic construct is that human–nature interactions produce both resources and hazards. The critical agent is human-induced change, which transforms nature into beneficial and threatening outcomes. Interactions lead to disasters when hazards are extreme in magnitude, exposed populations are large and the human use system is particularly vulnerable. This recognition that hazard is a joint product of events (or perturbations), degree of exposure and what now would be called sensitivity or susceptibility is a construct still very current, and which the IPCC efforts took many years to recognize. The approach to vulnerability in this early work on natural hazards failed to provide an in-depth conceptualization of vulnerability; but it did offer an extended treatment of coping, termed *adaptive capacity* (a combination of resistance and adaptive capacity), and social resilience was also very much part of the analysis (Burton et al, 1978, 1993). Social resilience, for its part, was seen as a mix of coping measures – individual and collective, incidental and purposeful – and adjustments and adaptations that make up society's interaction with the flow of hazards that confronts society. This work remains valuable for its formats for assessing human behaviour around coping actions and its attention to broadening the range of human choice in hazard management. What the body of work missed substantially was the social, economic and political structures that shape patterns of human vulnerability and constrain choice – issues taken up by the political economy school who focused on social vulnerability.

The several decades since the first edition of *The Environment as Hazard* have witnessed the evolution of a diverse body of theory and empirical analysis treating environmental and technological risk (Burton et al, 1978). Some of this work has probed aspects of vulnerability in great depth, such as the landmark assessment of differential susceptibility of humans to chemicals by Calabrese (1978), the role of organizational structure and technological complexity in the proneness of certain industrial systems to catastrophic failure (Perrow, 1984, 1999), the insecurity of energy systems to economic and political change (Khatib, 2000; UCS, 2002), the vulnerability of critical societal infrastructure to catastrophic natural events and terrorism (Haimes, 1990, 1998), and the obstacles facing communities in long-term recovery from industrial disasters (Erikson, 1994; Mitchell, 1996, 2000). Although not specifically focused on vulnerability, diverse theoretical work in hazards research provides important entry points into the cultural, social and political processes operating in society's encounters with environmental change. Cultural theory (Douglas and Wildavsky, 1982; Thompson et al, 1990; van Asselt and Rotmans, 1995), for example, offers a broad interpretation into how cultural biases enter into the types of hazards that are addressed and the types of coping and management systems that are employed. The 'social amplification of risk' (see Chapter 6, Volume I) provides a conceptual framework for examining the social and political processes by which different societies process threats and risks, allowing some (and their impacted peoples or ecosystems) to grow while assiduously reducing others. A large body of empirical work now is available documenting these 'amplification' and 'attenuation' processes (see Chapter 11, Volume I). Social processes generating 'hidden hazards' and vulnerable groups from society's scrutiny and action have also been delineated (see Chapter 7, Volume I). Gender issues associated with differential public concern and response to risk have been documented (Flynn et al, 1994), and broader social theory probing the splintering of societies and the world economy into risk winners and risk losers, and a dynamic leading to ever-increasing environmental degradation in the risk society have been set forth (Beck, 1992, 1995, 1999).

If hazard research was the focus of the first generation of natural hazards studies, critical theory and political ecology have received much attention over the past decade, and it is to this perspective we now turn.

Critical theory to political ecology

Vulnerability would appear to be a logical outgrowth of natural hazards approached through the lens of Marxian and, more broadly, critical approaches. Their intricacies and complexities notwithstanding, critical approaches are predicated on various assumptions that emphasize the socioeconomic structures that control individual and group action and that captured attention in early studies on natural hazards. Thus, natural hazards are to be understood in terms of the social conditions (political economy) that place people in harm's way (e.g. living on hillsides that are prone to landslides) and that reduce their coping capacity for managing the hazard (Hewitt, 1983). In this sense natural hazards (environmental perturbations and stressors) are a social construction in which different social units are differentially 'placed in harm's way' and have differential coping capacities. These approaches are grounded in the belief that no individual or group would voluntarily choose a more hazardous or risk-prone setting, and do so only when other options are unavailable or, in some cases, when social safety nets that provide a buffer to the risks taken disappear (e.g. insurance against floods). The more marginal economically and the weaker politically, the fewer the options, the more likely is the environment to be hazardous, and the greater are the difficulties in coping with stresses and perturbations (e.g. Wisner, 1988, 1993a, 1993b). Following this logic, vulnerability, or at least the vulnerability that warrants social concern, is linked to economic and political impoverishment, and so the research lens focuses on social units that are impoverished and what makes them this way (Waddell, 1983; Watts, 1983; Watts and Bohle, 1993).

The sources of impoverishments occur largely within the relations that a social unit maintains with other units, and tracing these relations may lead well beyond the geographical location of an impoverished group (e.g. Wisner, 1993a). Thus, critical approaches seek causal chains or networks of causes, often historical in nature, that more often than not lead to causes that are geographically exogenous to the social unit and linked to it through layers of connections that obfuscate the connectivity (Blaikie and Brookfield, 1987). For example, in south-central Mali, increased vulnerability of villages to catastrophic fires follows from state-directed fire policies that are, in turn, a response to international concerns about the deforestation and 'desertification' of the Sahelian fringe that find their way into various international accords and, ultimately, the conditions for international loans to the country in question (Laris, 2002). For much of the economically 'have-not' world, these relations are traceable to colonialism and are redirected and amplified under 'globalization'.

These kinds of orientations in vulnerability work join those on land change, sustainability, environmental justice, indigenous knowledge and rights and feminist concerns under the heading of 'political ecology' (Blaikie and Brookfield, 1987), an interdisciplinary field directed to human–environment relationships following a loose 'structural' approach that share a common suspicion of what they term 'mainstream' approaches (Turner, 1991). Vulnerability studies make up a prominent subset of research in this field (Wisner, 1993b; Blaikie et al, 1994; Dow and Downing, 1995; Ribot, 1995; Vogel, 1998, 2001). The name of the subfield notwithstanding, the intellectual origins of this line of vulnerability studies shape its questions and focus its analysis primarily on 'social vulnerability' (Ribot, 1995, 1996). Rarely does it address ecological or environmental vulnerability independent of its resource implications for the human occupants, despite complementary work on environmental entitlements (Leach et al, 1999).

Bohle's revised framework (Bohle, 2001; see also Watts and Bohle, 1993) illuminates these characteristics. For Bohle, three main approaches – internal–external, coping assets and conflict crisis – are central for understanding vulnerability. This last approach, of course, epitomizes critical and Marxian perspectives. Bohle (2001) credits Chambers (1989) and Giddens (1996) with enlarging vulnerability studies in political ecology beyond external and structural considerations to include internal and agent-based ones (but does not access the extensive literature outside political ecology on these themes). Likewise, vulnerability assessment is rich in treating coping capacity, although an in-depth conceptualization awaited Sen's entitlement theory (Drèze and Sen, 1989; Sen, 1981, 1990), which is consistent with the base precepts of political ecologists in its emphasis on the role of social structures in defining entitlements (Watts and Bohle, 1993; Ribot, 1995). These issues notwithstanding, the latest work by Bohle (2001) reworks the three-pronged configuration of vulnerability – entitlements, empowerment and political economy – provided by Watts and Bohle (1993) to a more integrative approach, as discussed in the following section. The focus of the political economy/political ecology school, however, remains strongly aimed at vulnerability of the social unit (e.g. Blaikie et al, 1994; Bohle et al, 1994; Ribot, 1995), and it is not clear where critical theory/political ecology approaches rest with regard to coupled human–environment systems or ecological systems without human considerations.

INTEGRATIVE ANALYSES

An encouraging sign in vulnerability research and assessment is the gradual emergence of more integrative approaches. Bohle (2001), drawing upon an impressive array of research by his graduate students over a decade, has proposed a research strategy that structures analysis, following Chambers, to focus on both the 'external' and 'internal' dimensions of vulnerability. The work of Downing (Downing, 1991a, 1991b; Bohle et al, 1994) stands

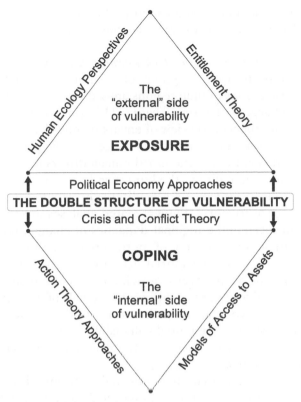

Source: adapted from Bohle (2001, p4)

Figure 14.1 *A conceptual model for analysing vulnerability*

out in regard to the high degree of integration that it achieves across ideological and theoretical perspectives. In addition, several recent analyses have sought to bridge the divide between ecological and social analysis and to suggest more integrative pathways for assessing vulnerability. Here we note in brief several of the more prominent of these.

Bohle (2001) has set forth a template – less than a model – that covers both external and internal dimensions of vulnerability, and the interactions between them (see Figure 14.1). The first theoretical strand focuses upon the interactions between 'structure' and 'agency', with the question still open as to the explanatory power of each. A second strand focuses upon the assets that people control that allow them to buffer themselves from stresses and perturbations. Such control is presumably closely linked with the political economy of the region and the ways in which various populations are 'embedded' in the basic structures and dynamics of economy and polity. The third strand – crisis and conflict theory – treats tensions over risk and criticality and seeks to solve conflicts in coping actions. This approach takes important strides to identify and illuminate the social structural component of vulnerability, but accords relatively little attention to the human component or the coping potentials of those affected and largely externalizes the environment (WBGU/GACGC, 2000, p184).

In *Regions at Risk: Comparisons of Threatened Environments* (Kasperson et al, 1995), an interdisciplinary group of natural and social scientists examines the trajectories of nine regions around the world with regard to vulnerability and degradation (see Chapter 11 in this volume). Eschewing a geocentric or anthropocentric view of environmental change, the authors centre their analysis on environmental change over the past 50–100-year period, in each region linking the broad human drivers of environmental change and natural variability to the human and ecological outcomes that they produce. These outcomes are qualitatively classified into levels of endangerment or 'criticality'. Disaggregated analyses are then used to identify the socioeconomic regional dynamics of change, ranging from social polarization, exploitation of peripheries (marginalization), to 'trickle-down' processes. Substantial attention is given to contextual effects and what the authors term 'a rich tapestry of human causation'. Ample evidence is offered of the differential speeds of ecological change and human response systems, propelling many of these systems towards trajectories of overshoot. Differential vulnerability is heavily implicated in the nature of human driving forces and the structure and effectiveness of human response systems.

The authors conclude that environmental 'criticality' is a function of the speed and intensity of environmental degradation, the vulnerability of people and ecosystems affected, and coping capacities and resilience (see Chapter 11). They also conclude that environmental criticality emerges historically through a series of stages in which the decisive attributes are the regenerative capacities of affected ecosystems and the buffering and mitigative costs incurred by affected societies. They assign particular importance to the speed of progression through these stages and the extent to which non-linearities, punctuated surprises and threshold effects are present.

A highly ambitious analysis that internalizes vulnerability into a broader set of dynamics, linking stresses and effects across ecological and social science domains, is the work on *syndromes of global change* undertaken at the Potsdam Institute for Climate Impact Research (www.pik_potsdam.de; Schellnhuber et al, 1997; Petschel-Held et al, 1999; WBGU/GACGC, 2000, pp176–185). This approach at understanding global change seeks to be interdisciplinary, comprehensive and integrative. The basic intent is to describe global change by assessing 'archetypical, dynamic, co-evolution patterns of civilization–nature interactions', termed *syndromes* (Schellnhuber et al, 1997, p23). These patterns represent different sub-dynamics of global change, which are modelled by the use of qualitative differential equations. The syndromes exhibit various *symptoms*, those qualities of global change that figure prominently in the ongoing problematic developments worldwide both in the natural environment and in society. Some 80 symptoms (e.g. urban sprawl, the growing significance of non-governmental organizations, or NGOs, terrestrial run-off and increasing mobility) make up 16 syndromes (see Box 14.3). A major goal is to detect geographical patchworks that sufficiently characterize syndromes on a global scale. The early analyses suggest promise not only for interrelating stresses, dynamics and effects, but also suggest how key vulnerabilities enter. To reach the 'syndrome' level, however, requires significant aggregation of highly complex components and processes operating in diverse ways. A critical issue for this approach is that inherent in all global aggregation: how to reconcile generic properties or propositions with sufficient specificity so that they hold true at the regional scale.

Finally, a nearly decadal effort of work by ecological and social scientists under the auspices of the Resilience Alliance (www.resalliance.org) has generated a significant body of integrated work focused on coupled social–ecological systems. Enlisting theories of resilience and the evolution of ecosystems, this group has constructed a model of system co-evolution. This conceptual model and resilience theory together form the foundations of Holling's extended theory of 'panarchy', a framework for understanding the co-evolution of ecological and social systems (Gunderson and Holling, 2001). This model envisions a nested set of adaptive cycles, arranged as a dynamic hierarchy in space and time, which provide the novelty and persistence required for the sustainability of both human and ecological systems. According to the concept of panarchy, an ecological community passes through four phases:

1   the *conservation phase*, with its great connectedness, energy stored in biomass and low leakage of nutrients;
2   the *release phase*, when some dramatic event disturbs the consolidated biomass, food web and nutrient cycles;
3   the *reorganization phase*, which is the most formless, with free-floating nutrients, less trapped energy and many open niches; and
4   the *exploitation phase*, in which species that can exploit these opportunities invade and increase, and, over time, lock in new food webs and nutrient cycles (Holling, 1998, p33).

Figure 14.2 depicts a stylized representation of the four phases.

The exit from the cycle is the stage where the potential can leak away and where a flip into a less productive and less organized system is most likely. The model, furthermore, contributes to Holling's five paradigms of nature: *nature cornucopian*; *nature anarchic*; *nature balanced*; *nature resilient*; and *nature evolving* (Holling, 1998, p34).

Members of the Resilience Alliance have drawn upon studies that have explored alternative stability domains in different ecosystems, including

Box 14.3 Syndromes of global change

Utilization syndromes

Sahel syndrome	Overuse of marginal land
Overexploitation syndrome	Overexploitation of natural ecosystems
Rural exodus syndrome	Degradation through abandonment of traditional agricultural practices
Dustbowl syndrome	Non-sustainable agro-industrial use of soils and bodies of water
Katanga syndrome	Degradation through depletion of non-renewable resources
Mass tourism syndrome	Development and destruction of nature for recreational ends
Scorched Earth syndrome	Environmental destruction through war and military action

Development syndromes

Aral Sea syndrome	Damage of landscapes as a result of large-scale projects
Green Revolution syndrome	Degradation through the transfer and introduction of inappropriate farming methods
Asian Tiger syndrome	Disregard for environmental standards in the course of rapid economic growth
Favela syndrome	Socioecological degradation through uncontrolled urban growth
Urban sprawl syndrome	Destruction of landscapes through planned expansion of urban infrastructures
Disaster syndrome	Singular anthropogenic environmental disasters with long-term impacts

Sink syndromes

Smokestack syndrome	Environmental degradation through large-scale diffusion of long-lived substances
Waste dumping syndrome	Environmental degradation through controlled and uncontrolled disposal of waste
Contaminated land syndrome	Local contamination of environmental assets at industrial locations

The names of the syndromes refer to functional patterns or characteristic features. These patterns of non-sustainable development can be grouped according to basic human use (and misuse) of nature: as a source for production, as a medium for socioeconomic development and as a sink for civilization's outputs.

Source: Schellnhuber et al (1997, p23)

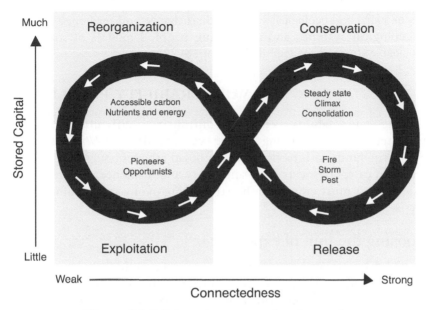

Figure 14.2 *Schematic representation of panarchy*

lakes, coral reefs, woodlands, deserts and oceans, to assess large shifts that are apparent, feedback mechanisms, the role of stochastic events and implications for management strategies (Scheffer et al, 2001). Among the more important findings are that contrasts among states in ecosystems are usually due to a shift in dominance among organisms with different life forms, that stochastic events (outbreaks of pathogens, fires or climate extremes) usually trigger state shifts, and that feedbacks that stabilize different states involve biological, physical and chemical mechanisms. The authors infer a number of management implications. Since systems often lack early-warning signals of massive change, societal attention tends to focus on precipitating events rather than on underlying loss of resilience. But disturbance is a natural part of ecosystem change that promotes diversity and renewal processes. The authors conclude that management efforts to reduce unwanted state shifts need to focus much more on the slow and gradual changes (e.g. land use, nutrient stocks, soil processes and the biomass of long-lived organisms) than on the more dramatic events (e.g. hurricanes, droughts and disease outbreaks) that are inherently more difficult to predict or control.

Finally, it is worth noting the workshop of 44 distinguished scientists convened in 2000 in Tempe, Arizona, in the US to consider the needs and priorities for interdisciplinary environmental research. The results accorded particular attention to the issues surrounding the evolution and resilience of coupled social and ecological systems, and identified five principal research needs in this area: the evolution of social norms regarding the environment; past and future land-use change; feedback loops in social and ecological systems; disturbance and resilience in social and ecological systems; and developing coupled models of social and ecological systems (Kinzig et al, 2000, pp18–22).

STRUCTURING THE VULNERABILITY PROBLEM

Analysing vulnerability entails confronting difficult issues in conceptualizing the framing of the vulnerability *problematique*. The literature abounds with treatments of a broad array of issues that fall into three main categories: exposure, sensitivity and resilience. A series of key considerations must enter into the structuring of any sound vulnerability analysis.

Choosing the unit of exposure

An initial decision involves selecting the exposure unit – who or what is at risk (e.g. a social group, ecosystem or place) – that is to be examined. In reality, many different entities may be at risk; therefore, defining the purposes and scope of the analysis is a critical decision in structuring the vulnerability problem. If the problem is defined principally in terms of understanding the effects of a specific perturbation (e.g. climate change or economic downturn), selection among the wide array of affected systems and causal factors has already narrowed the scope of analysis to a single perturbation or set of perturbations. If the intent is to understand the vulnerability of a particular group, region or ecosystem, then that unit or system can be the starting point; but the analysis can be framed as a multi-stress or multi-perturbation situation. A major pitfall lies in selecting the unit of exposure a priori, assuming that the social groups or ecosystems most at risk are known, and so the analysis ends up missing other highly vulnerable components or exposure units.

The work on the International Decade for Natural Disaster Reduction (UN, 1988; USNRC, 1991; Mileti, 1999) and the early IPCC assessments (Tegart et al, 1993; Watson et al, 1998) amply demonstrates that impact analyses have focused strongly on the vulnerability of natural systems, with some attention to economic implications. By contrast, environment–development approaches concentrated almost exclusively on 'marginalized' human groups and the political economy involved in the 'production' of 'social vulnerability' (Blaikie et al, 1994), particularly influencing the World Food Programme (FIVIMS, 1999, 2000, 2001), the United Nations Environment Programme (UNEP, 1999) and the United Nations Development Programme (UNDP, 2000).

The effect of these different foci has been to drive a wedge between natural and social systems and to conduct segmentary analyses of what are, in essence, *coupled human–environment systems*. The major, and a highly noteworthy, exception to this trend has been the work of the Resilience Alliance, which starts with the essential hypotheses that resilience is important for both ecological systems and social institutions, that the well-being of social and ecological systems are closely linked, and that resilience and flexibility are essential for forging capability to respond over time to surprises and crises (e.g. Berkes and Folke, 1998). The focus on coupled human–environment systems is only now beginning to attract wider attention, but holds promise for more integrated assessments that unite rather than separate the work of social and ecological scientists (Kinzig et al, 2000).

Multiple stresses and vulnerabilities

Much of the past treatment of vulnerability has proceeded using established methods of environmental impact and natural hazards analysis in which the assessment begins with a particular developmental project (e.g. a new technology or facility) or a natural hazard event (e.g. a flood) and traces through the effects, positive and negative, likely to result. Accordingly, the analysis of vulnerability has largely been predicated upon a major type of perturbation or disturbance. Such structuring of the vulnerability analysis has occluded or concealed causal agents that are actually shaping the vulnerability of high-risk social groups, ecosystems and places via multiple socioeconomic stresses and perturbations acting on the unit and the 'normal' or 'everyday' processes that are causing the unit to become more exposed to threats, more defenceless in dealing with the threats, and more prone to damage from the accumulating multiple stresses, perturbations and shocks.

No matter what the unit of analysis, a complex of perturbations and stresses, both biophysical (e.g. climate change and sea-level rises) and socioeconomic and political (e.g. social inequalities, population increases and civil strife), threatens the unit of exposure. An early example to capture such suites in a 'multiple stress model of decision making' at the municipal level is available in Kasperson (1969). These *suites of stresses and perturbations* may interact, register their toll across differing exposure units, and call into play different types and levels of vulnerability. Moreover, the sequencing and temporal character of these suites of stresses shape the interactions and their ultimate impacts. At root, vulnerability assessment needs to grapple with the highly dynamic and interactive flow of multiple stresses and perturbations that can undergo analysis as complexes or constellations of threats. It is an imposing task for current assessment methodologies, which usually involve the typical 'divide and conquer' procedures of the social sciences or risk analysis. In particular, a multiple-perturbation/multiple-vulnerability framing suggests that starting with the vulnerable people and systems and the outcomes to be avoided, working *backwards* through major sources of vulnerability and *towards* suites of stresses may shed more, and perhaps different, light on vulnerability than the working-forward strategy of typical environmental impact assessments.

Cross-scale interactions

Much of the existing body of vulnerability research focuses on specific spatial and temporal scales of analysis, geared to the particular perturbation under study – for example, impacts on agriculture in South Africa from a three-year El Niño Southern Oscillation (ENSO) event (Anyamba, 1997) – or to the availability of data on national-level 'hot spots' (Brklacich and Leybourne, 1999; Lonergan et al, 2000). Meanwhile, a significant dearth of information on effects at other scales presents a significant impediment to more sub-regional and localized analyses and applications.

The analyses in *Regions at Risk* (Kasperson et al, 1995; and see Chapter 11 in this volume), for example, found that in the shift from the global to small regional scales, the impact of population, affluence and technology (IPAT) construct for explaining impacts gave way to a much richer and diverse set of explanatory variables, including multiple types of vulnerability and a greater role for technology (Ausubel, 1996; Chertow, 2001). Other studies of driving forces at finer local scales have found similar results (Dietz and Rosa, 1994; Angel et al, 1998; DeHart and Soulé, 2000; York et al, 2002), and Easterling (1997) has argued at length why regional-scale studies are essential. For political ecologists (Blaikie and Brookfield, 1987, p27), the analysis starts with the 'land or resource managers' but follows the chain of causal connections that invariably leads to larger-scale drivers and processes (e.g. international timber companies operating through state agencies to create pressures on local land). The net effect of pre-selecting particular levels of scale, particular scale interactions or particular 'chains of explanation' is to assume away much of the potential richness of the operation of cross-scale dynamics in the production and amelioration of vulnerabilities. As Wilbanks and Kates (1999, p608) observe: 'Where global change is concerned, it can be argued that a focus on a single scale tends to emphasize processes operating at that scale, information collected at that scale and parties influential at that scale – raising the possibility of misunderstanding cause and effect by missing the relevance of processes that operate at a different scale.' In short, the structuring of vulnerability analysis needs to take account of the dynamics of *cross-scale interactions* – those in which events or phenomena at one scale influence phenomena at other scales (Holling, 1978, 1995; Turner et al, 1990b; Cutter et al, 1996; Gibson et al, 1998; Wilbanks and Kates, 1999; Cash and Moser, 2000).

Causal statements or 'maps' of the sources of sensitivity and resilience invariably are predicated on phenomena at certain levels or scales, or are typically focused on particular scale interactions (usually dyads). The crucial issue in linking scale to causal interpretations is identifying where the variables that explain a pattern or relationship are, in fact, located (rather than assumed) and whether they are fully captured in the scale chosen for the vulnerability analysis. And analysis needs to be sensitive to accelerated scale interactions which, as Jodha (1995, 2001a, 2001b) persuasively argues, are likely to distance resource users from the resource base, to disconnect production from consumption, and to separate the production of knowledge from its applications. Cash and Moser (2000, p113) suggest the use of hierarchy theory to capture scale issues in social–ecological interactions. Such theory views phenomena at a particular scale as the result of both the smaller/faster dynamics of system components at the next lower scale and the constraints imposed by the larger/slower system dynamics at the next higher scale. Understanding cross-scale dynamics, they argue, requires simultaneously capturing *both* the driving and constraining forces at higher and lower scales.

Scale mismatches, such as the mismatch between the environmental system and the jurisdictional scope of the political authority, are a central issue in vulnerability analysis. Driving forces often emanate from macro-forces, institutions or policies set at higher-level scales – land tenure regimes, technological change, international financial institutions and government policy – and are articulated through a finer pattern of local scales with highly variable local resource and ecological settings (Geoghegan et al, 1998; Pritchard et al, 1998). Political jurisdiction links to management by way of projects and 'hot spots,' not via ecosystems or populations. Similarly, the timescale of political institutions stresses business cycles, electoral terms of office and budget processes, rather than spans of biological generation or ecosystem change (Lee, 1993, p63). As a result, many environmental problems become exports to distant places (e.g. deforestation in Indonesia and the international timber trade) or to distant generations (e.g. the disposal of hazardous chemical or radioactive materials). Moreover, the scope of political authority often matches poorly the scope of impacts and vulnerabilities, so that 'transboundary' effects on vulnerable people and places are increasingly common (Linnerooth-Bayer et al, 2001).

Internalizing such exports to the source and scale of problem generation is particularly difficult because vulnerable groups or groups controlling vulnerable ecosystems are commonly economically and politically marginalized. Meanwhile, the lack of cross-scale regimes and institutions encourages *policy pathologies* in which environmental and human systems proceed at difficult paces and at different scales, creating serious mismatches, and conflict, among ecosystems, social processes and institutions (Holling, 1995; Cash and Moser, 2000).

Finally, a word about conjunctions that occur among processes that are operating at different time and spatial scales. Social–ecological systems have a cadence as they pass through seasonal, annual and longer-term variability and fluctuations. They typically maintain rather different rhythms and patterns; but at certain critical times, conjunctions among processes occur and *windows of vulnerability* appear (Dow, 1992, pp432–433). Multiple stresses may then generate high-intensity perturbations and coincide with peaks of vulnerability, creating crises of damage and disturbance in the social–ecological system. Often these emerge as major 'surprises' and as 'shocks' to the institutions and managers of environmental systems (Lawless, 1977; Brooks, 1986; Kates and Clark, 1996; Schneider et al, 1998).

Endogenous perturbations

Studies of natural hazards have mostly focused on the perturbations and stresses that develop outside the affected exposure unit or the human–environment system, as in the case of tornadoes or drought (Glantz, 1988; Cutter, 1996; Wilhite, 2000). Vulnerability assessment has maintained this emphasis inasmuch as the critical stressors identified operate in distant (distal) processes that play out in a particular locale (e.g. international factors drawing down local production systems; Blaikie and Brookfield, 1987; Kasperson et al, 1995). Political-economic and entitlement perspectives that favour themes that locate the plight of the human system (environmental systems receive short shrift in this literature) on social structures beyond the influence of those affected have particularly influenced this direction of work. This emphasis has deflected attention away from those individual and societal actions that, no matter how limited, may exacerbate the perturbations and stressors, and even create new ones (Glantz, 1988).

Endogenous stressors are those emanating within the coupled human–environment system that, with the help of exogenous forces, can accumulate to the tipping point of a perturbation. Salinization in the Tigris-Euphrates (Jacobsen and Adams, 1958), deforestation in the Maya lowlands (Turner, 1983) and on Easter Island (Acharya, 1995), and the Great Plains' Dust Bowl (Worster, 1979; Brooks et al, 2000, pp74-77), for example, are cases in which local land-use decisions apparently precipitated a 'bite-back' on the system. The environmental system, given sufficient time, often recovers (e.g. Maya lowlands; see Turner, 1983), but not without significant societal repercussions. Chapter 11 in this volume indicates that endogenous stressors may become less important as the human wealth of a coupled system provides substitutes for the drawdown of nature, new technologies reduce the impacts of effluents and wastes from high-consumption societies, or both. Nevertheless, a full vulnerability treatment needs to treat endogenous linkages of potential perturbations and stressors.

Cumulative (iterative) vulnerability

These cadences and linkages over time and space suggest that vulnerability is better thought of and structured as an evolving set of processes than as a condition or state, or, as so often happens, a snapshot of an evolving process. Put slightly differently, vulnerability has its own history and its own trajectory, both of which must be understood. Rangasami (1985) has emphasized the problem in entitlement theory of depicting events when social groups collapse into starvation as 'entitlement failures'. She argues instead for viewing vulnerability in terms of long-term socioeconomic and political processes that conspire to keep vulnerable members of the society prone to disaster when perturbations or accumulating stresses register their effects. Bohle et al (1994) have described the sequence of events involved in the occurrence of a food crisis in which a concatenation of events combines with structural vulnerability to produce an emergency or disaster. Building upon their portrayal of interacting forces, they depict vulnerability as a sequential or iterative process in which shifting suites of stresses intersect periodically with evolving vulnerability to draw down coping resources and adaptive capacity at intervals of major perturbations, making the social–ecological system increasingly prone to disaster over time. The coping resources enlisted to recover from crisis events are able to be partially replenished through, for example, the changes in anticipatory behaviour and adaptive strategies or through the acquisition (perhaps across scale) of new entitlements; but the total stock of coping resources in this case is also partially depleted. The production and amelioration of vulnerability, through both 'stores' of coping resources and through social learning from experience, may be viewed as dynamic, uncertain and cumulative in nature.

But stresses are also multiple, complex and cumulative. Consider the case of chemicals in the environment. Some 20,000 pesticide products are on the market in the Toxic Substances Control Act (TSCA) inventory in the US. In addition, some 80,000 chemicals now on the market are in the inventory, and some 2000 chemicals make their way to the market annually. Examining the cumulative ecological and health effects of this mounting chemical inventory is an overwhelming task. Even assessing the effects of the much more limited set of chemicals at a particular site is daunting. Such an analysis needs to take account of the *accumulation* of chemicals in the environment over time, across sources, across multiple routes and pathways of exposure, to people and environmental components with widely differing susceptibilities to damage. Then potential effects from the multiple stressors and exposures must be integrated for the stressors acting together within a coherent picture of risk for the different exposure units (people and ecosystems), and for chemicals whose toxicities we often know little about and even then only at high levels of uncertainty. Still the job is not done, as the chemical threats then need to be related to other substantially different social and economic stresses, and finally to often widely differing vulnerabilities to individual stresses. Recognizing cumulative risk/vulnerability in logical fashion aside, the development of methodologies and approaches required to achieve a sound picture of differential cumulative vulnerability is no easy task.

Causal 'maps' of vulnerability

To understand the incidence and depth of vulnerability is to understand the sources of vulnerability and their causal linkages to the proximate factors that shape the three dimensions of vulnerability – exposure, sensitivity and resilience. In some cases, these roots are exogenous, lying in the basic social and economic structures that shape the differential endowments and access to entitlements of different populations. In other cases, the roots are endogenous in the social–environment system in which internal structures and processes generate both stresses and vulnerabilities that threaten the security of the system. In other cases, the roots lie in management systems and institutions. Management solutions intended to control near-term perturbations, as conventional natural resources management tends to do, may increase the 'brittleness' of institutions and the potential for larger-scale and more devastating disturbances (Holling, 1995). In still other cases, the behaviour of the local resource manager and the capacity to learn and fashion effective adaptive strategies is the key. Most typically, multiple sources of vulnerability exist, and it is necessary to demonstrate and to model – not to assume – analytic approaches to uncover the causal factors and the linkages in enlarging or reducing risk.

Such causal maps of vulnerability, and particularly those that capture social–ecological interactions, are rare and at an early stage of development. Bohle et al (1994, pp39–44) have described the causal structure of vulnerability as involving the intersection of three axes – *human ecology*, the way in which labour transforms nature and generates hazards; *expanded entitlements*, including endowments, social entitlements and empowerment; and *political economy*, the national regimes of accumulation influenced by transnational processes. The triangle represented by connecting these axes is taken to be the 'space of vulnerability' in which particular groups can be located. The framework is really an icon of vulnerability, however, rather than a causal framework as causal linkages are not shown. Downing (1991a), drawing on the work of Hohenemser et al (1985), has developed a causal structure of hunger that, in particular, illustrates connections across scale, domains (scales) ranging from national to individual levels, a simplified causal model of hazard (causes and consequences) and a set of directional linkages (see Figure 14.3).

Turning to energy as a key life-support system, the recent *World Energy Assessment* (Goldemberg, 2000) suggests how extensively energy system-related vulnerabilities are intertwined with a host of social, democratic and economic factors. Worldwide, 2 billion people lack access to electricity and another 2 billion people use traditional solid fuels for cooking. Burning solid fuels in poorly ventilated spaces is estimated to cause about 2 million deaths per year, disproportionately concentrated among women and children in developing countries (Smith, 1993). Particulate matter and hydrocarbons are a growing serious global hazard. Hundreds of millions of people – mainly women and children – spend several hours daily gathering and transporting firewood and water to meet household needs. Affordable

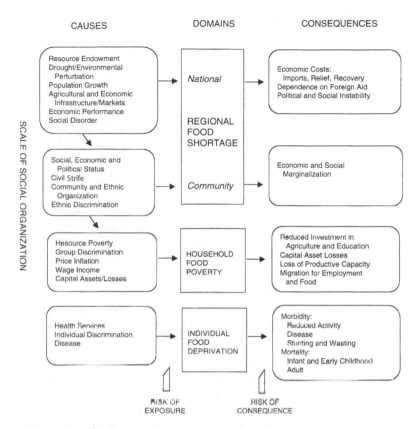

Source: Downing (1991a, p6)

Figure 14.3 *A causal structure of hunger*

energy is a key lever in increasing household productivity and breaking out of the poverty cycle. Yet poor households throughout the world, who pay a larger fraction of their income for energy than the rich, are highly vulnerable to potential increases in the price of energy. Table 14.1 suggests some of the principal social, economic and institutional interventions by which changed energy systems could reduce the vulnerabilities of many populations globally.

These various examples point to the need to develop a causal 'map' of the factors, structures and processes that produce differential vulnerability to stresses and perturbations. Such maps should consider both the proximate determinants of the three key dimensions of vulnerability – exposure, sensitivity and resilience – but also the extent to which these are linked to underlying social and economic structures and to ecosystem components and dynamics. In short, the causal maps need to capture cross-scale dynamics. Causal linkages among factors need to be carefully established and documented empirically. Sensitivity analyses will then be required to assess where interventions may yield the greatest gain in vulnerability reduction.

Table 14.1 *Energy-related options to address social issues*

Social challenge	Energy linkages and interventions
Alleviating poverty in developing countries	• Improve health and increase productivity by providing universal access to adequate energy services – particularly for cooking, lighting and transport – through affordable, high-quality, safe and environmentally acceptable energy carriers and end-use devices. • Make commercial energy available to increase income-generating opportunities.
Increasing opportunities for women	• Encourage the use of improved stoves and liquid or gaseous fuels to reduce indoor air pollution and improve women's health. • Support the use of affordable commercial energy to minimize arduous and time-consuming physical labour at home and at work. • Use women's managerial and entrepreneurial skills to develop, run and profit from decentralized energy systems.
Speeding the demographic transition (to low mortality and low fertility)	• Reduce child mortality by introducing cleaner fuels and cooking devices and providing safe, potable water. • Use energy initiatives to shift the relative benefits and costs of fertility – for example, adequate energy services can reduce the need for children's physical labour for household chores. • Influence attitudes about family size and opportunities for women through communications made accessible through modern energy carriers.
Mitigating the problems associated with rapid urbanization	• Reduce the 'push' factor in rural–urban migration by improving the energy services in rural areas. • Exploit the advantages of high-density settlements through land planning. • Provide universal access to affordable multi-modal transport services and public transportation. • Take advantage of new technologies to avoid energy-intensive, environmentally unsound development paths.

Source: Goldemberg (2000, p9)

Coping 'pools' and strategies

Since we have argued that vulnerability is best conceived as a process, a set of cross-scale dynamics and historical trajectories, it follows that learning and coping by the exposed system are essential ingredients in vulnerability. *Coping,* in our usage, refers to *the wide-ranging set of mechanisms used and actions undertaken to reduce or ameliorate threats and potential adverse impacts*. Human coping may take various forms. Some coping measures involve *anticipatory actions* by which people seek to avoid exposure to threats or to increase their buffering or resistance to future stresses (and thereby lower their sensitivity to the hazards) that may occur. Coping also occurs as stresses unfold, accumulate and (possibly) concatenate, and the exposed people or systems undertake short-term coping measures, referred to here as *adjustments*, which while adding to buffering capacity do not usually alter the social–environmental systems in fundamental ways. Finally, more fundamental interventions, referred to here as *adaptations*, typically occur after the perturbation or shock has registered its effects, as the coupled social ecosystem seeks to reduce its vulnerability to future perturbations and stresses.

Thus, it is essential in a vulnerability analysis to define the coping pool and resources available to a social–ecological system at risk. In his seminal piece on 'Choice of use in resource management', White (1961) set forth a template for assessing the range of choices open to an individual, group or institution managing hazards. His particular interest was enquiring into the adequacy of decision processes in processing such choices and in identifying ways of enlarging the range of choice. In this decision paradigm, attention was given to various behaviour constraints after Simon (1957, 1979), such as *bounded rationality* and *compliant behaviour*, which led to suboptimal resource choices. This framework was later extended to hazard and vulnerability analysis. We know, for example, that households and other exposed systems have portfolios of investments and other stores of coping resources that can be drawn down during periods of perturbation and stress. In such cases more vulnerable units are forced to expend important assets earlier and more completely than resilient units (Vogel, 1998). Some years later, Brooks (1986) took up the question of *response pools*, arguing that surprises and shocks are inevitable in most systems and that the size and variegation of the human responses available to a threatened system are critical to its ability to respond to such events with minimal damage and disruption.

As with Holling, Brooks was particularly concerned with the tendency to create technological monocultures by focusing too heavily upon efficiency at the expense of variety. He also observed that pruning the tree of technology, based solely on the first reasonably successful partial solution to a problem, is a common tendency in society – one that can reduce significantly the response pools available for dealing with surprise and increase the overall vulnerability of society to technological and environmental surprises. Learning curves, as a result, can actually contract response pools as society embraces near-term solutions at the expense of longer-term learning. Unfortunately, this promising line of enquiry into vulnerability has not been taken up as vulnerability analyses have focused on other issues.

One of those 'other' issues has been how the broader social and economic structures have differentially shaped the magnitude of coping resources across society, empowering and endowing some members of society while weakening and depleting others. It has also been argued that more vulnerable people become marginalized to more hazardous environments simultaneously as their endowments and coping capacities are diminished and they are disenfranchised politically. This has led to considerable interest in *social vulnerability*, conceived as 'a multi-layered and multi-dimensional social space defined by determinate political, economic and institutional capabilities of people in specific places and at specific times' (Watts and Bohle, 1993, p46). Various agencies have identified major types of 'assets' that contribute to buffering against perturbations, increasing resilience and providing livelihood security (Carney et al, 1999). Kelly and Adger (2000, p326) have argued for an analysis of the 'architecture of entitlements', which they describe as 'the influence levels of vulnerability within a community or nation that promote or constrain options for adaptation'.

Finally, renewed interest in adaptive capacity and strategies of adaptation in relation to vulnerability is plentiful. Like other forms of coping, adaptation is iterative, dynamic and processual. Interactions occur throughout the coupled social and environmental system, involving an adaptation cycle that occurs in space and time (see Figure 14.4).

These stages and questions replicate those that occur in coping geared to sensitivity issues. It is useful, however, to distinguish between *adaptive capacity*, referring here to the potential to adapt to new circumstances and surprises, and the *adaptation that actually occurs* (it should be noted that some ecologists would use a more specific definition, such as 'system robustness to changes in resilience'; Gunderson, 2000, p435). As we have noted in Chapter 11 of this volume, societal responses that take place in the face of growing environmental degradation and threat typically lag seriously behind and fall well short of potential effectiveness across diverse threatened regions throughout the globe. Analyses are needed to estimate and profile adaptive capacity and to explain the large gap that often exists between mitigative and adaptive capacity and actions actually undertaken.

With this discussion of the structuring of vulnerability analysis in hand, we now turn to a new conceptual framework, designed to guide a more integrative analysis of vulnerability.

A NEW CONCEPTUAL FRAMEWORK

The authors, with input and advice from others in the Research and Assessment Systems for Sustainability Program (www.sust.harvard.edu) have created a framework for vulnerability that addresses coupled human–environment systems and their interactive vulnerabilities. It seeks to capture as much as possible of the 'totality' of the different elements

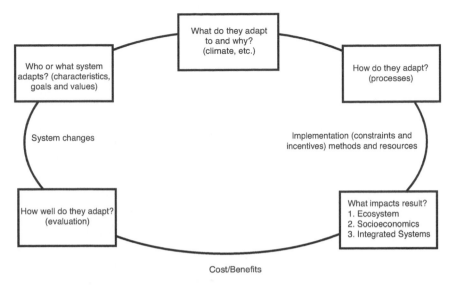

Cost/Benefits

Source: adapted from Wheaton and MacIver (1999, p218)

Figure 14.4 *The adaptation cycle through space and time*

that have been identified and demonstrated in risk, hazards and vulnerability studies and to frame them with regard to their complex linkages. It is not an explanatory framework; rather, it provides the components and linkages that must be addressed to capture the phenomena and processes that give rise to vulnerability. In this regard, it also serves as a definitional framework, identifying which component and linkages fall into the core arena of vulnerability. The framework recognizes that the components and linkages in question vary by the scale of analysis undertaken, and that the scale of the assessment may change the specific components but not the overall structure of the framework. As with complex agent-based modelling, full implementation of the framework will exceed the capacity and, in some instances, the needs of most vulnerability practitioners (i.e. agency use), who may focus on the development and simplification of the framework's subsystems. The framework, however, serves as a reminder of what is missing in assessments based on such simplifications.

The framework (see Figure 14.5) recognizes two basic parts to the problem and assessment: perturbation stresses and the coupled human–environment system. Of the two, vulnerability emerges from the attributes of the coupled system, comprising sets of components that fall largely with the categories of exposure, sensitivity and resilience. In this section we briefly describe each part and provide examples.

Perturbations and *stresses* are both human and natural in kind and are affected by processes that often operate at scales larger than the event in question (e.g. local drought) and, in many cases, whose origins are

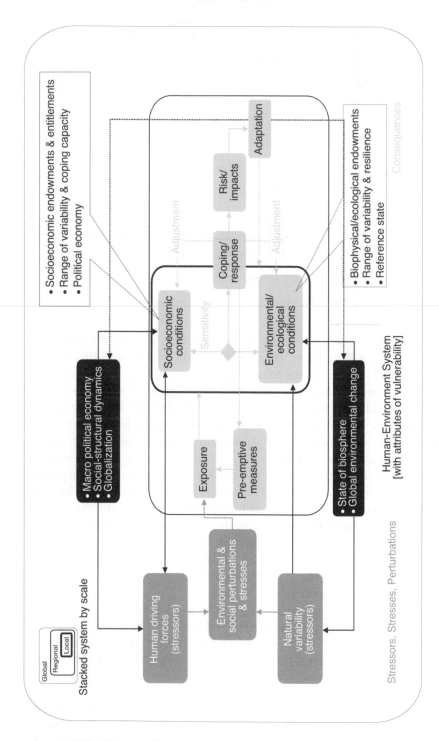

Figure 14.5 *A framework for analysing vulnerability*

exogenous to the ultimate location of the event. Importantly, the social–environment system, in turn, can exacerbate the perturbations and stresses. For example, globally induced climate warming triggers increased variation in precipitation in a tropical forest frontier, while political strife elsewhere drives large numbers of immigrants to the frontier. These exogenous forces create new levels of land pressures that stress forest recovery, make cropping riskier, and increase the amount of forest cut and the local dependency on cultivation. The cropping strategies undertaken by economically marginal immigrants may intensify land degradation, which becomes an endogenous force that increases per capita land clearance and reduces the capacity of the forest to recover.

The *coupled human–environment system* maintains some level of vulnerability to these perturbations and stresses, related to the manner in which they are experienced. This experience is registered, first, in terms of the nature of the *exposure* (e.g. intensity, frequency and duration), and involves measures that the human and environment subsystems may take to reduce the exposure. Take the case of slash-and-burn tropical farmers exposed to increasing climatic aridity. Given the convective nature of tropical precipitation, they may take pre-emptive measures to reduce exposure to drought by scattering cultivated plots in different niches across the landscape in the hope that some plots will gain more precipitation or take advantage of different soil moisture regimes.

The coupled system experiences some level of harm to exposure (i.e. risk and impacts), determined by its sensitivity. Little work has been directed to the determinants and measures of *sensitivity* (at least in human subsystems), and the level of harm experienced (e.g. human deaths incurred, cost of material damages and loss of ecosystem function) is usually the measure. The linkage between exposure and impact is not necessarily direct, however, because the coupled system maintains coping mechanisms that permit immediate or near-term adjustments that reduce the harm experienced and, in some cases, change the sensitivity of the system itself. Sensitivity is therefore a dynamic quality of the coupled system. Drought and land degradation, for example, reduce yields and household consumption, raising the potential for malnutrition in the human subsystem. To cope, a cash crop may be planted with low-yielding but drought-resistant crop varieties, producing food that becomes available as the regular harvest stocks dwindle. Where the opportunity exists, some segment of the household may seek off-farm employment, using the wages gained to purchase food. Likewise, the biophysical subsystem adjusts to the changing precipitation regimes by favouring more xeric species, changing the character of forest succession and recovery rates of soils whose nutrients have been reduced by cultivation.

If specific perturbation stress persists over the long run (i.e. change in the external conditions in which the coupled system operates), the kind and quality of *system resilience* changes. This last change demands *adaptation* (fundamental change) in the coupled system. The human subsystem must be altered or it ceases to function (e.g. abandonment of a place or region), and the environmental subsystem changes by definition (e.g. climate and vegetation change). Consider the case of farmers responding to increasing aridity and a shift from tropical forests to savannah woodlands by banding into communities that develop the infrastructure and institutions for small-scale irrigation. With improved water sources, a shift occurs to the commercial production of orchard products. Alternatively, the migrants move elsewhere, abandoning the area, which is subsequently taken over by commercial cattle enterprises. Both of these adaptations, of course, carry with them implications for the environmental subsystem.

By definition, no part of a system in this framework is unimportant, and work directed specifically to them and their linkages is needed. Emergent vulnerable communities, however, signal the need for devoting much more attention to the base characteristics of coupled systems because they hold the clues to the root causes of vulnerability and, hence, those elements that must be addressed in order to institute measures that reduce vulnerability.

PRIORITIES FOR NEXT STEPS

Vulnerability research has foundered amid the abstract theoretical discussion and acrimonious debate over alternative conceptual approaches and basic theory. The most pressing need for the next stage of work is for rigorous testing and concrete applications of the various conceptual frameworks, including the framework presented above. This empirical work should particularly emphasize comparative case studies, for most extant evidence comes from single case studies that proceed from a particular conceptual approach. Vulnerability research has tended to be overly context or case specific and, as a result, the development and the impacts of empirical findings have been distressingly non-cumulative. The time is ripe for systematic cross-case and cross-theory analysis, designed to glean what a particular conceptual framework can and cannot deliver, and what aspects of our understanding are best established and supported by empirical evidence. A model meta-analysis (Geist and Lambin, 2001) provides a useful guide for such comparative work.

Once comparative case studies yield lessons or general rules, a larger portion of vulnerability research must subject those lessons to strict 'hypothesis' testing, replete with quantitative analyses. Such an orientation is imperative, if applied and human-dimensions work on vulnerability is to link more usefully to research on the ecological and biophysical dimensions of vulnerability. Indeed, a major impediment to incorporating work on human vulnerability within research on global environmental change has been the paucity of research designs capable of contributing to model development and empirical tests, regardless of place-based orientation.

Applying and testing the framework set forth above, or any alternative framework, will, at first glance, be daunting, owing to the setting of multiple stresses and multiple vulnerabilities and the complexity of linkages, scale interactions and dynamics. Full application of the framework will exceed the resources and available time of most analysts and practitioners; therefore, comprehensive assessments or applications will be rare. Yet, even when the objectives of the study may be highly specific or the scope of analysis quite narrow, a major value of such a framework is to assist in structuring the problem and analysis. In the case of climate change, for example, where interest focuses on a particular set of perturbations or stresses, it will be useful to construct the analysis as a multiple-stress problem and to focus on climate perturbations, but also to consider other stresses in the analysis. Or, alternatively, the concern may well be with marginalized, indigenous groups in society; but the analysis can be structured so that the unit of exposure is indigenous group–ecological interactions. Yet another case might involve a focus upon local preparedness and coping; but the linkages with structural factors at higher scales (macro-political economy, globalization) can still be considered as context. So the broad conceptual framework can play a significant role in framing studies and assessment in a more sophisticated form.

Another possibility is that analysis can proceed by making judgements on the question: what are the essentials that should exist in any sound vulnerability analysis? Particularly for those seeking to use the conceptual framework outlined above, if the full range of complexity is beyond the capability and resources of the analyst, what are the essential elements of the framework that should certainly be included? Appreciating that the answer will ultimately depend upon the purposes of the study or assessment, we urge consideration of a minimal set of essentials that would include:

- a multiple stress/multiple vulnerability framing of the problem, including potential interactions among the stresses;
- inclusion of exposure, sensitivity and resilience as major dimensions to be analysed;
- a structuring of the exposure unit as a coupled system of ecological–social interactions;
- a causal mapping of vulnerability, its social and ecological roots, and causal linkages;
- cross-scale interactions (what is the architecture of stresses, responses, and coping resources and strategies over multiple scales?);
- vulnerability as process: the history, iteration and cumulative nature of vulnerability as evolutionary process;
- coping and adaptation, by which human societies and ecosystems buffer themselves against perturbations, and their ability to recover or reorganize in modified or new systems.

In any given case, of course, the list of 'essentials' will differ; but we suggest the foregoing as a starting set for consideration.

It is also important to appreciate that the existing knowledge base of vulnerability is highly uneven. Since conceptual approaches have been very general in construction and case studies have typically been quite specific in focus and approach, it is scarcely surprising that major gaps exist in the components of the conceptual framework elaborated above. We have yet to determine, for example, how best to structure and implement multiple-stress types of analysis. We lack detailed causal 'maps' of the factors and processes that shape vulnerabilities in a particular place. Few studies exist that detail the dynamics of human–environment interactions as systems degrade over time under stress. We know far too little about coping resources, adaptive capacities, and how they are mobilized and used in different places. We need to target these voids in knowledge and systematically begin to fill the empirical gaps and to grow and deepen our conceptual structures.

Meanwhile, assessment and practice cannot wait; it is not possible to delay until the next generation of vulnerability research to develop a strategic knowledge base. It is quite apparent that traditional environmental impact analysis lacks the conceptual and methodological capability to assess vulnerability issues; indeed, such analysis typically obscures the situation of vulnerable ecosystems and human groups within the mass of aggregated data and categories of effects derived from top-down assessments. Just as vulnerability assessments raise new questions and concepts, they also will require new assessments processes and procedures. Working backwards from most vulnerable groups and from those outcomes we seek most to avoid (from right to left in Figure 14.5) to vulnerabilities and, eventually, suites of stresses may yield new knowledge and insights. Entering assessments at different scales of analysis, and particularly the local scales of place-based assessments, promises greater depth and texture – and, perhaps, a fundamentally different format and presentation – than existing analyses. These more bottom-up assessments raise new possibilities for collaborative data gathering, analysis and entrance of value issues. And since the local people at risk will also be partners in these assessments, results will matter more in the policy and decision process.

Going forward with vulnerability analysis is part of the larger task of creating a sustainability science, a science dedicated to supporting the transition to a more sustainable world, a science that forges new interactions between science and the humanity it serves.

ACKNOWLEDGEMENTS

The writing of this chapter has relied on contributions from many quarters. Support for the transatlantic collaboration among the authors, based at the

George Perkins Marsh Institute at Clark University and the Stockholm Environment Institute (SEI), comes largely from the Research and Assessment Systems for Sustainability Program (www.sust.harvard.edu). Coordinated by William C. Clark and Nancy M. Dickson at Harvard University, the programme is supported by a core grant from the US National Science Foundation (NSF award BCS-0004236) with contributions from the Office of Global Programs at the National Oceanic and Atmospheric Administration (NOAA). Under its auspices, several workshops have served to keep the topic of vulnerability high on the research and political agenda. We are immensely grateful to the participants at these workshops (for full lists of these participants, see Clark et al, 2000; Research Assessment and Systems for Sustainability Program, 2001).We owe a particular debt of gratitude to Robert W. Kates for his evaluation of an early draft of this chapter. Also, we thank Jill Jäger, executive director of the International Human Dimensions Programme on Global Environmental Change (IHDP), for unwavering support of our efforts to advance the concept of vulnerability.

Many individuals merit special mention for their unflagging promotion of the various incarnations of our proposed framework for analysing vulnerability. Pamela Matson and her team at Stanford University have been endeavouring to test the applicability of the framework to an ongoing case study in the Yaqui Valley in Mexico, and, at Harvard, Robert Corell and James McCarthy are doing likewise for a case study in the Arctic. Meanwhile, a group at Clark University's Graduate School of Geography is working closely with two of us (Turner and Schiller) to apply the framework to another Mexican case, the southern Yucatán peninsular region. A generous gift from the Howard and Leah Green Fund, the George Perkins Marsh Institute, Clark University, provided the initial support for Andrew Schiller's post-doctoral position.

Also at Clark University, Ke Chen and Alex Pulsipher have enlivened our discussions and challenged our conceptual lapses. A sterling assemblage of two dozen researchers at SEI's workshop (Kasperson and Kasperson, 2001a) helped to sharpen our thinking. At SEI, Teresa Ogenstad cheerfully produced numerous early drafts of our manuscript. Erik Willis, in SEI's centre at York, in the UK, performed magic on Figure 14.5. Finally, we salute Clark University's indefatigable Mimi Berberian and Lu Ann Pacenka for transforming what appeared to be an irretrievably mangled penultimate draft into a real chapter.

References

Abraham, C. M. and S. Abraham (1991) The Bhopal case and the development of environmental law in India. *International and Comparative Law Quarterly*, vol 40, no 2 (April), pp334–365

Abramovitz, J. N. (2001) *Unnatural Disasters*. Worldwatch Paper 158. Washington, DC: Worldwatch Institute

ACGIH (American Conference of Governmental Industrial Hygienists) (1984) *TLVs: Threshold Limit Values for Chemical Substances and Physical Agents in the Work Environment and Biological Exposure Indices with Intended Changes for 1984–1985*. Cincinnati: ACGIH.

Acharya, A. (1995) Small islands awash in a sea of trouble. *World Watch*, Vol 8, no 6, pp24–33

Adger, W. N. (2000) Institutional adaptation to environmental risk under the transition in Vietnam. *Annals of the Association of American Geographers*, vol 90, pp738–758

Adger, W. N. and P. M. Kelly (1999) Social vulnerability to climate change and the architecture of entitlements. *Mitigation and Adaptation Strategies for Global Change*, vol 4, nos 3–4, pp253–266

Adler, N. J. (1986) *International Dimensions of Organizational Behaviour*. Boston: Kent

Agarwal, A. and S. Narain (1999) Kyoto Protocol in a unique world: The imperative of equity in climate negotiations. In K. Hultcrantz (ed) *Towards Equity and Sustainability in the Kyoto Protocol*. Papers presented at a seminar during the Fourth Conference of the Parties to the United Nations Framework Convention on Climate Change, Buenos Aires, 8 November 1988. Stockholm: Stockholm Environment Institute, pp17–30

Agrawal, Y. K. and K. P. S. Raju (1980) Effects of lead from motor vehicle exhausts in Baroda City. *International Journal of Environmental Studies*, vol 14, pp313–316

Aguilar, A. G., E. Ezcurra, T. Garcia, M. Mazari Hiriart and I. Pisanty (1995) The Basin of Mexico. In J. X. Kasperson, R. E. Kasperson and B. L. Turner, II (eds) *Regions at Risk: Comparisons of Threatened Environments*. Tokyo: United Nations University Press, pp304–366

Aiken, S. R. and C. H. Leigh (1985) On the declining fauna of Malaysia. *Ambio*, vol 14, pp15–22

Alexander, D. (1993) *Natural Disasters*. New York: Chapman and Hall

All India Reporter (1988a) M. C. Mehta v. Union of India. Supreme Court, pp1037–1048

All India Reporter (1988b) UP Pollution Control Board vs M/s Modi Distillery. Supreme Court, pp1128–1133

Aluma, R. J. W. (1979) Uganda: A damage report. *Unasylva*, vol 31, no 126, pp20–24

Angel, D. P., S. Attoh, D. Kromm, J. DeHart, R. Slocum and S. White (1998) The drivers of greenhouse gas emissions: What do we learn from local case studies? *Local Environment*, vol 3, no 3, pp263–278

Anyamba, A. (1997) *Interannual Variations of NDVI over Africa and their Relationship to ENSO: 1982–1983.* PhD thesis, Worcester, MA: Graduate School of Geography, Clark University

Argent, J. and T. O'Riordan (1995) The North Sea. In J. X. Kasperson, R. E. Kasperson, and B. L. Turner, II (eds) *Regions at Risk: Comparisons of Threatened Environments.* Tokyo: United Nations University Press, pp367–419

Ashford, N. A. (1984) Control the transfer of technology. *New York Times*, 9 December, section 3, p2

Ashford, N. A., and C. Ayers (1985) Policy issues in transferring technology to developing countries. *Ecology Law Quarterly*, vol 12, no 4, pp871–905

Austin, J. E. (1990) *Managing in Developing Countries: Strategic Analysis and Operating Techniques.* New York: Free Press

Ausubel, J H. (1996) Can technology spare the earth? *American Scientist*, vol 84, pp166–178

Baidhya, B. N. (1982) Alarming status of great Indian rhinoceros. *Environmental Conservation*, vol 9, pp346–347

Baram, M. (1985) Chemical industry hazards: Liability, insurance and the role of risk analysis. Unpublished paper presented to the Joint Conference on the Transportation, Storage and Disposal of Hazardous Materials. Laxenburg, Austria, International Institute for Applied Systems Analysis, 1–6 July (for a slightly different version of this paper, see Baram, 1987)

Baram, M. (1987) Chemical industry hazards: Liability, insurance and the role of risk analysis. In P. R. Kleindorfer and H. C. Kunreuther (eds) *Insuring and Managing Hazardous Risks: From Seveco to Bhopal and Beyond.* New York: Springer-Verlag, pp415–441

Baum, A., R. Fleming and J. Singer (1983) Coping with victimization by technological disaster. *Journal of Social Issues*, vol 39, no 2, pp117–138

Bebbington, A. (1999) Capitals and capabilities: A framework for measuring peasant viability, rural livelihoods and poverty. *World Development*, vol 27, no 12, pp2021–2044

Beck, U. (1992) *Risk Society: Toward a New Modernity.* London: Sage

Beck, U. (1995) *Ecological Enlightenment: Essays on the Politics of the Risk Society.* Atlantic Highlands, NJ: Humanities Press

Beck, U. (1999) *World Risk Society.* London: Polity

Beierle, T. C. and D. M. Konisky (1999) *Public Participation in Environmental Planning in the Great Lakes Region.* Discussion Paper 99–50, September. Washington, DC: Resources for the Future

Berkes, F., and C. Folke (1998) Linking social and ecological systems for resilience and sustainability. In F. Berkes and C. Folke, with J. Colding (eds) *Linking Social and Ecological Systems: Management Practices and Social Mechanisms for Building Resilience*, Cambridge: Cambridge University Press, pp1–25

Bhagavan, M. R. (1986) The wood fuel crisis in SADCC countries. *Ambio*, vol 13, pp25–27

Bhargava, D. S. (1985) Matching river quality to use. *Environmental Professional*, vol 7, pp240–247

Bhushan, B. and A. Subramaniam (1985) Bhopal: What really happened? Special Report 1, *BusinessIndia*, vol 7, no 182, March, p109

Biswas, M. R. and A. K. Biswas (1978) Loss of productive soil. *International Journal of Environmental Studies*, vol 12, pp189–197

Biswas, M. R. and A. K. Biswas (1984) Complementarity between environment and development. *Environmental Conservation*, vol 11, pp35–44

Blaikie, P. (1985) *The Political Economy of Soil Erosion in Developing Countries*. Harlow, UK: Longman

Blaikie, P. M. and H. C. Brookfield (1987) *Land Degradation and Society*. London: Methuen

Blaikie, P., T. Cannon, I. Davies and B. Wisner (1994) *At Risk: Natural Hazards, People's Vulnerability and Disaster*, London: Routledge

Blanchard, E. P., Jr. (1990) Remarks delivered to members of the Norwegian parliament. Wilmington, DE, 13 February

Blassing, R. (2000) Countries agree on Biosafety Protocol regulating transboundary movements of GMOs. *International Environment Reporter*, vol 23, no 3, 2 February, pp71–72

Blau, P. M. (1964) *Exchange and Power in Social Life*. New York: Wiley

Boffey, P. (1984) Bhopal's doctors given high praise. *New York Times*, 18 December, pA8

Bogard, W. C. (1989) Bringing social theory to hazards research: Conditions and consequences of the mitigation of environmental hazards. *Sociological Perspectives*, vol 31, no 2, pp147–168

Bohle, H.-G. (2001) Vulnerability and criticality: Perspectives from social geography. *IHDP Update*, vol 2, pp1, 3–5

Bohle, H.-G., T. E. Downing and M. J. Watts (1994) Climate change and social vulnerability: Towards a sociology and geography of food insecurity. *Global Environmental Change*, vol 4, pp37–48

Bolin, R. (1982) *Long-term Family Recovery from Natural Disaster*. Boulder, CO: Westview

Bolin, R. and P. Bolton (1986) *Race, Religion and Ethnicity in Disaster Recovery*. Program on Environment and Behavior Monograph no 42. Boulder, CO: Institute of Behavioral Science, University of Colorado

Bolin, R. and D. Klenow (1983) Response of the elderly to disaster: An age-stratified analysis. *Journal of Aging and Human Development*, vol 16, pp283–296

Bolin, R. and L. Stanford (1991) Shelter, housing, and recovery: A comparison of US disasters. *Disasters*, vol 15, pp24–34

Bowonder, B. (1980) Issues in environmental risk assessment. *Journal of Environmental Systems*, vol 10, pp307–333

Bowonder, B. (1981a) Environmental risk assessment issues in the Third World. *Technological Forecasting and Social Change*, vol 19, no 1, February, pp99–127

Bowonder, B. (1981b) Environmental risk management in the Third World. *International Journal of Environmental Studies*, vol 18, pp223–226

Bowonder, B. (1982) Deforestation in India. *Journal of Environmental Systems*, vol 12, pp199–217

Bowonder, B. (1983a) Environmental management conflicts in developing countries. *Environmental Management*, vol 7, pp211–222

Bowonder, B. (1983b) Environmental quality and water resources in India. *International Journal of Water Resources Development*, vol 1, pp253–267

Bowonder, B. (1983c) Management of urban environment in India. *Journal of Environmental Systems*, vol 12, pp199–217

Bowonder, B. (1985a) Environmental perceptions and management. *Mazingira*, vol 8, pp14–15

Bowonder, B. (1985b) Strategies for managing environmental problems in developing countries. *Environmental Professional*, vol 7, pp108–115

Bowonder, B. (1986a) Catastrophe theory and environmental changes. *Science and Public Policy*, vol 11, pp94–99

Bowonder, B. (1986b) Environmental management problems in India. *Environmental Management*, vol 10, pp599–609

Bowonder, B. (1987) Global forests. *Futures*, vol 19, pp43–63

Bowonder, B. (1988) *Implementing Environmental Policy in India*. New Delhi: Friedrich Ebert Stiftung

Bowonder, B. and S. S. Arvind (1989) Environmental regulations and litigation, India. *Project Appraisal*, vol 4, pp182–196

Bowonder, B. and R. Chettri (1984) Urban water supply in India. *Urban Ecology*, vol 8, pp295–311

Bowonder, B. and T. Miyake (1988) Managing hazardous facilities: Lessons from the Bhopal accident. *Journal of Hazardous Materials*, vol 19, pp237–269

Bowonder, B., P. V. Muralikrishna, M. R. Reddy, A. S. Ramasastry, S. Singh, E. Srinivas and S. S. Arvind (1988) *Corporate Responses to Environmental Policies*. Hyderabad: Centre for Energy, Environment and Technology, Administrative Staff College of India

Bowonder, B., K. V. Ramana and T. H. Rao (1987) Sedimentation of reservoirs. *Water Resources Development*, vol 2, pp11–28

Brickman, R., S. Jasanoff and T. Ilgen (1985) *Controlling Chemicals: The Politics of Regulation in Europe and the United States*. Ithaca, NY: Cornell University Press

Brklacich, M. and S. Leybourne (1999) Food security in a changing world. *AVISO*, vol 4, September, pp1–9, www.gechs.org

Broad, W. C. (1987) Does the fear of litigation dampen the drive to innovate? *New York Times*, 12 May, ppC1 and C9

Bromet, E. (1980) *Three Mile Island: Mental Health Findings*. Pittsburgh: Western Psychiatric Institute and Clinic

Brookfield, H., I. Potter and Y. Byron (1995) *In Place of the Forest: Environmental and Socio-economic Transformation in Borneo and the Eastern Malay Peninsula*. Tokyo: United Nations University Press

Brooks, E. and J. Emel (1995) The Llano Estacado of the American Southern High Plains. In J. X. Kasperson, R. E. Kasperson and B. L. Turner, II (eds) *Regions at Risk: Comparisons of Threatened Environments*. Tokyo: United Nations University Press, pp255–303

Brooks, E. and J. Emel, with B. Jokisch and P. Robbins (2000) *The Llano Estacado of the US Southern High Plains: Environmental Transformation and the Prospect for Sustainability*. Tokyo: United Nations University Press

Brooks, H. (1986) The typology of surprises in technology, institutions, and development. In W. C. Clark and R. E. Munn (eds) *Sustainable Development of the Biosphere*. Cambridge: Cambridge University Press, pp325–348

Brown, L. R. (1983) Soils and civilization. *Third World Quarterly*, vol 5, pp103–119

Brown, L. R. (1984) Global 1055 of top soil. *Journal of Soil and Water Conservation*, vol 39, pp162–165

Browning, J. B. (1984) Director of Health, Safety and Environmental Affairs, Union Carbide Corporation. In *News for Release*, vol P–0082–84, 6 December

Bruce, J. P., H. Lee and E. F. Haites (eds) (1996) *Climate Change 1995: Economic and Social Dimensions of Climate Change. Contributions of Working Group III to the Second Assessment Report of the Intergovernmental Panel on Climate Change*. New York: Cambridge University Press

Buchholz, R. A., W. D. Evans and R. A. Wagley (1985) Dow Chemical and product stewardship. In R. A. Buchholz (ed) *Management Response to Public Issues: Concepts and Cases in Strategy Formulation*. Englewood Cliffs, NJ: Prentice Hall, pp99–111

Budnitz, R. (1984) 'External initiators in probabilistic reactor accident analysis: Earthquakes, fires, floods, winds', *Risk Analysis*, vol 4, pp313–322

Bunker, S. G. (1980) Forces of destruction in Amazonia. *Environment*, vol 22, no 7, pp14–24

Burton, I., R. W. Kates and G. F. White (1978) *The Environment as Hazard*. New York: Oxford University Press

Burton, I., R. W. Kates and G. F. White (1993) *The Environment as Hazard*, second edition. New York: Guilford Press

Business and Political Observer (1991a) Five major amendments to EPA on Anvil. 12 November, p5

Business and Political Observer (1991b) MPCB well within powers to give NOC to NOCIL. 22 November, p3

Calabrese, E. T. (1978) *Pollutants and High-risk Groups: The Biological Basis of Increased Human Susceptibility to Environmental and Occupational Pollutants.* New York: Wiley

Camerer, C. and A. Vepsalainen (1988) The economic efficiency of corporate culture. *Strategic Management Journal*, vol 9, pp115–126

Cannon, T. (1994) Vulnerability analysis and the explanation of 'natural' disasters. In Ann Varley (ed) *Disasters, Development and the Environment*. Chichester: Wiley, pp13–30

Carney, D. (ed) (1998) *Sustainable Rural Livelihoods: What Contributions Can We Make.* Papers presented at DFID's Natural Resources Advisers' Conference, July 1998. London: Department for International Development (DFID)

Carney, D., M. Drinkwater, T. Rusinow, K. Neefjes, S. Wanmali and N. Singh (1999) *Livelihood Approaches Compared: A Brief Comparison of the Livelihoods Approaches of the UK Department for International Development (DFID), CARE, Oxfam and the United Nations Development Programme (UNDP).* London: DFID

Cash, D. (1997) Local response to global change: Creating effective bridges between science, policy and action. Paper prepared for the workshop on Regional Climate Change Impacts on Great Plains Ecosystems, 27–29 May 1997, Fort Collins, CO

Cash, D. W. and S. C. Moser (2000) Linking local and global scales: Designing dynamic assessment and management processes. *Global Environmental Change*, vol 10, pp109–120

Central Pollution Control Board (1991) *Annual Report, 1989–1990.* Delhi: Central Pollution Control Board

Centre for Science and Environment (1985) *The State of India's Environment 1984–1985: The Second Citizens' Report.* New Delhi: Centre for Science and Environment

Chalat, S. (1989) Transcript of taped interview with Halina S. Brown and Allen V. White, 6 November 1989. Halina S. Brown (Clark University, Worcester, MA, 01610) provided the authors with a copy of the transcript

Chambers, R. (1989) Editorial introduction: Vulnerability, coping and policy. *IDS Bulletin*, vol 20, no 3, pp1–8

Chambers, R. (1997) *Whose Reality Counts? Putting the First Last.* London: Intermediate Technology Publications

Chambers, R. and G. Conway (1992) *Sustainable Rural Livelihoods: Practical Concepts for the 21st Century.* IDS Discussion Paper 296. Brighton, UK: Institute of Development Studies, University of Sussex

Chambers, R., A. Pacey and L. A. Thrupp (eds) (1989) *Farmer First: Farmer Innovation and Agricultural Research*. London: Intermediate Technology Publications

Charnes, A., W. W. Cooper and E. Rhodes (1978) Measuring the efficiency of decisionmaking units. *European Journal of Operational Research*, vol 2, pp429–444.

Chemical and Engineering News (1985) Industry safety reforms: Bhopal inspires new initiatives. *Chemical and Engineering News*, vol 63, no 13, 1 April, p6

Chemocology (1983) Toxicology testing helps company assure chemical, product safety. *Chemocology*, March, p7

Chen, R. S. (1994) The human dimensions of vulnerability. In R. H. Socolow, C. Andrews, F. Berkhout and V. Thomas (eds) *Industrial Ecology and Global Change*. Cambridge: Cambridge University Press, pp85–105

Chertow, M. R. (2001) The IPAT equation and its variants: Changing views of technology and environmental impact. *Journal of Industrial Ecology*, vol 4, no 4, pp13–29

Chitnis, V. S. (1987) Environment Protection Act, 1986: A critique. In P. Diwan (ed) *Environment Protection: Problems, Policy Administration, Law*. New Delhi: Deep and Deep, pp152–156

Choy, C. L. (1987) History and managerial culture in Singapore: Pragmatism, openness, and paternalism. *Asia Pacific Journal of Management*, vol 4, May, p139

Citizens Fund (1990) *Manufacturing Pollution: A Survey of the Nation's Toxic Polluters*. Washington, DC: Citizens Fund

Clark, T. N. (ed) (1974) *Comparative Community Politics*. Beverly Hills, CA: Sage

Clark, W. C. (1986) Sustainable development of the biosphere: Themes for a research program. In W. C. Clark and R. E. Munn (eds) *Sustainable Development of the Biosphere*. New York: Cambridge University Press for IIASA, pp5–48

Clark, W. C., J. Jäger, R. Corell, R. E. Kasperson, J. J. McCarthy, D. Cash et al (2000) *Assessing Vulnerability to Global Environmental Risks: Report of the Workshop on Vulnerability to Global Environmental Change: Challenges for Research, Assessment and Decision Making*, 22–25 May, Airlie House, Warrenton, Virginia. Research and Assessment Systems for Sustainability Program Discussion Paper 2000–12. Cambridge, MA, Environmental and Natural Resources Program, Belfer Center for Science and International Affairs (BCSIA), Kennedy School of Government, Harvard University; available on line at www.sust.harvard.edu

Clark, W. C., J. Jäger and J. van Eijndhoven (eds) (2001) *Learning to Manage Global Environmental Risks: Volume 1, A Comparative History of Social Responses to Climate Change, Ozone Depletion and Acid Rain*. Cambridge, MA: MIT Press

CMA News (1983) CMA unveils position paper on environmental auditing: Majority of member companies use audits or intend to. *CMA News*, vol 11, no 8, p12

Cochran, C. E. (1974) Political science and the public interest. *Journal of Politics*, vol 36, pp327–355

Corps of Engineers (1979) *A Report on the Assessment of Flood Damages Resulting from the Storm of 6–7 February 1978 along the Coastline from Orleans, Massachusetts, to New Castle, New Hampshire*. Waltham, MA: US Army Corps of Engineers, New England Division

Cox, G.V., T. F. O'Leary and G. D. Strickland (1985) The chemical industry's view of risk assessment. In C. Whipple and V. T. Covello (eds) *Risk Analysis in the Private Sector*. New York: Plenum, pp271–284

Cutter, S. L. (1984) Risk cognition and the public: The case of Three Mile Island. *Environmental Management*, vol 8, no 1, pp15–20

Cutter, S. L. (1993) *Living with Risk: The Geography of Technological Hazard*. London: Edward Arnold

Cutter, S. L. (1996) Vulnerability to environmental hazards. *Progress in Human Geography*, vol 20, pp529–539

Cutter, S. L., D. Holm and L. Clark (1996) The role of geographic scale in monitoring environmental justice. *Risk Analysis*, vol 16, no 4, pp517–525

Cutter, S. L., J. T. Mitchell and M. S. Scott (2000) Revealing the vulnerability of people and places: A case study of Georgetown County, South Carolina. *Annals of the Association of American Geographers*, vol 90, no 4, pp713–737

Da Fonseca, G. A. B. (1985) Vanishing Brazilian Atlantic forest. *Biological Conservation*, vol 34, pp17–34

Dagani, R. (1985) Data on MIC's toxicity are scant, leave much to be learned. *Chemical and Engineering News*, vol 63, no 6, 11 February, pp37–40

Darkoh, M. B. K. (1982) Desertification in Tanzania. *Geography*, vol 67, pp220–331

Davidar, D. J. (1982) India: Every river polluted and few effective controls. *Ambio*, vol 11, pp63–64

Davidar, D. J. (1985) Beyond Bhopal: The toxic waste hazard in India. *Ambio*, vol 14, no 2, pp112–116

De la Torre, J. and B. Toyne. (1978) Cross national managerial interaction: A conceptual model. *Academy of Management Review*, vol 3, July, pp462–474

Deal, T. E. and A. A. Kennedy (1982) *Corporate Cultures: The Rites and Rituals of Corporate Life.* Reading, MA: Addison-Wesley

Dean, N. L., J. Paje and R. J. Burke (1989) *The Toxic SOD: The SOD Largest Releases of Toxic Chemicals in the United States, 1987.* Washington, DC: National Wildlife Federation

DeHart, J. L. and P. T. Soulé (2000) Does I = PAT work in local places? *Professional Geographer*, vol 52, no 1, pp1–10

Delhi Science Forum (1985) Bhopal gas tragedy. *Social Scientist* (Kerala), vol 13, no 140, p36

Denison, D. R. (1990) *Corporate Culture and Organizational Effectiveness.* New York: Wiley

Dewalt, G. R. (1983) Cattle are eating the forests. *Bulletin of the Atomic Scientists*, vol 39, no 1, pp18–23

Diamond, S. (1985a) The disaster in Bhopal: Lessons for the future. *New York Times*, 3 February, p8

Diamond, S. (1985b) India, Carbide trade charges. *New York Times*, 20 June, pD5

Dietz, T. and E. A. Rosa (1994) Rethinking the environmental impacts of population, affluence and technology. *Human Ecology Review*, vol 1, pp277–300

Dissanayake, C. B. (1982) The environmental pollution of Kandy Lake: A case study from Sri Lanka. *Environment International*, vol 7, pp343–351

D'Monte, D. (1984) India's environment: Pollution and poverty. *Ambio*, vol 13, no 4, pp272–273

Dooley, J. E., B. Hanson, J. X. Kasperson, R. E. Kasperson, T. O'Riordan and H. Paschen (1987) The emergence of risk analysis on the international scene. In R. E. Kasperson and J. X. Kasperson (eds) *Nuclear Risk Analysis in Comparative Perspective: The Impacts of Large-scale Risk Assessment in Five Countries.* Boston: Allen & Unwin, pp1–26

Douglas, M. and A. Wildavsky (1982) *Risk and Culture: An Essay on the Selection of Technological and Environmental Dangers.* Berkeley: University of California Press

Dow, K. (1992) Exploring differences in our common future(s): The meaning of vulnerability to global environmental change. *Geoforum*, vol 23, pp417–436

Dow, K. (1993) Unpublished literature review on the 'concept of vulnerability' and the 'factors contributing to vulnerability'. Worcester, MA: Jeanne X. Kasperson Research Library, Clark University

Dow, K. and T. E. Downing (1995) Vulnerability research: Where things stand. *Human Dimensions Quarterly*, vol 1, pp3–5

Downing, T. E. (1991a) *Assessing Socioeconomic Vulnerability to Famine: Frameworks, Concepts and Applications.* Research Report 91–1, April. Providence, RI: Alan Shawn Feinstein World Hunger Program, Brown University

Downing, T. E. (1991b) Vulnerability to hunger and coping with climate change in Africa. *Global Environmental Change*, vol 1, pp365–380

Doyle, J. (1992) Hold the applause: A case study of environmentalism. *The Ecologist*, vol 22, no 3, May/June, pp84–90

Drabek, T. E. and W. H. Key (1984) *Conquering Disaster: Family Recovery and Long-term Consequences*. New York: Irvington

Drèze, J. and A. K. Sen (1989) *Hunger and Public Action*. Oxford: Clarendon Press

Dumaine, B. (1990) Creating a new company culture. *Fortune*, vol 121, 15 January, pp127–131

Dunlap, R. E., G. H. Gallup, Jr. and A. M. Gallup (1993) *Health of the Planet: Results of a 1992 National Opinion Survey of Citizens in 24 Nations*. Princeton, NJ: George H. Gallup International Institute

DuPont (1989) *This Is DuPont*. Wilmington, DE: DuPont

DuPont (1990) Personal communication during visit to Clark University, 23 May

Easterling, W. E. (1997) Why regional studies are needed in the development of full-scale integrated assessment modelling of global change processes. *Global Environmental Change*, vol 7, pp337–356

Economic Times (1992a) PM assures green channel for delayed power projects. 5 April, p1

Economic Times (1992b) Prosecution of pollution units strongly resented. 8 February, p2

Ehrlich, P. R. (1980) Variety is the key to life. *Technology Review*, vol 28, no 5, pp2–14

Ehrlich, P. R. and J. P. Holdren (1971) Impact of population growth. *Science*, vol 171, pp1212–1217

EK–A (Energikommissionens Expertgrupp für Sakerhat och Miljo) (1978) *Miljvesfekter och risker vid utnyttjandet av energi* (*Environmental Effects and Hazards of Energy Exploitation*). Stockholm: Allmänna Förlagg

Ember, L. R. (1985) Technology in India: An uneasy balance of progress and tradition. *Chemical and Engineering News*, vol 63, no 6, 11 February, pp61–65

Erikson, K. T. (1976) *Everything In Its Path*. New York: Simon & Schuster

Erikson, K. T. (1994) *A New Species of Trouble: The Human Experience of Modern Disasters*. New York: Norton

ESRC (Economic and Social Research Council) Global Environmental Change Research Programme (1999) *The Politics of GM Food: Risk, Science and Public Trust*. Special Briefing no 5. Falmer, Brighton, UK: University of Sussex

Ezcurra, E., M. Mazari Hiriart, I. Pisanty and A. G. Aguilar (1999) *The Basin of Mexico: Critical Environmental Issues and Sustainability*. Tokyo: United Nations University Press

Fare, F. K. (1986) Climate, drought and desertification. *Nature and Resources*, vol 20, no 1, pp2–8

Farmer, F. R. (1967) Reactor safety analysis as related to reactor siting. In *Containment and Siting of Nuclear Power Plants*, IAEA Proceedings Series. Vienna: Austrian International Atomic Energy Agency

Farnesworth, C. H. (1981) US proposes eased car standards. *New York Times*, 7 April, ppA1, D7

Financial Times (London) (1984) Union Carbide halts production of pesticide gas. 5 December, p1

Finn, D. (1983) Land use and abuse in the East African region. *Ambio*, vol 12, pp296–301

Fischhoff, B. (1977) Cost–benefit analysis and the art of motorcycle maintenance. *Policy Sciences*, vol 8, pp177–202

Fischhoff, B. (1979) Behavioral aspects of cost–benefit analysis. In T. Goodman and W. D. Row (eds) *Energy Risk Management*. London: Academic Press, pp269–283

Fischhoff, B., S. Lichtenstein, P. Slovic, S. Derby and R. Keeney (1982) *Acceptable Risk*. New York: Cambridge University Press

Fischhoff, B., S. Lichtenstein, P. Slovic, R. Keeney and S. Derby (1980) *Approaches to Acceptable Risk: A Critical Guide*. NUREG/CR-1614; ORNL sub-7656/1. Oak Ridge, TN: Oak Ridge National Laboratory

Fitzmaurice, J. (1996) *Damming the Danube: Gabcikovo and Post-Communist Politics in Europe*. Boulder, CO: Westview

FIVIMS (Food Insecurity and Vulnerability Information Mapping System) (1999) *The State of Food Insecurity in the World 1999*. Rome: Food and Agriculture Organization of the United Nations

FIVIMS (2000) *The State of Food Insecurity in the World 2000*. Rome: Food and Agriculture Organization of the United Nations

FIVIMS (2001) *The State of Food Insecurity in the World 2001*. Rome: Food and Agriculture Organization of the United Nations

Flynn, C. B. (1982) Reactions of local residents to the accident at Three Mile Island. In D. L. Sills, C. P. Wolf and V. B. Shelanski (eds) *Accident at Three Mile Island: The Human Dimensions*. Boulder, CO: Westview, pp49–63

Flynn, C. B. and J. A. Chalmers (1980) *The Social and Economic Effects of the Accident at Three Mile Island*. NUREG/CR–1215. Washington, DC: US Nuclear Regulatory Commission

Flynn, J., P. Slovic and H. Kunreuther (eds) (2001) *Risk, Media and Stigma: Understanding Public Challenges to Modern Science and Technology*. London: Earthscan

Flynn, J., P. Slovic and C. K. Mertz (1994) Gender, race, and perception of environmental health risks. *Risk Analysis*, vol 14, no 6, pp1101–1108

Folke, C., J. Colding and F. Berkes (2003) Synthesis: Building resilience and adaptive capacity in social-ecological systems. In F. Berkes, J. Colding and C. Folke (eds) *Navigating Social-ecological Systems: Building Resilience for Complexity and Change*. Cambridge: Cambridge University Press, pp352–387

Foster, H. D. (1980) *Disaster Planning: The Preservation of Life and Property*. New York: Springer-Verlag

French, H. F. (2000) *Vanishing Borders: Protecting the Planet in an Age of Globalization*. London: Earthscan

Freudenburg, W. R. (1988) Perceived risk, real risk: Social science and the art of probabilistic risk assessment. *Science*, vol 242, pp44–49

Freudenburg, W. R. (1992) Nothing recedes like success? Risk analysis and the organizational amplification of risk. *Risk: Issues in Health and Safety*, vol 3, no 3, winter, pp1–35

Fuchs, R. (1999) Viewpoint: START and the IHDP. *IHDP Update*, no 1, pp2–3

Gabor, T. and T. K. Griffith (1980) The assessment of community vulnerability to acute hazardous materials incidents. *Journal of Hazardous Materials*, vol 8, pp323–333

Garrick, B. J. (1984) Recent case studies and advancements in probabilistic risk assessment. *Risk Analysis*, vol 4, pp267–279

Geertz, C. (1973) *The Interpretation of Cultures*. New York: Basic Books

Geist, H. J. and E. F. Lambin (2001) *What Drives Tropical Deforestation?: A Meta-analysis of Proximate and Underlying Causes of Deforestation Based on Subnational Scale Case Study Evidence*. LUCC Report Series no 4. Louvain-la Neuve, Belgium: University of Louvain

Geoghegan, J., L. Pritchard, Jr., Y. Ogneva-Himmelberger, R. R. Chowdhury, S. Sanderson and B.L. Turner, II (1998) 'Socializing the pixel' and 'pixelizing the social' in land-use and land-cover change. In D. Liverman, E. F. Moran, R. R. Rindfuss and P. C. Stern (eds) *People and Pixels: Linking Remote Sensing and Social Science*. Washington, DC: National Academy Press, pp51–69

Ghafourian, H. (1983) An investigation of the amount of lead in blood of Tehran citizens. *International Journal of Environmental Studies*, vol 21, pp309–316

Ghinure, G. P. S. (1985) Water pollution: A major crisis in Nepal. *Environmentalist*, vol 5, pp193–196

Gibson, C., E. Ostrom and T. K. Ahn (1998) *Scaling Issues in the Social Sciences*. IHDP Working Paper no 1. Bonn: IHDP

Giddens, A. (1979) *Central Problems in Social Theory: Action, Structure and Contradiction in Social Analysis*. Berkeley, CA: University of California Press

Giddens, A. (1996) Affluence, poverty and the idea of a post-scarcity society. *Development and Change*, vol 27, no 2, April, pp365–377

Gladwin, T. N. (1985) The Bhopal tragedy: Lessons for management. *NYU Business*, vol 5, spring/summer, pp17–21

Gladwin, T. N. and V. Terpstra (1978) Introduction. In V. Terpstra (ed) *The Cultural Environment of International Business*. Cincinnati, OH: South-Western

Glantz, M. H. (ed) (1988) *Societal Responses to Regional Climate Change: Forecasting by Analogy*. Boulder, CO: Westview

Glazovsky, N. F. (1995) The Aral Sea basin. In J. X. Kasperson, R. E. Kasperson and B. L. Turner, II (eds) *Regions at Risk: Comparisons of Threatened Environments*. Tokyo: United Nations University Press, pp92–139

Gleick, P. H. (1998) *The World's Water 1998–1999. The Biennial Report on Freshwater Resources*. Washington, DC: Island Press

Global Environmental Assessment Project (1997) *A Critical Evaluation of Global Environmental Assessments: The Climate Experience*. Calverton, MD: CARE

Glynn, P. W. (1983) Extensive bleaching and death of reef corals of the Pacific coast of Panama. *Environmental Conservation*, vol 10, pp149–154

Godschalk, D. R., D. J. Brower and T. Beatley (1989) *Catastrophic Coastal Storms*. Durham, NC: Duke University Press

Goldemberg, J. (ed) (2000) *World Energy Assessment: Energy and the Challenge of Sustainability*. New York: United Nations Development Programme (UNDP), United Nations Department of Economic and Social Affairs and World Energy Council

Gore, C. (1993) Entitlement relations and unruly social practices: A comment on the work of Amartya Sen. *Journal of Development Studies*, vol 29, no 3, pp429–460

Gori, G. B. (1982) The regulation of carcinogenic hazards. In C. Hohenemser and J. X. Kasperson (eds) *Risk in the Technological Society*. AAAS Symposium Series. Boulder, CO: Westview, pp167–187

Government of Karnataka (1987) *Report of Task Force on Safety in Hazardous Industries in Karnataka*. Bangalore: Government of Karnataka

Grainger, A. (1986) Deforestation and progress in afforestation in Africa. *International Tree Crops Journal*, vol 4, pp33–48

Granot, H. (1999) Facing catastrophe: Mad cows and emergency policy-making. *International Journal of Mass Emergencies and Disasters*, vol 17, no 2, August, pp161–184

Greene, O. (1998) Implementation review and the Baltic Sea regime. In D. Victor, K. Raustiala and E. B. Skolnikoff (eds) *The Implementation and Effectiveness of International Environmental Commitments: Theory and Practice*. Cambridge, MA: MIT Press, pp177–220

Gregory, K. L. (1983) Native-view paradigms: Multiple cultures and cultural conflicts in organizations. *Administrative Science Quarterly*, vol 28, pp359–376

Gunderson, L. H. (2000 Ecological resilience: In theory and application. *Annual Review of Ecology and Systematics*, vol 31, pp425–439

Gunderson, L. H. and C. S. Holling (eds) (2001) *Panarchy: Understanding Transformations in Systems of Humans and Nature*. Washington, DC: Island Press

Gunderson, L. H., C. S. Holling and S. S. Light (eds) (1995) *Barriers and Bridges to the Renewal of Ecosystems and Institutions*. New York: Columbia University Press

Hadden, S. (1987) Statutes and standards for pollution control in India. *Economic and Political Weekly*, vol 22, no 16, 18 April, pp709–720

Hadden, S. G. (1994) Citizen participation in environmental policy making. In S. Jasanoff (ed) *Learning from Disaster*. Philadelphia: University of Pennsylvania Press, pp91–112

Hagman, G. (1984) *Prevention Better than Cure: Report on Human and Environmental Disasters in the Third World*, second edition. Stockholm: Swedish Red Cross

Haimes, Y. Y. (1990) *Hierarchical Multiobjective Analysis of Large-scale Systems*. New York: Hemisphere

Haimes, Y. Y. (1998) *Risk Modelling, Assessment and Management*. New York: Wiley

Hansson, B. (1987) Major energy risk assessments in Sweden: Information flow and impacts. In R. E. Kasperson and J. X. Kasperson (eds) *Nuclear Risk Analysis in Comparative Perspective: The Impacts of Large-scale Risk Assessment in Five Countries*. Boston: Allen & Unwin, pp50–85

Harriss, R. C., C. Hohenemser and R. W. Kates (1978) Our hazardous environment. *Environment*, vol 20, September, pp6–15, 38–41

Harriss, R. C., C. Hohenemser and R. W. Kates (1985) Human and nonhuman mortality. In R. W. Kates, C. Hohenemser and J. X. Kasperson (eds) *Perilous Progress: Managing the Hazards of Technology*. Boulder, CO: Westview, pp129–135

Hayes, J. and C. W. Allinson (1988) Cultural differences in the learning style of managers. *Management International Review*, vol 28, no 3, pp75–80

Haynes, K., S. Ratick, W. Bowen and J. Cummings-Saxton (1993) Environmental decision models: US experiences and new approaches to pollution management. *Environment International*, vol 19, pp200–220

Helms, J. (1981) Threat perceptions in acute chemical disasters. In J. K. Gray and E. L. Quarantelli (eds) *Social Aspects of Acute Chemical Emergencies, Journal of Hazardous Materials*, vol 4, no 4, pp321–329

Hewitt, K.(ed) (1983) *Interpretations of Calamity from the Viewpoint of Human Ecology*. Boston: Allen & Unwin

Hewitt, K. (1997) *Regions of Risk: A Geographical Introduction to Disasters*. London: Longman

Hewitt, K. and I. Burton (1971) *The Hazardousness of a Place: A Regional Ecology of Damaging Events*. Department of Geography Research Paper no 6. Toronto, Canada: University of Toronto Press for the Department of Geography

Hinrichsen, D. (1983) Saving Sagarmartha. *Ambio*, vol 12, pp203–205

Hofstede, G. (1980) *Culture's Consequences: International Differences in Work-related Values*. Beverly Hills, CA: Sage

Hohenemser, C., R. E. Kasperson and R. W. Kates (1977) The distrust of nuclear power. *Science*, vol 196, 1 April, pp25–34

Hohenemser, C., R. E. Kasperson and R. W. Kates (1982) Causal structure: A framework for policy formulation. In C. Hohenemser and R. E. Kasperson (eds) *Risk in the Technological Society*. AAAS Selected Symposium no 5. Boulder, CO: Westview, pp109–139

Hohenemser, C., R. E. Kasperson and R. W. Kates (1985) Causal structure. In R. W. Kates, C. Hohenemser and J. X. Kasperson (eds) *Perilous Progress: Managing the Hazards of Technology*. Boulder, CO: Westview, pp25–42

Hohenemser, C. and O. Renn (1988) Chernobyl's other legacy: Shifting public perceptions of nuclear risk. *Environment*, vol 30, April, pp4–11, 40–45

Holcomb, J. (1990) How greens have grown. *Business and Society Review*, vol 75, fall, pp20–25

Holdren, J. P. and P. R. Ehrlich (1974) Human population and the global environment. *American Scientist*, vol 62, pp282–292

Holling, C. S. (1978) Adaptive environmental assessment and management. *EIA [Environmental Impact Assessment] Review*, vol 2, pp24–25

Holling, C. S. (1986) The resilience of terrestrial ecosystems: Local surprise and global change. In W. C. Clark and R. E. Munn (eds) *Sustainable Development of the Biosphere*. Cambridge: Cambridge University Press, pp292–317

Holling, C. S. (1995) Sustainability: The cross-scale dimension. In M. Munasinghe and W. Shearer (eds) *Defining and Measuring Sustainability: The Biophysical Dimensions*. Washington, DC: Distributed for the United Nations University by the World Bank, pp65–75

Holling, C. S. (1996a) Engineering resilience versus ecological resilience. In P. C. Schulze (ed) *Engineering within Ecological Constraints*. Washington, DC: National Academy Press, pp31–44

Holling, C. S. (1996b) Surprise for science, resilience for ecosystems, and incentives for people. *Ecological Applications*, vol 6, August, pp733–735

Holling, C. S. (1997) Regional responses to global change. *Conservation Ecology*, vol 1, no 2, p3, www.consecol.org/journal

Holling, C. S. (1998) The renewal, growth, birth, and death of ecological communities: *Whole Earth*, vol 93, summer, pp32–35

Holling, C. S. and G. K. Meffe (1996) Command and control and the pathology of natural resource management. *Conservation Biology*, vol 10, no 2, pp328–337

Holt, R. R. (1982) A selective review of research on health effects of the accident at Three Mile Island. Draft paper, New York University

Holusha, J. (1987) US fines Chrysler $1.5 million, citing workers' exposure to peril. *New York Times*, 7 July, ppA1, A17

Holzheu, F. and P. Wiedemann (1993) Introduction: Perspectives on risk perception. In B. Rück (ed) *Risk is a Construct*. Münich: Knesebeck, pp9–20

Homans, G. C. (1958) Human behavior as exchange. *American Journal of Sociology*, vol 63, pp597–606

Houghton, J. T., L. G. Meira Filho, B. A. Callander, N. Harris, A. Kattenburg and K. Maskell (eds) (1996) *Climate Change 1995: The Science of Climate Change*. Cambridge: Cambridge University Press

Howard-Clinton, E. G. (1984) Emerging concepts of environmental issues in Africa. *Environmental Management*, vol 8, pp187–190

Huber, P. (1986) The Bhopalization of American tort law. In National Academy of Engineering (ed) *Hazards: Technology and Fairness*. Washington, DC: National Academy Press, pp89–110

Hunt, R. (1990) *Saugus River and Tributaries Flood Damage Reduction Study, Lynn, Malden, Revere, and Saugus, Massachusetts*. Waltham, MA: US Army Corps of Engineers, New England Division

IER (International Environmental Reporter) (2000) Text: Final draft of Biosafety Protocol approved at Montreal meeting on Biological Diversity Convention, 29 January. *IER*, vol 23, no 3, 2 February, pp125–134

IFRC (International Federation of Red Cross and Red Crescent Societies) (1999a) *Vulnerability and Capacity Assessment: An International Federation Guide*. Geneva: IFRC

IFRC (1999b) *Work Plan for the Nineties*. Geneva: IFRC

IFRC (2001) *World Disasters Report 2001: Focus on Recovery*. Geneva: IFRC

IHDP Update (2001) Special issue on vulnerability, vol 2, pp1–16, www.ihdp.org

IIED (International Institute for Environment and Development) and WRI (World Resources Institute) (1987) *World Resources 1987*. New York: Basic Books

ILO (International Labour Office) (1977) *Occupational Exposure Limits to Airborne Toxic Substances*. Geneva: ILO

Industrial Union Department (1980) AFL–CIO v. American Petroleum Institute et al, US Supreme Court, 2 July 1980

Ingersoll, V. H. and G. B. Adams (1983) Beyond organizational boundaries: Exploring the organizational metamyth. Paper presented at the Organizational Folklore Conference, Los Angeles, March

Inhaber, H. (1979) Risk with energy from conventional and nonconventional sources. *Science*, vol 203, pp718–723

Inhaber, H. (1982) *Energy Risk Assessment*. New York: Gordon and Breach

Inhaber, H. (1985) Risk in developing countries. *Risk Analysis*, vol 5, no 2, June, p87

IPCC (Intergovernmental Panel on Climate Change) (2001) *Climate Change 2001: Synthesis Report*. Cambridge: Cambridge University Press

Isabella, L. A. (1986) Culture, key events, and corporate social responsibility. *Research in Corporate Social Performance and Policy*, vol 8, pp175–192

Jackson, J. B. C., M. X. Kirby, W. H. Berger, K. A. Bjorndal, L. W. Botsford, B. J. Bourque et al (2001) Historical overfishing and the recent collapse of coastal ecosystems. *Science*, vol 293, pp629–637

Jackson, P. (1982) Future of elephants and rhinos in Africa. *Ambio*, vol 11, pp202–205

Jacobsen, T. and R. M. Adams (1958) Salt and silt in Ancient Mesopotamian agriculture. *Science*, vol 128, pp1251–1258

Jalees, K. and K. Vemuri (1980) Pesticide pollution in India. *International Journal of Environmental Studies*, vol 15, pp49–54

Jaques, E. (1951) *The Changing Culture of a Factory*. New York: Dryden

Jelinek, M. (1979) *Institutionalizing Innovation: A Study of Organizational Learning Systems*. New York: Praeger

Jiang, H., P. Zhang, D. Zheng and F. Wang (1995) The Ordos Plateau of China. In J. X. Kasperson, R. E. Kasperson and B. L. Turner, II (eds) *Regions at Risk: Comparisons of Threatened Environments*. Tokyo: United Nations University Press, pp420–459

Jiang, H. (1999) *The Ordos Plateau of China: An Endangered Environment*. Tokyo: United Nations University Press

Jodha, N. S. (1980) The process of desertification and the choice of interventions. *Economic and Political Weekly*, vol 15, pp1351–1356

Jodha, N. S. (1995) The Nepal middle mountains. In J. X. Kasperson, R. E. Kasperson, and B. L. Turner, II (eds) *Regions at Risk: Comparisons of Threatened Environments*. Tokyo: United Nations University Press, pp140–185

Jodha, N. S. (2001a) Interacting processes of environmental vulnerabilities in mountain areas. *Issues in Mountain Development*, vol 2001/2, September, pp1–6

Jodha, N. S. (2001b) *Life on the Edge. Sustaining Agriculture and Community Resources in Fragile Environments*. Oxford: Oxford University Press

Johnson, B. B. and V. T. Covello (eds) (1987) *The Social and Cultural Construction of Risk*. Dordrecht: Reidel

Johnson, D. L. and L. A. Lewis (1995) *Land Degradation: Creation and Destruction*. Oxford: Blackwell

Johnson, J. H., Jr. and D. J. Zeigler (1984) A spatial analysis of evacuation intentions at the Shoreham Nuclear Power Station. In M. J. Pasqualetti and D. Pijawka (eds) *Nuclear Power: Assessing and Managing Hazardous Technology*. Boulder, CO: Westview, pp279–301

Joksimovich, V. (1984) A review of plant specific PKAS. *Risk Analysis*, vol 4, pp255–266

Joshi, V. T. (1991) Madhya Pradesh: Not learning from experience. In *The Hindu Survey of the Environment*, pp59, 61. Madras: Kasturi and Sons Ltd

Kadry, A. (1983) Salvaging Egypt's Nubian monuments. *Ambio*, vol 12, pp206–209

Kail, L. K., J. M. Poulson and C. S. Tyson (1982) *Operational Safety Survey*. Danbury, CT: Union Carbide Corporation

Karan, P. P., W. A. Bladen and J. R. Wilson (1986) Technological hazards in The Third World. *Geographical Review*, vol 76, no 2, April, pp195–208

Karrh, B. W. (1984) An illustration of voluntary actions to reduce carcinogenic risks in the workplace. In Paul F. Deisler, Jr (ed) *Reducing the Carcinogenic Risks in Industry*. New York: Marcel Dekker, pp121–134

Karrh, B. W. (1989) Remarks at the DuPont International Safety Symposium. Houston, 16 January

Kasperson, J. X. and R. E. Kasperson (1987a) Priorities in profile: Managing risks in developing countries. *Risk Abstracts*, vol 4, no 3, July, pp113–118

Kasperson, J. X. and R. E. Kasperson (1993) Corporate culture and technology transfer. In H. S. Brown, P. Derr, O. Renn, and A. White (eds) *Corporate Environmentalism in a Global Economy*. Westport, CT: Quorum, pp149–177

Kasperson, J. X. and R. E. Kasperson (2001a) *International Workshop on Vulnerability and Global Environmental Change, 17–19 May 2001, Stockholm Environment Institute (SEI), Stockholm, Sweden: A Workshop Summary*. SEI Risk and Vulnerability Programme Report 2001–2001. Stockholm: SEI

Kasperson, J. X. and R. E. Kasperson (eds) (2001b) *Global Environmental Risk*. Tokyo: United Nations University Press, and London: Earthscan

Kasperson, J. X., R. E. Kasperson and B. L. Turner, II (eds) (1995) *Regions at Risk: Comparisons of Threatened Environments*. Tokyo: United Nations University Press

Kasperson, R. E. (1969) Environmental stress and the political system. In R. E. Kasperson and J. V. Minghi (eds) *The Structure of Political Geography*. Chicago: Aldine, pp481–496

Kasperson, R. E. (1980) Public opposition to nuclear power: Retrospect and prospect. *Science, Technology and Human Values*, vol 5, pp11–23

Kasperson, R. E. (ed) (1983a) *Equity Issues in Radioactive Waste Management*. Cambridge, MA: Oelgeschlager, Gunn & Hain

Kasperson, R. E. (1983b) Worker participation in protection: The Swedish alternative. *Environment*, vol 25, no 4, May, pp13–20, 40–43

Kasperson, R. E. (1987) The Kemeny Commission and the accident at Three Mile Island. In *Nuclear Risk Analysis in Comparative Perspective: The Impacts of Large-scale Risk Assessment in Five Countries*. Boston: Allen & Unwin, pp163–196

Kasperson, R. E. (1992) The social amplification of risk: Progress in developing an integrative framework of risk. In S. Krimsky and D. Golding (eds) *Social Theories of Risk*. Westport, CT: Praeger, pp153–178

Kasperson, R. E. and M. Breitbart (1974) *Participation, Decentralization and Community Planning*. Commission on College Geography, Resource Paper no 25. Washington, DC: Association of American Geographers

Kasperson, R. E., J. E. Dooley, B. Hanson, J. X. Kasperson, T. O'Riordan and H. Paschen (1987) Large-scale nuclear risk analysis: Its impacts and future. In *Nuclear Risk Analysis in Comparative Perspective: The Impacts of Large-scale Risk Assessment in Five Countries*. Risks and Hazards Series, vol 4. Boston: Allen & Unwin, pp219–236

Kasperson, R. E. and J. X. Kasperson (1983) Determining the acceptability of human risk: Ethical and policy issues. In J. T. Rogers and D. V. Bates (eds) *Risk: A Symposium on the Assessment and Perception of Risk to Human Health in Canada*. Ottawa: The Royal Society of Canada, pp135–155

Kasperson, R. E. and J. X. Kasperson (eds) (1987b) *Nuclear Risk Analysis in Comparative Perspective: The Impacts of Large-scale Risk Assessment in Five Countries*. Risks and Hazards Series, vol 4. Boston: Allen & Unwin

Kasperson, R. E., J. X. Kasperson, C. Hohenemser and R. W. Kates (1988a) *Corporate Management of Health and Safety Hazards: A Comparison of Current Practice*. Boulder, CO: Westview

Kasperson, R. E., J. X. Kasperson, C. Hohenemser and R. W. Kates (1988b) Managing hazards at PETROCHEM Corporation. In *Corporate Management of Health and Safety Hazards: A Comparison of Current Practice*. Boulder, CO: Westview, pp15–41

Kasperson, R. E., J. X. Kasperson, C. Hohenemser and R. W. Kates (1988c) Managing occupational and catastrophic hazards at the Rocky Flats nuclear arsenal plant. In *Corporate Management of Health and Safety Hazards: A Comparison of Current Practice*. Boulder, CO: Westview, pp79–99

Kasperson, R. E., J. X. Kasperson, C. Hohenemser and R. W. Kates (1988d) Managing occupational hazards at a PHARMACHEM Corporation plant. In *Corporate Management of Health and Safety Hazards: A Comparison of Current Practice*. Boulder, CO: Westview, pp43–56

Kasperson, R. E. and M. Morrison (1982) A proposal for international risk management research. In C. Hohenemser and J. X. Kasperson (eds) *Risk in the Technological Society*. AAAS Selected Symposium, vol 65. Boulder, CO: Westview, pp303–331

Kasperson, R. E. and P.-J. M. Stallen (eds) (1991) *Communicating Risks to the Public: International Perspectives*. Dordrecht: Kluwer

Kates, R. W. (1981) Drought in the Sahel. *Mazingira*, vol 5, no 2, pp72–83

Kates, R. W. (1985) The interaction of climate and society. In R. W. Kates, J. H. Ausubel and M. Berberian (eds) *Climate Impact Assessment*. SCOPE 27. Chichester: Wiley, pp3–36

Kates, R. W. and B. Braine (1983) Locus, equity and the West Valley nuclear wastes. In R. E. Kasperson (ed) *Equity Issues in Radioactive Waste Management*. Cambridge, MA: Oelgeschlager, Gunn & Hain, pp94–117

Kates, R. W. and W. C. Clark (1996) Environmental surprise: Expecting the unexpected. *Environment*, vol 38, no 2, March, pp6–11, 28–34

Kates, R. W., W. C. Clark, R. Corell, J. M. Hall, C. C. Jaeger, I. Lowe et al (2000) *Sustainability Science*. Research and Assessment Systems for Sustainability Program Discussion Paper 2000–33. Cambridge, MA: Environment and Natural Resources Program, Belfer Center for Science and International Affairs (BCSIA), Kennedy School of Government, Harvard University. Modified version published in *Science*, vol 292, pp641–642, 27 April 2001. Available online at www.sust.harvard.edu

Kates, R. W., C. Hohenemser and J. X. Kasperson (eds) (1985) *Perilous Progress: Managing the Hazards of Technology*. Boulder, CO: Westview

KBS (Kärn-Bränsle-Säkerhet) (1977) *Handling of Spent Nuclear Fuel and Final Storage of Vitrified High Level Reprocessing Waste*, vols 1–4. Solna, Sweden: AB Teleplan

KBS (1979) *Handing and Final Storage of Unprocessed Spent Nuclear Fuel*, vols 1–2. Solna, Sweden: AB Teleplan

Kelly, P. M. and W. N. Adger (2000) Theory and practice in assessing vulnerability to climate change and facilitating adaptation. *Climatic Change*, vol 47, pp325–352

Kelman, S. (1981) *Regulating America, Regulating Sweden: A Comparative Study of Occupational Safety and Health Policy*. Cambridge: MIT Press

Kennedy, D. (2002) Science, terrorism, and natural disasters (editorial). *Science*, vol 295, p405

Keswani, R. (1982a) Sage, please save this city. *Septahik Report*, 17 September

Keswani, R. (1982b) Bhopal on the mouth of a volcano. *Septahik Report*, 1 October

Keswani, R. (1982c) If you don't understand, you will be wiped out. *Septahik Report*, 8 October

Keswani, R. (1984) Bhopal's killer plant. *Indian Express* (Delhi), 9 December

Khandekar, S. (1985) An area of darkness. *India Today*, 30 June, p134

Khandekar, S. and S. Dubey (1984) City of death. *India Today*, 10 December, p11

Khatib, H. (2000) Energy security. In J. Goldemberg (ed) *World Energy Assessment: Energy and the Challenge of Sustainability*. New York: United Nations Development Programme (UNDP), United Nations Department of Economic and Social Affairs, and World Energy Council, pp113–114

Khator, R. (1988) Organizational response to the environmental crisis in India. *Indian Journal of Political Science*, vol 49, no 1, January–May, pp14–39

Kimmerle, G. and A. Iben (1964) Toxicitat von Methylisocyanat und dessen quantitativer Bestimmung in der Luft, *Arkhiv fur Toxicologie*, vol 20, pp235–241 (An English translation, Toxicity of methyl isocycanate and its quantitative determination in air' appears on pp500–510 of *Hazardous Air Pollutants: Hearing*, 14 December 1984, 98th Congress, 2nd session, 1984, Serial 192

Kinzig, A. P., J. Antle, W. Ascher, W. Brock, S. Carpenter, F. S. Chapin, III, et al (2000) *Nature and Society: An Imperative for Integrated Environmental Research*. A report from a workshop, Developing a Research Agenda for Linking Biogeophysical and Socioeconomic Systems, 5–8 June 2000, Tempe, Arizona. www.lsweb.la.asa.edu/akinzig/report.htm

Kletz, T. A. (1978) What you don't have can't leak. *Chemistry and Industry*, vol 9, 6 May, pp287–292

Kletz, T. A. (1985) *What Went Wrong? Case Histories of Process Plant Disasters*. Houston: Gulf Publishing Company

Kletz, T. A. (1988) *Learning from Accidents in Industry*. London: Butterworths

Kletz, T. A. (1993) *Lessons from Disasters: How Organizations Have No Memory and Accidents Recur*. Rugby, Warwickshire, UK: Institution of Chemical Engineers

Kluckhohn, C. (1951) The concept of culture. In D. Lerner and H. D. Lasswell (eds) *The Policy Sciences*, Stanford, CA: Stanford University Press

Kniffin, F. R. (1987) Corporate crisis management. *Industrial Crisis Quarterly*, vol 1, spring, pp19–23

Koenigsberger, M. D. (1986) Paper presented at Governor's Conference on Pollution Prevention Pays. Nashville, TN, March

Kottary, S. (1984–1985) Whose life is it anyway? *The Illustrated Weekly of India*, 30 December 1984–5 January 1985, p8

Kramer, P. (1983) Galapagos: Islands under Siege. *Ambio*, vol 12, pp174–179

La Porte, T. R. (1987) *High Reliability Organizations: The Research Challenge*. Working paper. Berkeley, CA: University of California, Department of Political Science

La Porte, T. R. (1996) High reliability organizations: Unlikely, demanding and at risk. *Journal of Contingencies and Crisis Management*, vol 4, no 6, pp60–70

La Porte, T. R. and A. Keller (1996) Assuring institutional constancy. *Public Administration Review*, vol 56, no 6, pp535–544

Lagadec, P. (1982) *Major Technological Risk: An Assessment of Industrial Disasters*. New York: Pergamon

Lagadec, P. (1987) From Seveso to Mexico and Bhopal: Learning to cope with crises. In P. R. Kleindorfer and H. C. Kunreuther (eds) *Insuring and Managing Hazardous Risks: From Seveso to Bhopal and Beyond*. Berlin: Springer-Verlag, pp13–46

Lalvani, G. H. (1985) Law and pollution control. In J. Bandyopadhyay, N. D. Jayal, U. Schoettli and C. Singh (eds) *India's Environment: Crisis and Responses*. Dehra Dun: Natraj Publishers, pp284–290

Lamb, R. (1983) Soil erosion: The real cause of the Ethiopian famine. *Environmental Conservation*, vol 10, pp157–159

Lamotte, M. (1983) Undermining Mount Nimba. *Ambio*, vol 12, pp174–179

Lanza, G. R. (1985) Blind technology transfer: The Bhopal example. *Environmental Science and Technology*, vol 19, pp581–586

Laris, P. (2002) Burning the seasonal mosaic: Preventative burning strategies in the wooded savanna of southern Mali. *Human Ecology*, vol 30, no 2, June, pp155–186

Laurent, A. (1983) The cultural diversity of Western conceptions of management. *International Studies of Management and Organization*, vol 13, spring/summer, pp75–96

Lawless, E. W. (1977) *Technology and Social Shock*. New Brunswick, NJ: Rutgers University Press

Le Houerou, H. N. (1977) Man and desertification in the Mediterranean region. *Ambio*, vol 6, pp363–365

Leach, M., R. Mearns and I. Scoones (1999) Environmental entitlements: Dynamics and institutions in community-based natural resource management. *World Development*, vol 27, no 2, pp225–247

Lee, K. N. (1993) *Compass and Gyroscope: Integrating Science and Politics for the Environment*. Washington, DC: Island Press

Leonard, H. J. (1985) Confronting industrial pollution in rapidly industrializing countries: Myths, pitfalls, and opportunities. *Ecology Law Quarterly*, vol 12, no 4, pp779–816

Lepkowski, W. (1985) People of India struggle toward appropriate response to tragedy. *Chemical and Engineering News*, vol 63, no 6, 11 February, pp16–26

Levidow, L. (2001) Genetically modified crops: What transboundary harmonization in Europe? In J. Linnerooth-Bayer, R. Löfstedt and G. Sjöstedt (eds) *Transboundary Risk Management*. London: Earthscan, pp59–90

Levi-Strauss, C. (1963) *Structural Anthropology*. Translated by C. Jacobson and B. Schoeph. New York: Basic Books

Liedtka, J. (1988) Managerial values and corporate decision-making: An empirical analysis of value congruence in two organizations. *Research in Corporate Social Performance and Policy*, vol 11, pp55–91

Lindblom, C. (1959) The science of muddling through. *Public Administration Review*, vol 19, spring, pp79–87

Lindell, M. K. and R. W. Perry (1980) Evaluation criteria for emergency response plans in radiological transportation. *Journal of Hazardous Materials*, vol 3, pp335–348

Linnerooth-Bayer, J. (1996) Fairness in dealing with transboundary risks. In R. E. Löfstedt and G. Sjöstedt (eds) *Environmental Aid Programmes to Eastern Europe*. Ashgate, Aldershot, UK: Avebury Studies in Green Research, pp167–184

Linnerooth-Bayer, J., R. Löfstedt and G. Sjöstedt (eds) (2001) *Transboundary Risk Management*. London: Earthscan

Liverman, D. M. (1990) Vulnerability to global environmental change. In R. E. Kasperson, K. Dow, D. Golding and J. X. Kasperson (eds) *Understanding Global Environmental Change: The Contributions of Risk Analysis and Risk Management*. Worcester, MA: The Earth Transformed Program, Clark University, pp27–44

Löfstedt, R. E. (1996a) Fairness across borders: The Bärseback nuclear power plant. *Risk: Health, Safety & Environment*, vol 7, no 2, spring, pp134–144

Löfstedt, R. E. (1996b) Risk communication: The Bärseback nuclear power plant case. *Energy Policy*, vol 24, no 8, pp684–686

Löfstedt, R. E. and V. Jankauskas (2001) Swedish aid and the Ignalina power plant. In J. Linnerooth-Bayer, R. E. Löfstedt and G. Sjöstedt (eds) *Transboundary Risk Management*. London: Earthscan, pp33–58

Lonergan, S. (1999) *GECHS Science Plan*. IHDP Report no 11. Bonn: International Human Dimensions Programme on Global Environmental Change

Lonergan, S., K. Gustavson and B. Carter (2000) The index of human security. *Aviso*, vol 6, January, pp1–11, www.gechs.org/aviso

Lonergan, S. and A. Swain (1999) Environmental degradation and population displacement. *AVISO*, vol 2 (entire issue)

Lowrance, W. W. (1976) *Of Acceptable Risk. Science and the Determination of Safety*. Los Altos, CA: W. Kaufmann

Luck, S. H. (1983) Recent trends of desertification in the Maowusu Desert: China. *Environmental Conservation*, vol 10, pp213–224

Luhmann, N. (1986) *Ökologische Kommunikation*. Opladen: Westdeutscher Verlag

Luhmann, N. (1993) *Risk: A Sociological Theory*. New York: Aldine de Gruyter

Lundgren, B. (1985) Global deforestation. *Agroforestry Systems*, vol 3, pp91–95

March, J. G. and H. A. Simon (1958) *Organizations*. New York: Wiley

Martin, J. (1982) Stories and scripts in organizational settings. In A. Hastrof and A. Isen (eds) *Cognitive Social Psychology*. New York: Elsevier–North Holland, pp155–304

Mather, J. T. and G. V. Sdasyuk (eds) (1991) *Global Change: Geographic Approaches*. Tucson: University of Arizona

McCarthy, J. J., O. F. Canziani, N. A. Leary, D. J. Dokken and K. S. White (eds) (2001) *Climate Change 2001: Impacts, Adaptation and Vulnerability*. Cambridge: Cambridge University Press for the Intergovernmental Panel on Climate Change (IPCC)

Meadows, D. H., D. L. Meadows and J. Randers (1992) *Beyond the Limits: Confronting Global Collapse, Envisioning a Sustainable Future*. Post Mills, VT: Chelsea Green

Meadows, D. H., D. L. Meadows, J. Randers and W. W. Behrens, II (1972) *The Limits to Growth: A Report for the Club of Rome's Project on the Predicament of Mankind*. New York: Universe Books

Medawar, C. (1979) *Insult or Injury? An Enquiry into the Marketing and Advertising of British Food and Drug Products in the Third World*. London: London Social Audit

MEF (Ministry of Environment and Forests) (1991) *Annual Report, 1990–1991*. Delhi: MEF

MEF (1992) *Policy Statement for Abatement of Pollution*. No H. 11013(2)/90-CPW. New Delhi: MEF

Melville, M. (1981). Risks on the job: The worker's right to know. *Environment*, vol 23, November, pp12–20, 42–45

Mendeloff, J. (1980) *Regulating Safety: An Economic and Political Analysis of Occupational Safety and Health Policy*. Cambridge, MA: MIT Press

Meyers, P. W. (1990) Nonlinear learning in large technological firms. *Research Policy*, vol 19, pp97–115

Middlebrooks, E. J. M., P. M. Armenante and J. B. Carmichael (1981) Industrial pollution discharges from the West African region. *Environment International*, vol 5, pp177–191

Mileti, D. S. (1975) *Natural Hazard Warning Systems in the United States: A Research Assessment*. Program on Technology, Environment and Man, Monograph 13. Boulder, CO: Institute of Behavioral Science, University of Colorado

Mileti, D. S. (1999) *Disasters by Design: A Reassessment of Natural Hazards in the United States*. Washington, DC: Joseph Henry Press

Millennium Ecosystem Assessment (2003) *Ecosystems and Human well-being: A Framework for Advancement*. Washington, DC: Island Press

Minor, M., K. Kawamura and F. F. Lynes (1986) Risk in developing countries: Comment. *Risk Analysis*, vol 6, no 1, March, pp1–2

Mitchell, J. K. (1984) Hazard perception studies: Convergent concerns and divergent approaches during the past decade. In T. F. Saarinen, D. Seamon and J. L. Sell (eds) *Environmental Perception and Behavior: An Inventory and Prospect*. Research Paper no 209. Chicago: Department of Geography, University of Chicago, pp33–59

Mitchell, J. K. (1989) Hazards research. In G. L. Gaile and C. J. Wilmott (eds) *Geography in America*. Columbus, OH: Merrill Publishing, pp410–424

Mitchell, J. K (1996) *The Long Road to Recovery: Community Response to Industrial Disaster*. Tokyo: United Nations University Press

Mitchell, J. K. (2000) Urban metabolism and disaster vulnerability in an era. In H.-J. Schellnhuber and V. Wenzel (eds) *Earth System Analysis: Integrating Science for Sustainability, Complemented Results of a Symposium Organized by the Potsdam Institute (PIK)*. Berlin: Springer-Verlag, pp259–377

Mitchell, J. K. (2001) What's in a name? Issues of terminology and language in hazards research. *Environmental Hazards*, vol 2, no 3, September, pp87–88

Mitroff, I. I., T. Pauchant, M. Finney and C. Pearson (1989) Do (some) organizations cause their own crises?: The cultural profiles of crisis-prone vs. crisis-prepared organizations. *Industrial Crisis Quarterly*, vol 3, pp269–283

Morehouse, W. and A. Subramaniam (1986) *The Bhopal Tragedy*. New York: Council of International and Public Affairs

Moser, C. O. M. (1998) The asset vulnerability framework: Reassessing urban poverty reduction strategies. *World Development*, vol 26, no 1, pp1–19

Moser, S. C. (1997) *Mapping the Territory of Uncertainly and Ignorance: Broadening Current Assessment and Policy Approaches to Sea-level Rise*. PhD thesis. Worcester, MA: Graduate School of Geography, Clark University

Moursy, S. (1986) Oil pollution studies in Nile water. *Environment International*, vol 9, pp103–106

Munich Re (2001) *Annual Review: Natural Catastrophes 2001*. Münich, Germany: Münchener Rückversicherungs–Gesellschaft.

Muralidharan, S. (1985) The Bhopal tragedy: Carbide's counter-offensive. Special report. *The Herald Review* (Bangalore), 7 April, p35

Myers, N. (1988) Threatened biotas: 'Hot spots' in tropical forests. *The Environmentalist*, vol 8, no 3, pp187–208

Nash, J. R. (1976) *Darkest Hours: A Narrative Encyclopedia of Worldwide Disasters from Ancient Times to the Present*. Chicago: Nelson-Hall

National Geographic Society (1989) *Endangered Earth* (map). Washington, DC: National Geographic Society

National Safety Council (1984) *Accident Facts, 1984 Edition*. Chicago: National Safety Council

Nations, J. D. and D. I. Komer (1983) Central America's tropical forests. *Ambio*, vol 12, pp232–238

Nautiyal, J. C. and P. S. Babor (1985) How to avert an environmental disaster. *Interdisciplinary Science Reviews*, vol 10, pp27–41

Nealon, P. and B. Brelis (1996) The eastcoaster of '96: N. E. digs out; forecasters say to dig in for more. *The Boston Globe*, 9 January, pp1, 21

Nelkin, D. M. (1981) Nuclear power as a feminist issue. *Environment*, vol 23, pp14–20, 38–39

Nelkin, D. and M. S. Brown (1984) *Workers at Risk: Voices from the Workplace*. Chicago: University of Chicago Press

Nepstad, D. C., A. Verissimo, A. Alencar, C. Nobre, E. Lima, P. Lefebvre, P. Schlesinger, C. Potter, P. Moutinho, E. Mendoza, M. Cochrane and V. Brooks (1999) Large-scale impoverishment of Amazonian forests by logging and fire. *Nature*, vol 398, pp505–508

New Jersey Radioactive Waste Advisory Committee (1994) *Proposed Voluntary Siting Plan for Locating a Low-level Radioactive Waste Disposal Facility in New Jersey.*

New York Times (1985) Many at plant thought MIC was chiefly a skin–eye irritant. 30 January, pA6

Newman, L. F. (ed) (1990) *Hunger in History: Food Shortage, Poverty, and Deprivation*. Cambridge, MA: Blackwell

Nigg, J. M. (1987) Communication and behavior: Organizational and individual response to warnings. In R. R. Dynes, B. de Marchi and C. Pelanda (eds) *Sociology of Disasters: Contributions of Sociology to Disaster Research*. Milan: Franco Angeli, pp103–117

Nyström, M. and C. Folke (2001) Spatial resilience of ecosystems. *Ecosystems*, vol 4, no 5, pp406–417

Nyström, M., C. Folke and F. Moberg (2000) Coral reef disturbance and resilience in a human dominated environment. *Trends in Ecological Evolution*, vol 15, pp413–417

Obeng, L. E. (1978) Starvation or bilharziasis. *Water Supply and Management*, vol 2, pp343–350

OECD (Organisation for Economic Co-operation and Development) (1972) *Problems of Trans-frontier Pollution: An OECD Record of a Seminar on the Economic and Legal Aspects of Trans-frontier Pollution*. Paris: OECD

OFDA (Office of US Foreign Disaster Assistance) (1988) *Disaster History: Significant Data on Major Disasters Worldwide, 1900–Present*. Washington, DC: OFDA

O'Keefe, P. (1983a) The causes, consequences, and remedies of soil erosion in Kenya. *Ambio*, vol 12, pp302–305

O'Keefe, P. (1983b) Natural wood supplies in Kenya. *Environmental Monitoring and Assessment*, vol 3, pp109–110

Oliver-Smith, A. (1986) Introduction: Disaster context and causation: An overview of changing perspectives in disaster research. *Studies in Third World Societies*, Publication no 36, June, pp1–34

Oluwande, P. A. (1983) Pollution levels in some Nigerian rivers. *Water Research*, vol 17, pp957–964

Ong, J. E. (1982) Mangroves and aquaculture in Malaysia. *Ambio*, vol 11, pp252–257

O'Riordan, T. (1984) The Sizewell B inquiry and a national energy strategy. *The Geographical Journal*, vol 150, pp171–182

OSHA (Occupational Safety and Health Administration) (1983) Hazard communication: Final Rule. *Federal Register,* Vol 48, no 228, 25 November, pp53280–53348

Osore, H. (1983) Pollution and public health in East Africa. *Ambio,* vol 12, no 6, pp316–321

Otway, H., P. Haastrup and W. Cannell (1987) *An Analysis of the Print Media in Europe Following the Chernobyl Accident.* Ispra, Italy: Joint Research Centre of the Commission of the European Community

Paine, R. T., M. J. Tegner and E. A. Johnson (1998) Compounded perturbations yield ecological surprises. *Ecosystems,* vol 1, pp535–545

Park, C. (1991) Trans-frontier air pollution: Some geographical issues. *Geography,* vol 76, no 335 (1), pp21–35

Parr, A. R. (1987) Disasters and disabled persons: An examination of the safety needs of a neglected minority. *Disasters,* vol 11, pp148–159

Paschen, H., G. Beckmann, G. Frederichs, F. Gloede, F.-W. Heuser and H. Hörtner (1987) The German risk study for nuclear power plants. In R. E. Kasperson, and J. X. Kasperson (eds) *Nuclear Risk Analysis in Comparative Perspective: The Impacts of Large-scale Risk Assessment in Five Countries.* Boston: Allen & Unwin, pp86–125

Patarasuk, W. (1991) The role of transnational corporations in Thailand's manufacturing industries. *Regional Development Dialogue,* vol 12, spring, pp92–114

Peat, Marwick, Mitchell and Co (1983) *An Industry Survey of Chemical Company Activities to Reduce Unreasonable Risk.* Final report, prepared for the Chemical Manufacturers Association, 11 February. Washington, DC: Peat, Marwick, Mitchell and Co

Perrow, C. (1984) *Normal Accidents: Living with High-risk Technologies.* New York: Basic Books (second edition, 1999, Princeton University Press)

Perrow, C. (1985) *Normal Accidents: Living with High-risk.* New York: Basic Books

Perrow, C. (1999) *Normal Accidents: Living with High-risk Technologies, with a New Afterword and a Postscript on the Y2K Problem.* Princeton, NJ: Princeton University Press

Perry, R. W. (1983) Environmental hazards and psychopathology: Linking natural disasters with mental health. *Environmental Management,* vol 7, no 6, pp543–552

Perry, R. W. (1985) *Comprehensive Emergency Management: Evacuating Threatened Populations.* Greenwich, CT: JAI Press

Perry, R. W., M. Greene and A. Mushkatel (1983) *American Minority Citizens in Disaster.* Seattle: Battelle

Perry, R. W. and M. K. Lindell (1991) The effects of ethnicity on evacuation decisionmaking. *International Journal of Mass Emergencies and Disasters,* vol 9, pp47–68

Persley, G. J. and J. N. Siedow (1999) *Applications of Biotechnology to Crops: Benefits and Risks.* Issue Paper no 12, December. Ames, IA: Council for Agricultural Science and Technology

Peters, H. P., G. Albrecht and L. Hennen (1987) *Reactions of the German Population to the Chernobyl Accident: Results of a Survey.* Jül-Spez-400/Translation (May). Jülich, Germany: The Nuclear Research Centre

Peters, T. J. (1978) Symbols, patterns, and settings: An optimistic case for getting things done. *Organizational Dynamics,* vol 7, no 2, pp3–23

Petschel-Held, G., A. Block, M. Cassel-Gintz, J. Kropp, M. K. B. Lüdeke, O. Moldenhauer, F. Reusswig and H-J. Schellnhuber (1999) Syndromes of global change: A qualitative modelling approach to assist global environmental management. *Environmental Modeling and Assessment,* vol 4, pp295–314

Pezzoli, K. (1998) *Human Settlements and Planning for Ecological Sustainability: The Case of Mexico City.* Cambridge, MA: MIT Press

Phung, D. L. (1984) *Assessment of Light Water Reactor Safety since the Three Mile Island Accident.* ORAU/IEA-84-3(M). Oak Ridge: Institute for Energy Analysis

Pidgeon, N., R. E. Kasperson and P. Slovic (eds) (2003) *The Social Amplification of Risk.* Cambridge: Cambridge University Press

Pijawka, K. D. (1983) *A Comparative Study of the Regulation of Pesticide Hazards in Canada and the United States.* PhD thesis. Worcester, MA: Clark University

Pijawka, K. D. (1984) *Assessment of Hazardous Materials Management in the State of Arizona.* Tempe: Arizona State University for Arizona Department of Energy Services

Pijawka, K. D. and A. E. Radwan (1985) The transportation of hazardous materials: Risk assessment and hazard management. *Dangerous Properties of Industrial Materials Report* 5, September/October, pp2–11

Pinsdorf, M. (1987) *Communicating When Your Company is Under Siege: Surviving Public Crisis.* Lexington, MA: Lexington Books

Platt, R. (1991) Lifelines: An emergency management priority for the United States in the 1990s. *Disasters*, vol 15, pp172–176

Pondy, L. R., P. J. Frost, G. Margan and T. C. Dandridge (1983) *Organizational Symbolism.* Greenwich, CT: JAI

Potter, L., H. Brookfield and Y. Byron (1995) The eastern Sundaland region of South-East Asia. In J. X. Kasperson, R. E. Kasperson and B. L. Turner, II (eds) *Regions at risk: Comparisons of Threatened Environments.* Tokyo: United Nations University Press, pp460–518

Powell, D. and W. Leiss with A. Whitfield (1997) Mad cows or crazy communication? In *Mad Cows and Mother's Milk: The Perils of Poor Risk Communication.* Montreal, Quebec, Canada: McGill–Queen's University Press, pp3–25

Pritchard, L., Jr, J. Colding, F. Berkes, U. Svedin and C. Folke (1998) *The Problem of Fit between Ecosystems and Institutions.* IHDP Working Paper no 2. Bonn: International Human Dimensions Programme on Global Environmental Change (IHDP)

Project 88 (1988) *Project 88: Harnessing Market Forces to Protect the Environment.* Washington, DC: US Government Printing Office

Puchachenko, Y. G. (1989) *Ekoystemy v kriticheskikh sostoyanikh* (*Ecosystems in Critical Stages*). Moscow: Nauka

Quaraishi, T. A. (1985) Residential woodburning and air pollution. *International Journal of Environmental Studies*, vol 24, pp19–34

Quarantelli, E. L. (1979) *Consequences of Disasters for Mental Health: Conflicting Views.* Preliminary Paper 62. Columbus, OH: Disaster Research Center, Ohio State University

Quarantelli, E. L. (1981) *Socio-behavioral Responses to Chemical Hazards.* Columbus, OH: Disaster Research Center, Ohio State University

Quarantelli, E. L. (1988) Assessing disaster preparedness planning: A set of criteria and their applicability to developing countries. *Regional Development Dialogue*, vol 9, no 1, spring, pp48–69

Quarantelli, E. L. (1991) *Patterns of Sheltering and Housing in American Disasters.* Preliminary Paper no 170. Newark, DE: University of Delaware, Disaster Research Center

Qureshi, A. (1985) Did Buch okay Carbide plant site? *Hindustan Times,* 13 February, pp1, 8

Rajendran, S. and R. Reigh (1981) Environmental health in Malaysia. *Bulletin of the Atomic Scientists*, vol 37, no 6, pp30–35

Rajitsinh, M. K. (1979) Forest destruction in Asia and the South Pacific. *Ambio*, vol 8, pp192–201

Ramaseshan, R. (1984) Government responsibility for Bhopal gas tragedy. *Economic and Political Weekly*, vol 19, 15 December, p2109

Ramaseshan, R. (1985) Bhopal gas tragedy: Callousness abounding. *Economic and Political Weekly*, vol 20, 12 January, pp56–57

Randrianarijaona, P. (1983) Erosion of Madagascar. *Ambio*, vol 12, pp308–311

Rangasami, A. (1985) Failure of exchange entitlements, theory of famine: A response. *Economic and Political Weekly*, vol 21, pp1741–1801

Rapp, A. (1986) Erosion and sedimentation by water. *Ambio*, vol 15, pp215–225

Raskin, P., M. Chadwick, T. Jackson and G. Leach (1966) *The Sustainability Transition: Beyond Conventional Development*. Polestar Series Report no 1. Stockholm: Stockholm Environment Institute

Reich, M. (1994) Toxic politics and pollution victims in the Third World. In S. Jasanoff (ed) *Learning from Disaster*. Philadelphia: University of Pennsylvania Press, pp180–203

Reissland, J. and V. Harries (1979) A scale for measuring risks. *New Scientist*, vol 83, pp809–811

Renn, O. (1988) Public responses to Chernobyl: Lessons for risk management and communication. In *Uranium and Nuclear Energy: 1987*, pp53–66. London: The Uranium Institute

Renn, O. (1992) Concepts of risk: A classification. In S. Krimsky and D. Golding (eds) *Social Theories of Risk*. Westport, CT: Praeger, pp53–79

Repetto, R. (1985) *Paying the Price: Pesticide Subsidies in Developing Countries*. Research Report no 2. Washington, DC: World Resources Institute

Research and Assessment Systems for Sustainability Program (2001) *Vulnerability and Resilence for Coupled Human–Environment Systems: Report of the Research and Assessment Systems for Sustainability Program 2001 Summer Study, 29 May–1 June, Airlie House, Warrenton, Virginia*. Research and Assessment Systems for Sustainability Program Discussion Paper 2001–17. Cambridge, MA: Environment and Natural Resources Program, Belfer Center for Science and International Affairs, Kennedy School of Government, Harvard University. Available online at www.sust.harvard.edu

Ribot, J. C. (1995) The causal structure of vulnerability: Its application to climate impact analysis. *Geo Journal*, vol 35, pp199–122

Ribot, J. C. (1996) Introduction. Climate variability, climate change and vulnerability: Moving forward by looking back. In J. C. Ribot, A. R. Magalhäes and S. S. Panagides (eds) *Climate Variability, Climate Change and Social Vulnerability in the Semi-arid Tropics*. Cambridge: Cambridge University Press, pp1–10

Rijnmond Public Authority (1982) *Risk Analysis of Six Potentially Hazardous Industrial Objects in the Rijnmond Area: A Pilot Study*. Boston: Reidel

Roberts, K. H. (1989) New challenges in organizational research: High reliability organizations. *Industrial Crisis Quarterly*, vol 3, no 2, pp111–125

Roberts, K. H. and G. Gargano (1990) Managing a high-reliability organization: A case for interdependence. In M. A. Glinow and S. A. Mohrman (eds) *Managing Complexity in High Technology Organizations*. New York: Oxford University Press, pp146–159

Roberts, K. H., T. R. La Porte and D. M. Rousseau (1994) The cultures of high reliability: Quantitative and qualitative assessment aboard nuclear-powered aircraft carriers. *The Journal of High Technology Management*, vol 5, no 1, spring, pp141–161

Rocheleau, D., P. Benjamin and A. Diang'a (1995) The Ukambani region of Kenya. In J. X. Kasperson, R. E. Kasperson and B. L. Turner, II (eds) *Regions at Risk:*

Comparisons of Threatened Environments. Tokyo: United Nations University Press, pp186–254

Rochlin, G. I. (1989) Informal organizational networking as a crisis-avoidance strategy: US Naval flight operations as a case study. *Industrial Crisis Quarterly*, vol 3, pp159–176

Rolén, M. (1996) *Culture, Perceptions and Environmental Problems: Interscientific Communication on Environmental Issues*. Stockholm: Swedish Council for Planning and Coordination of Research

Ronen, S. (1986) *Comparative and Multinational Management*. New York: Wiley

Ronen, S. and O. Shenkar (1985) Clustering countries on attitudinal dimensions: A review and synthesis. *Academy of Management Review*, vol 10, pp435–454

Rosen, M. (1985) Breakfast at Spiro's: The dramaturgy of dominance. *Journal of Management*, vol 11, pp31–48

Rosencranz, A., S. Divan and A. Scott (1994) Legal and political repercussions in India. In S. Jasanoff (ed) *Learning from Disaster*, pp44–65. Philadelphia: University of Pennsylvania Press

Rossi, P. H., J. D. Wright, E. Weber-Burdin and J. Pereira (1983) *Victims of the Environment: Loss from Natural Hazards in the United States, 1970–1980*. New York: Plenum

Rousseau, D. M. (1989) The price of success?: Security-oriented cultures and high reliability organizations. *Industrial Crisis Quarterly*, vol 3, pp285–302

Rout, M. K. (1985) *The Bhopal Tragedy: Analysis of Related Issues*. Bhubaneswar: Orissa State Prevention and Control of Pollution Board

Rowe, W. D. (1977) *An Anatomy of Risk*. New York: Wiley

Rowe, W. D. (1982) *Corporate Risk Assessment*. New York: Marcel Dekker

Runglertkrengkrai, S. and S. Engkaninan (1987) The pattern of managerial behaviour in Thai culture. *Asia Pacific Journal of Management*, vol 5, September, pp8–15

Salem, S. L., K. A. Solomon and M. S. Yesley (1980) *Issues and Problems in Inferring a Level of Acceptable Risk*. RAND/ R-2561-DOE. Santa Monica, CA: Rand Corporation

Sanchez, P. A. and S. W. Buol (1975) Soils of the tropics and the world food crisis. *Science*, vol 188, pp598–603

Sanchez, R. A. (1990) Health and environmental risks of the *maquiladora* in Mexicali. *Natural Resources Journal*, vol 30, no 1, pp163–186

Sani, S. (1985) Air pollution and air quality management in Malaysia. *Environmental Professional*, vol 7, pp168–177

Sankar, N. (1992) Industry alone not to blame. In *The Hindu Survey of the Environment, 1992*. Madras: *The Hindu Survey of the Environment*, pp137, 139, 141

Sapulkas, A. (1984) A Three Mile Island for chemicals. *New York Times*, 16 December, section 3, pp1, 30

Schaumberg, F. D. (1980) Environmental management in developing countries of Latin America. *Journal of Environmental Systems*, vol 9, pp89–97

Scheffer, M., S. Carpenter, J. A. Foley, C. Folke and B. Walker (2001) Catastrophic shifts in ecosystems. *Nature*, vol 413, 11 October, pp591–596

Schein, E. H. (1980) *Organizational Psychology*, third edition. Englewood Cliffs, NJ: Prentice-Hall

Schein, E. H. (1985) *Organizational Culture and Leadership*. San Francisco: Jossey-Bass

Schellnhuber, H. J., A. Block, M. Cassel-Gintz, J. Kropp, G. Lammel, W. Lass et al (1997) Syndromes of global change. *GAIA*, vol 6, pp19–34

Schneider, S. (1989) The greenhouse effect: Science and policy. *Science*, vol 243, pp771–781

Schneider, S. H., B. L. Turner, II and H. Morehouse-Garriga (1998) Imaginable surprise in global change science. *Journal of Risk Research*, vol 1, no 2, April, pp165–185

Science (1987) Risk assessment issue, vol 236, 17 April, pp241, 267–300

Scoones, I. (1996) *Hazards and Opportunities: Farming Livelihoods in Dryland Africa. Lessons from Zimbabwe*. London: Zed Books

Scoones, I. and J. Chambers (eds) (1994) *Beyond Farmer First: Rural People's Knowledge, Agricultural Research and Extension Practice*. London: Intermediate Technology Publications

SEI (Stockholm Environment Institute) (1996) *Baltic 21: Creating an Agenda 21 for the Baltic Sea Region, Main Report*. Stockholm: SEI.

Sen, A. K. (1977) Starvation and exchange entitlements: A general approach and its application to the great Bengal famine. *Cambridge Journal of Economics*, vol 1, pp33–59

Sen, A. K. (1981) *Poverty and Famines: An Essay on Entitlements and Deprivation*. Oxford: Oxford University Press

Sen, A. K. (1984) Rights and capabilities. In A. K. Sen (ed) *Resources, Values and Development*, pp307–324. Oxford: Blackwell

Sen, A. K. (1990) Food entitlements and economic chains. In L. F. Newman, W. Crossgrove, R. W. Kates, R. Matthews and S. Millman (eds) *Hunger in History: Food Shortage, Poverty and Deprivation*. Oxford: Blackwell, pp374–386

Sewell, W. R. D. and H. D. Foster (1976) Environmental risk: Management strategies in the developing world. *Environmental Management*, vol 1, no 1, pp49–59

Shastri, S. (1988) Public interest litigation and environmental pollution. In G. S. Nathawat, S. Shastri and J. P. Vyas (eds) *Man, Nature and Environmental Law*. Jaipur: RBSA, pp131–150

Shell Internationale Petroleum and B. V. Maatschappif (1987) Shell Moerdujk: A modern facility. *Industry and Environment*, vol 10, no 1, January/February/March, pp11–14

Sheppard, C. R. C. (1977) Effect of Athens pollution outfalls on marine fauna of the Saronikos Gulf. *International Journal of Environmental Studies*, vol 11, pp39–43

Short, J. F., Jr. (1992) Defining, explaining, and managing risks. In J. F. Short and L. Clarke (eds) *Organizations, Uncertainties and Risk*. Boulder, CO: Westview, pp3–23

Shrivastava, P. (1987a) *Bhopal: Anatomy of a Crisis*. Cambridge, MA: Ballinger

Shrivastava, P. (1987b) Crisis communications. *Industrial Crisis Quarterly*, vol 1, pp2

Shrivastava, P. (1987c) Preventing industrial crises: The challenges of Bhopal. *International Journal of Mass Emergencies and Disasters*, vol 5, no 3, November, pp199–221

Shrivastava, P. (1992) *Bhopal: Anatomy of a Crisis*, second edition. London: Chapman

Simon, H. (1951) A formal theory of the employment relationship. *Econometrics*, vol 9, pp293–305

Simon, H. A. (1957) *Models of Man*. New York: Wiley

Simon, H. A. (1979) *Models of Thought*. New Haven: Yale University Press

Simonis, U. E. (1987) *Ecological Modernisation: New Perspectives for Industrial Societies*. New Delhi: Friedrich Ebert Stiftung

Simons, M. (1985) Some smell disaster in a Brazilian industry zone. *New York Times*, 18 May, p2

Sinclair, C., L. Marstrand and P. Newick (1972) *Innovation and Human Risk: The Evolution of Human Life and Safety in Relation to Technical Change*. London: Center for the Study of Industrial Innovation

Singh, C. (1984) Legal policy for the control of environmental pollution. In P. Leelakrishnan (ed) *Law and Environment*. Cochin, India: University of Cochin, Department of Law, pp1–27

Singh, J. C. (1985) Soil loss from Himalaya. *Environmental Conservation*, vol 10, pp343–345

Slovic, P. (1987). Perception of risk. *Science*, vol 236, pp280–285

Slovic, P., B. Fischhoff and S. Lichtenstein (1982) Rating the risks: The structure of expert and lay perceptions. In C. Hohenemser and J. X. Kasperson (eds) *Risk in the Technological Society*. AAAS Symposium Series. Boulder, CO: Westview, pp141–166

Slovic, P., B. Fischhoff and S. Lichtenstein (1985) Characterizing perceived risk. In R. W. Kates, C. Hohenemser and J. X. Kasperson (eds) *Perilous Progress: Managing the Hazards of Technology*. Boulder, CO: Westview, pp91–125

Smircich, L. and M. B. Calás (1987) Organizational culture: A critical assessment. In F. M. Jablin, L. Putnam, K. H. Roberts and L. W. Poner (eds) *Handbook of Organizational Communication*. Newbury Park, CA: Sage, pp228–263

Smith, K. (1992) *Environmental Hazards: Assessing Risk and Reducing Disaster*. London: Routledge

Smith, K. (1993) Fuel combustion, air pollution exposure, and health: The situation in developing countries. *Annual Review of Energy and Environment*, vol 18, pp529–566

Smith, K. R., A. L. Aggarwal and R. M. Dave (1983) Air pollution and rural biomass fuels in developing countries. *Atmospheric Environment*, vol 17, pp2343–2362

Smith, N. J. H., E. A. S. Serrão, P. T. Alvim and I. C. Falesi (1995) *Amazonia: Resiliency and Dynamism of the Land and Its People*. Tokyo: United Nations University Press

Smith, P. and A. George (1988) The dumping grounds. *South*, vol 94, August, pp37–39

Smith, R. R. (1986) Biomass combustion and indoor air pollution. *Environmental Management*, vol 10, pp61–74

Sorensen, J. H. (1984) Evaluating the effectiveness of warning systems for nuclear power plant emergencies: Criteria and application. In M. J. Pasqualetti and D. Pijawka (eds) *Nuclear Power: Assessing and Managing Hazardous Technology*, Boulder, CO: Westview

Sorensen, J. H. and D. S. Mileti (1995) Pre-emergency information programs for accidents at nuclear power plants. In D. Golding, J. X. Kasperson and R. E. Kasperson (eds) *Preparing for Nuclear Power Plant Accidents*. Boulder, CO: Westview, pp311–336

Sorensen, J., J. Soderstrom, R. Bolin, E. Copenhaver and S. Carnes (1983) *Restarting TMI Unit One: Social and Psychological Impacts*, ORNL–5891. Oak Ridge, TN: Oak Ridge National Laboratory

Sorensen, J., B. Vogt and D. Mileti (1987) Evacuation: An assessment of planning and research, ORNL–6376. Oak Ridge, TN: Oak Ridge National Laboratory

Sridhar, M. K. C., P. A. Oluwande and A. O. Okbadyo (1981) Health hazards from open drains in a Nigerian city. *Ambio*, vol 10, pp19–33

Stallen, P.-J. M. and A. Tomas (1981) Psychological aspects of risk: The assessment of threat and control. In H. Kunreuther (ed) *Risk: A Seminar Series*. Laxenburg, Austria: International Institute for Applied Systems Analysis, pp195–239

Starr, C. (1969) Social benefit versus technological risk. *Science*, vol 165, pp1232–1238

Stavins, R. H. (1989) Harnessing market forces to protect the environment. *Environment*, vol 31, no 1, January/February, pp5–7, 28–35

Stern, P. C. and H. V. Fineberg (eds) (1996) *Understanding Risk: Informing Decisions in a Democratic Society*. Report for National Research Council, Committee on Risk Characterization. Washington, DC: National Academy Press

Stirling, A. and S. Mayer (1999) *Rethinking Risk: A Pilot Multi-criteria Mapping of a Genetically Modified Crop in Agricultural Systems in the UK*. Falmer, Brighton, UK: SPRU, University of Sussex

Stobaugh, R. and D. Yergin (eds) (1979) *Energy Future: Report of the Energy Project at the Harvard Business School*. New York: Random House

Stover, W. (1985) A field day for the legislators. In *The Chemical Industry after Bhopal*. Proceedings of a two-day international symposium, 7–8 November 1985. London: Oyez IBC

Suess, M. J. and J. W. Huismans (eds) (1983) *Management of Hazardous Wastes Policy Guidelines and Code of Practice*. Copenhagen: WHO Regional Office for Europe

Sunday (1985) The crime continues, *Sunday* (Calcutta), 7–13 April, pp23–26

Susman, P., P. O'Keefe and B. Wisner (1983) Global disasters: A radical interpretation. In K. Hewitt (ed) *Interpretations of Calamity from the Viewpoint of Human Ecology*. Boston: Allen & Unwin, pp263–283

Suter, G. W., II (1990) Endpoints for regional ecological risk assessments. *Environmental Management*, vol 14, no 1, pp9–23

Suter, G. W., II (1993) *Ecological Risk Assessment*. Boca Raton, FL: Lewis Publishers

Svedin, U. and B. Aniansson (eds) (1987) *Surprising Futures: Notes from an International Workshop on Long-term World Development*, Fribergh Manor, Sweden, January 1986. Stockholm: Swedish Council for Planning and Economic Research

Svenson, O. (1988). Managing product hazards at Volvo Car Corporation. In R. E. Kasperson, J. X. Kasperson, C. Hohenemser and R. W. Kates (eds) *Corporate Management of Health and Safety Hazards: A Comparison of Current Practice*. Boulder, CO: Westview, pp57–78

Szekely, F. (1983) The chemical industry and its impact on Latin American environment. *Mazingira*, vol 7, pp26–37

Tait, J. (1999) *More Faust than Frankenstein: The European Debate about Risk Regulation for Genetically Modified Crops*. SUPRA Paper no 6, August. Edinburgh: Scottish Universities Policy Research and Advice Network (SUPRA), University of Edinburgh

Talbott, S. and N. Chanda (eds) (2001) *The Age of Terror: America and the World after September 11*. New York: Basic Books

Tam, S. W. (1981) Causes of environmental degradation in eastern Barbados. *Agriculture and Environment*, vol 5, pp285–308

Tangley, L. (1986) Saving tropical forests. *Bioscience*, vol 36, pp4–8

Taylor, G. D. and P. E. Sudnik (1984) *DuPont and the International Chemical Industry*. Boston: Twayne

Tegart, W. J. McG. and G. W. Sheldon (eds) (1993) *Climate Change 1992: The Supplementary Report to the IPCC Impacts Assessment*. Canberra, Australia: Australian Government Publishing Service

Texler, J. (1986) *Environmental Hazards in Third World Development*. Studies on Developing Countries, no 120. Budapest: Institute for World Economics, Hungarian Academy of Sciences

The Hindu (1992) Government norms for environmental clearance of projects. 20 February, p28

Thompson, K. (1969) Insalubrious California: Perception and Reality. *Annals of the Association of American Geographers*, vol 59, pp50–64

Thompson, M., R. Ellis and A. Wildavsky (1990) *Cultural Theory*. Boulder, CO: Westview

Tierney, K. J. (1981) Community and organizational awareness of and preparedness for acute chemical emergencies. *Journal of Hazardous Materials*, Special Issue, vol 4, pp333–342

Tiffen, M., M. Mortimore and F. Gichuki (1994) *More People, Less Erosion: Environmental Recovery in Kenya*. Chichester: Wiley

Times of India (1985) Probe report gathered dust for three years. 2 January

Timmerman, P. (1981) *Vulnerability, Resilience and the Collapse of Society: A Review of Models and Possible Climatic Applications*. Toronto: Institute of Environmental Studies, University of Toronto

Train, R. (1985) Foreword, special issue: Environmental law and policy in developing countries. *Ecology Law Quarterly*, vol 12, no 4, pp675–680

Trainer, P. and R. Bolin (1976) Persistent effects of disasters on daily activities: A cross-cultural comparison. *Mass Emergencies*, vol 1, pp279–290

Travis, C. C. (1985) Health risks of residential wood heat. *Environmental Management*, vol 9, pp209–216

Tucker, J. B. (1983) Schistosomiasis and water projects. *Environment*, vol 25, no 7, pp17–20

Turner, B. L., II (1983) *Once Beneath the Forest: Prehistoric Terracing in the Rio Bec Region of the Maya Lowlands*. Boulder, CO: Westview

Turner, B. L., II (1991) Thoughts on linking the physical and human sciences in the study of global environmental change. *Research and Exploration*, vol 7, no 2, spring, pp133–135

Turner, B. L., II, W. C. Clark, R. W. Kates, J. F. Richards, J. T. Mathews and W. B. Meyer (eds) (1990a) *The Earth as Transformed by Human Action: Global and Regional Changes in the Biosphere over the Past 300 years*. Cambridge: Cambridge University Press with Clark University

Turner, B. L., II, R. E. Kasperson, W. B. Meyer, K. M. Dow, D. Golding, J. X. Kasperson, R. C. Mitchell and S. J. Ratick (1990b) Two types of global environmental change: Definitional and spatial-scale issues in their human dimensions. *Global Environmental Change*, vol 1, no 1, December, pp14–22

UCS (Union of Concerned Scientists) (2002) *Energy Security Solutions to Protect America's Power Supply and Reduce Oil Dependence*. Cambridge, MA: UCS

UK Health and Safety Executive (1978) *Canvey: An Investigation of Potential Hazards from Operations in the Canvey Island/Thurrock Area*. London: HMSO

UK Health and Safety Executive (1981) *Canvey, A Second Report: A Review of Potential Hazards from Operations in the Canvey Island/Thurrock Area Three Years after Publication of The Canvey Report*. London: HMSO

UN (United Nations) (1988) *International Decade for Natural Disaster Reduction: Report of the Secretary General, United Nations*. New York: United Nations

UNDP (United Nations Development Programme) (2000) *Overcoming Human Poverty: UNDP Poverty Report 2000*. New York: UNDP

UNDRO (United Nations Disaster Relief Organization) (1988) *Natural Disasters and Vulnerability Analysis*. Geneva: Office of the United Nations Disaster Relief Coordinator

UNEP (United Nations Environment Programme) (1986) *The State of the Environment 1986: Environment and Health*. Nairobi: UNEP

UNEP (1987) *Environmental Data Report*. New York: Basil Blackwell

UNEP (1999) Vulnerability and adaptation to climate change impacts. In Rabi Sharma (ed) *Newsletter of the United Nations Environment Programme*, pp1–4

UNEP/IEO (United Nations Environment Programme/Industry and Environment Office) (1990) *Environmental Auditing: Report of a United Nations Environment Programme/Industry and Environment Office (UNEP/IEO) Workshop, Paris, 10–11 January 1989*. Paris: UNEP/IEO

Union Carbide Corporation (1976) Toxicological properties of an extremely dangerous chemical. *Methyl isocyanate F-41443-A-76 and Material Safety Data Sheet F–43458-A.* Danbury, CT: Union Carbide Corporation

Union Carbide Corporation (1984) *Operational Safety Health Survey: Institute MIC II Unit.* Charleston, WVA: Union Carbide Corporation

Union Carbide Corporation (1985) *Bhopal Methyl Isocyanite Incident Investigation Team Report.* Danbury, CT: Union Carbide Corporation

UNISDR (United Nations International Strategy for Disaster Reduction) (2002) *Natural Disasters and Sustainable Development: Understanding the Links between Development, Environment and Natural Disasters.* Background Paper no 5, DESA/DSD/PC2/BP5. New York: Department of Economic and Social Affairs, UNISDR

US Congress (1984) House Committee on Energy and Commerce, Subcommittee on Health and the Environment. *Hazardous Air Pollutants. Hearing,* 14 December 1984. 98th Congress, second session, 1984, Serial 192, pp458–476

USEPA (US Environmental Protection Agency) (1985) *A Strategy to Reduce Risks to Public Health from Air Toxics.* Washington, DC: EPA

USEPA (1986) Environmental auditing policy statement. *Federal Register*, vol 51, 9 July, pp25004–25009

USEPA (1987) *Unfinished Business: A Comparative Assessment of Environmental Problems.* Washington, DC: Office of Policy Analysis, Office of Policy Planning and Evaluation, EPA, February

USEPA (1989) *The Toxics-release Inventory: National and Local Perspectives.* EPA 560/4-89-005. Washington, DC: EPA

USGAO (US General Accounting Office) (1985) *Probabilistic Risk Assessment: An Emerging Aid to Nuclear Power Plant Safety Regulation.* GAO/RCED-85-11. Washington, DC: GAO

USGAO (1990) *Nuclear Safety: Concerns about Reactor Restart and Implications for DOE's Safety Culture.* GAO/RCED-90-104. Washington, DC: GAO

USNRC (US National Research Council) Committee on Environmental Decision Making (1977) *Decision Making in the Environmental Protection Agency: Analytical Studies for the US Environmental Protection Agency,* vol 2. Washington, DC: National Academy of Sciences

USNRC Committee on Risk and Decision Making (1982) *Risk and Decision Making: Perspectives and Research.* Washington, DC: National Academy Press

USNRC (1983) *Risk Assessment in the Federal Government: Managing the Process.* Washington, DC: National Academy Press

USNRC (1984a) *Hurricane Diana, North Carolina. 10–14 September 1984.* Washington, DC: National Academy Press

USNRC (1984b) *Social and Economic Aspects of Radioactive Waste Disposal.* Washington, DC: National Academy Press

USNRC (1984c) *Toxicity Testing: Strategies To Determine Needs and Priorities.* Washington, DC: National Academy Press

USNRC (1987) *Confronting Natural Disasters: An International Decade for Natural Hazard Reduction.* Washington, DC: National Academy Press

USNRC (1989) *Improving Risk Communication.* Washington, DC: National Academy Press

USNRC (1991) *A Safer Future: Reducing the Impacts of Natural Disasters.* Washington, DC: National Academy Press

USNRC (1999) *Our Common Journey: A Transition Toward Sustainability.* Washington, DC: National Academy Press

USNUREG (US Nuclear Regulatory Commission) (1975) *Reactor Safety Study.* WASH 1400, NUREG 75/014. Washington, DC: USNUREG

USNUREG (1984) *Probabilistic Risk Assessment (PRA) Reference Document: Final Report*. NUREG-1050. Washington, DC: USNUREG

USNUREG (1985) *NRC Policy on Future Reactor Designs: Decisions on Severe Accident Issues in Nuclear Plant Regulation*. NUREG-I070. Washington, DC: USNUREG

USNUREG (1987) *Reactor Risk Reference Document: Main Report, Draft for Comment*. NUREG-1150, vol 1. Washington, DC: USNUREG

USOMB (US Office of Management and Budget) (1996) *Statement of Administration Policy S. 1271 – Nuclear Waste Policy Act 1996*. Washington, DC: OMB

US President's Commission on the Accident at Three Mile Island (1979*) The Need for Change: The Legacy of TMI*. Washington, DC: US Government Printing Office

US Risk Assessment Review Group (1978) *Report to the US Nuclear Regulatory Commission*. NUREG/CR–0400. Washington, DC: US Nuclear Regulatory Commission

Vallura, M. (1987) A high-tech cure-all for India's climate? *Boston Sunday Globe*, 5 April, p90

van Asselt, M. and J. Rotmans (1995) *Uncertainty in Integrated Assessment Modelling: A Cultural-perspective Based Approach*. GLOBO Report Series no 9/RIVM, Report no 461502009. Bilthoven, The Netherlands: National Institute of Public Health and the Environment

van der Leeuw, S. E. (2000) Land degradation as a socionatural process. In R. J. McIntosh, J. A. Tainter and S. K. McIntosh (eds) *The Way the Wind Blows: Climate, History and Human Action*. New York: Columbia University Press, pp357–383

van Eijndhoven, J. (1994) Disaster prevention in Europe. In S. Jasanoff (ed) *Learning from Disaster*. Philadelphia: University of Pennsylvania Press, pp113–132

Van Gelder, A. and R. Hosier (1983) Fuelwood production in developing countries. *Proceedings, Indian Academy of Sciences*, vol 6, part 1 (Engineering Sciences), pp59–78

Vari, A. and J. Linnerooth-Bayer (2001) A transborder environmental controversy on the Danube: The Gabcikovo–Nagymaros dam system. In J. Linnerooth-Bayer, R. E. Löfstedt and Gunnar Sjöstedt (eds) *Transboundary Risk Management*. London: Earthscan, pp155–181

Vaughan, D. (1996) *The Challenger Launch Decision: Risky Technology, Culture and Deviance at NASA*. Chicago: University of Chicago Press

Victor, D. G., K. Raustiala and E. B. Skolnikoff (eds) (1998) *The Implementation and Effectiveness of International Environmental Commitments: Theory and Practice*. Cambridge, MA: MIT Press

Vigue, D. I. (1997) Some don't Revere the name. *Boston Globe*, 23 March, ppB1, B5

Viscusi, W. K. (1983) *Risk by Choice: Regulating Health and Safety in the Workplace*. Cambridge, MA: Harvard University Press

Viscusi, W. K. (1987) *Learning about Risk: Consumer and Worker Responses to Hazard Information*. Cambridge, MA: Harvard University Press

Vogel, C. (1998) Vulnerability and global environmental change. *LUCC Newsletter*, vol 3, March, pp15–19

Vogel, C. (2001) *Plenary: Vulnerability*. 2001 Open Meeting of the Human Dimensions of Global Environmental Change Research Community, 6–8 October 2001, Hotel Gloria, Rio de Janeiro, Brazil

Waddell, E. (1983) Coping with frosts, governments, and disaster experts: Some reflections based on a New Guinea experience and a perusal of the relevant literature. In K. Hewitt (ed) *Interpretations of Calamity from the Viewpoint of Human Ecology*. Boston: Allen & Unwin, pp33–43

Waldholz, M. (1985) Bhopal death toll, survivor problems still being debated. *Wall Street Journal*, 21 March, p22

Walker, A. S. (1982) Deserts of China. *American Scientist*, vol 70, pp366–376

Walter, I. and J. L. Ugelow (1979) Environmental policies in developing countries. *Ambio*, vol 8, pp102–109

Warrick. R. A., C. Le Provost, M. F. Meier, J. Oerlemans and P. L. Woodworth (1996) Changes in sea level. In J. T. Houghton, L. G. Meira Filho, B. A. Callander, N. Harris, A. Kattenberg and K. Maskell (eds) *Climate Change 1995: The Science of Climate Change*. Cambridge: Cambridge University Press, pp359–405

Watson, R. T., M. C. Zinyowera and R. H. Moss (eds) (1996) *Climate Change 1995: Impacts, Adaptations and Mitigation of Climate Change: Scientific–Technical Analyses*. Cambridge: Cambridge University Press

Watson, R. T., M. C. Zinyowera and R. H. Moss (eds) (1998) *Regional Impacts of Climate Change: An Assessment of Vulnerability*. Cambridge: Cambridge University Press

Watts, M. J. (1983) On the poverty of theory: Natural hazards research in context. In K. Hewitt (ed) *Interpretations of Calamity from the Viewpoint of Human Ecology*. Boston: Allen & Unwin, pp231–262

Watts, M. J. and H.-G. Bohle (1993) The space of vulnerability: The causal structure of hunger and famine. *Progress in Human Geography*, vol 17, pp43–67

WBGU/GACGC (Wissenschaftlicher Beirat Globale Umweltveränderungen/ German Advisory Council on Global Change) (2000) Specific vulnerabilities of regions and social groups. In *World in Transition: Strategies for Managing Global Environmental Risks*, pp176–185. Annual Report 1998. Berlin: Springer-Verlag, www.wbgu.de

WCED (World Commission on Environment and Development) (1987) *Our Common Future*. Oxford: Oxford University Press

Webber, D. (1985) Settlement or litigation? For Union Carbide, that is the question. *Chemical and Engineering News*, vol 63, no 6, 11 February, pp47–52

Weick, K. E. (1987) Organizational culture as a source of high reliability. *California Management Review*, vol 29, no 2, winter, pp112–127

Weick, K. E. (1989) Mental models of high-reliability systems. *Industrial Crisis Quarterly*, vol 3, pp127–142

Weinberg, A. (1977) Is nuclear energy acceptable? *Bulletin of the Atomic Scientists*, vol 33, no 4, April, pp54–60

Wescoat, J. L., Jr. (1987) The 'practical range of choice' in water resources geography. *Progress in Human Geography*, vol 11, pp41–59

Wescoat, J. L., Jr. (1993) Resource management: UNCED, GATT, and global change. *Progress in Human Geography*, vol 17, pp232–240

Wheaton, D. B. and D. C. MacIver (1999) A framework and key questions for adapting to climate variability and change. *Mitigation and Adaptation Strategies for Global Change*, vol 4, nos 3–4

White, G. F. (1945) *Human Adjustment to Floods: A Geographic Approach to the Flood Problem in the United States*. Research Paper no 29. Chicago: University of Chicago, Department of Geography

White, G. F. (1961) Choice of use in resource management. *Natural Resources Journal*, vol 1, March, pp23–40

White, G. F. (ed) (1974) *Natural Hazards: Local, National, Global*. New York: Oxford University Press

White, G. F. and J. E. Haas (1975) *An Assessment of Research Needs on Natural Hazards*. Cambridge, MA: MIT Press

WHO (World Health Organization) and UNICEF (United Nations International Children Emergency Fund) (1986) Basic principles for control of acute respiratory infections in children in developing countries: A joint WHO/UNICEF statement. Geneva: WHO

Whyte, A. V. and I. Burton (eds) (1980) *Environmental Risk Assessment*. SCOPE 15. Chichester: Wiley

WICEM (World Industry Conference on Environmental Management) (1984) WICEM: Outcome and reactions. *Industry and Environment*, Special Issue no 5

Wijkman, A. and L. Timberlake (1985) *Natural Disasters: Acts of God or Acts of Man?* London: Earthscan

Wilbanks, T. J. and R. W. Kates (1999) Global change in local places. *Climatic Change*, vol 43, pp601–628

Wilhite, D. A. (ed) (2000) *Drought: A Global Assessment*. 2 vols. London: Routledge

Wilson, R. (1975) The cost of safety. *New Scientist*, vol 68, pp274–275

Wilson, R. (1979) Analyzing the daily risks of life. *Technology Review*, vol 81, February, pp41–46

Wisner, B. (1988) *Power and Need in Africa*. London: Earthscan

Wisner, B. (1993a) Disaster vulnerability: Geographical scale and existential reality. In H.-G. Bohle (ed) *Worlds of Pain and Hunger: Geographical Perspectives on Disaster Vulnerability and Food Security*, pp13–52. Saarbrücken, Germany: Breitenbach

Wisner, B. (1993b) Disaster, vulnerability: Scale, power and daily life. *Geo Journal*, vol 30, pp127–140

Wolfe, D. A. (1983) Marine pollution in China. *Oceanus*, vol 26, no 4, pp40–47

Woolard, E. S. (1989a) The ethic of environmentalism. *Executive Excellence*, November, pp6–7

Woolard, E. S. (1989b) Remarks at the DuPont International Safety Symposium. Houston, TX, 16 January

Woolard, E. S. (1989c) Remarks before the American Chamber of Commerce of Japan. Tokyo, 6 October

Woolard, E. S. (1990) Our commitment to the environment: DuPont policy must meet public expectations and desires. *Plastics World*, 22 April, p20

Woolard, E. S. (1992) An industry approach to sustainable development. *Issues in Science and Technology*, vol 8, no 3, spring, pp29–33

World Bank (1986) *Poverty and Hunger: Issues and Options for Food Security in Developing Countries*. Washington, DC: World Bank

World Bank (1988) World Bank guidelines for identifying, analysing, and controlling major hazard installations in developing countries. Appendix B in Technica Ltd, *Techniques for Assessing Industrial Hazards*, pp126–138

World Environment Centre (1985) *Project Aftermath Information Sheet*. 27 February. New York: World Environment Centre

WRI (World Resources Institute) (1984) *Improving Environmental Cooperation: The Roles of Multinational Corporations and Developing Countries*. Washington, DC: WRI

WRI and IIED (World Resources Institute and International Institute for Environment and Development) (1986) *World Resources 1986*. New York: Basic Books

Worster, D. (1979) *Dust Bowl: The Southern Plains in the 1930s*. Oxford: Oxford University Press

Worthy, W. 1985. Methyl isocyanate: The chemistry of a hazard. *Chemical and Engineering News* Vol 63, no 6, 11 February, pp33–46.

Wynne, B. and K. Dressel (2001) Cultures of uncertainty: Transboundary risks and BSE in Europe. In J. Linnerooth-Bayer, R. E. Löfstedt and G. Sjöstedt (eds) *Transboundary Risk Management*. London: Earthscan, pp121–154

Yergin, D. (1991) *The Prize: The Epic Quest for Oil, Money and Power.* New York: Simon & Shuster

Yergin, D. and M. Hillenbrand (eds) (1982) *Global Insecurity: A Strategy for Energy and Economic Renewal.* Boston: Houghton Mifflin

York, R., E. A. Rosa and T. Dietz (2002) Bridging environmental science with environmental policy: Plasticity of population, affluence, and technology. *Social Science Quarterly*, vol 83, no 1, March, pp18–34

Zeigler, D. J., J. H. Johnson, Jr. and S. D. Brunn (1983) *Technological Hazards.* Washington, DC: Association of American Geographers

Zinyowera, M. C., B. P. Jallow, R. S. Maya and H. W. O. Okoth-Ogendo (1998) Africa. In R. T. Watson, M. C. Zinyowera and R. H. Moss (eds) *The Regional Impacts of Climate Change: An Assessment of Vulnerability.* Cambridge: Cambridge University Press, pp29–84

Index